Patrick Staib

Confronting the Coffee Crisis

Food, Health, and the Environment

Series Editor: Robert Gottlieb, Henry R. Luce Professor of Urban and Environmental Policy, Occidental College

Confronting the Coffee Crisis

Fair Trade, Sustainable Livelihoods and Ecosystems in Mexico and Central America

edited by Christopher M. Bacon, V. Ernesto Méndez, Stephen R. Gliessman, David Goodman, and Jonathan A. Fox

The MIT Press
Cambridge, Massachusetts
London, England

For information on quantity discounts, email special_sales@mitpress.mit.edu.

Set in Sabon by The MIT Press. Printed and bound in the United States of America.

Library of Congress Cataloging-in-Publication Data

Confronting the coffee crisis : Fair Trade, Sustainable Livelihoods and Ecosystems in Mexico and Central America / edited by Christopher M. Bacon . . . [et al.].
    p.   cm. — (Food, health, and the environment series)
Includes bibliographical references and index.
ISBN 978-0-262-02633-8 (hardcover : alk. paper)—ISBN 978-0-262-52480-3 (pbk. : alk. paper)
1. Coffee industry—Mexico.   2. Coffee industry—Central America.   3. Coffee industry—Mexico—Case studies.   4. Coffee industry—Central America—Case studies.   I. Bacon, Christopher M.

HD9199.M62C66   2007
338.1'73730972—dc22
                                                                        2007020846

10 9 8 7 6 5 4 3 2

We dedicate this book to the communities engaged in the struggle to sustain their ecosystems, cultures, knowledge systems, and sense of place in a complex world.

Contents

Foreword

Around 2001, I learned of an exciting collaborative at the University of California at Santa Cruz, consisting of faculty, graduate students, and other food- and social-justice-oriented researchers and activists who informally called themselves the "coffee mafia." The collaborative had an ambitious agenda. How, they asked themselves, could an ecological approach be woven together with a social and economic justice agenda that addressed the crisis among coffee producers—small farmers in Nicaragua and other countries—who saw the rapid and painful decline of the price paid to the farmers during the years 1985–2005? And how could the decline in the price paid to farmers (which led to the loss of numerous small coffee farms and the migration of farmer-producers to the cities and across the borders in search of work) be reconciled with the fact that the retail price of coffee has remained about the same? Moreover, consumption has been essentially flat.

The researchers were also interested in coffee producers' responses to this crisis. These included the use of ecological strategies (shade-grown, organic, and other agro-forestry and agroecology practices) as well as Fair Trade certification programs and the marketing of the Fair Trade label to consumers and retail outlets in the United States and in Europe. The researchers recognized that, although these approaches and strategies were still limited, they nevertheless represented important efforts designed to challenge and begin to undo what has become an "interrelated livelihood and ecological crisis," as David Goodman puts it in his introductory chapter. Thus, in pulling together their material, the "coffee mafia" researchers focused on two important connected goals: to identify and analyze the multiple aspects of the crisis and to elaborate case studies that could evaluate promising strategies and the role of farmer organizations who have connected ecological and social-justice goals (including many of the agro-ecological practices that pre-dated the current crisis) and that place the coffee farmer at the center of those strategies for change.

The name "coffee mafia" was abandoned as some of the researchers went on to other projects and new researchers joined the original group. Together, this renewed collaborative continued to develop their "action-research" agenda, exploring how the coffee farmers could achieve what the great English writer Raymond Williams called the critical need to develop "sustainable livelihoods" in an era of unjust and ecologically destructive practices spawned by globalization. This volume is a result of that exploration. With its "action-research" focus and its elaboration of how deeply the social and the ecological are intertwined, it joins other books in the Food, Health, and Environment series that explore the global and local dimensions of food systems and questions of access, justice, and environmental and community well-being.

The books in the Food, Health, and Environment series examine how food is grown, processed, manufactured, distributed, sold, and consumed. They seek to evaluate how and what type of food is produced and accessed by communities and individuals, and how health and environmental factors are embedded in food-system choices and outcomes. They focus on the food security and well being of communities and on policy decisions and economic and cultural forces at the regional, state, national, and international levels that affect community and environmental health. Food, Health and Environment books provide a window into the public debates, theoretical considerations, and multidisciplinary perspectives that have made food systems and their connections to health and environment one of the most important and consequential historical and contemporary subjects to address.

Robert Gottlieb, Occidental College

Acknowledgments

The inspiration for this book flows from two sources. The first source emerges from the long-term relationships among mountainside small-scale farmers, rural producer organizations and their managed shade coffee landscapes in Mexico and Central America. The second one starts closer to the ocean at the University of California at Santa Cruz (UCSC). Our group of editors has benefited from the innovative scholarship in agroecology, political ecology, and Latin American social movements carried out in the departments of Environmental Studies and Latin American and Latino Studies at UCSC. The Community Agroecology Network and the Program in Community and Agroecology provided a dynamic place to work on the book, especially during the final stretch of preparing the manuscript.

We are grateful to Robert Gottlieb and Clay Morgan for their diligent editorial work, which included providing insightful comments, finding critical and constructive external reviews, and moving the project along in a timely manner.

The community of scholars we would like to thank includes the past and present members of the UCSC Agro-Food Studies Research Group, coordinated by David Goodman and Margaret FitzSimmons, and especially Melanie DuPuis and Eric Holt-Jimenez for their contribution to many stimulating discussions. Ravi Rajan always offered creative and engaged scholarship. We are especially grateful to the Agroecology Research group coordinated by Professor Gliessman, including Francisco Rosado-May, Rose Ann Cohen, Ariane DeBremond, Joji Muramoto, Wes Colvin, Hillary Melcareck, and Nicholas Babin. We also would like to send a word of appreciation to Suzanne Langridge. We have shared ongoing conversations with leaders of the specialty coffee industry, including Paul Katzeff and Mark Inman. Chris Bacon is deeply grateful to Maria Eugenia Flores Gómez for her support, which included driving through the worst snowstorm of 2007 to deliver the manuscript to Clay Morgan at The MIT Press.

Finally, we are indebted to foundations and specific granting agencies that provided financial support throughout the several years it took to do the research and prepare the manuscript. The Alfred E. Heller Endowed Chair in Agroecology provided valuable support for this work. Chris Bacon is grateful to the Switzer Foundation for providing financial support and an engaging network of scholars and practitioners. Ernesto Méndez gratefully acknowledges the support he received from the Rainer Fellows Program.

Confronting the Coffee Crisis

I

Context and Analytical Framework

1

The International Coffee Crisis: A Review of the Issues

David Goodman

Since 1999, small-scale coffee producers in Mexico and in Central America have confronted severely depressed export markets that are destroying their livelihoods, mortgaging their children's future, undermining the cohesion of families and communities, and threatening ecosystems. In the years 1999–2004, international coffee prices in real terms plumbed depths not experienced for a century, throwing thousands of family producers and rural workers into desperate struggles to hold on to their way of life and provoking others to abandon their holdings and migrate to an uncertain proletarian future in urban centers and in El Norte.[1] Such dramatic individual and collective experiences are at the heart of the dry prose of "chronic over-supply" and "structural crisis."

Abandonment of coffee farms and out-migration are major, harrowing consequences of the crisis, often represented in cold, abstract terms as "classic adjustment mechanisms." However, this book focuses primarily on collective responses to the collapse of export earnings, notably the mobilization of producer cooperatives to gain access to international networks for Fair Trade and organic coffee. Contributors to this volume have long and continuing research experience of social change in Meso-American coffee communities, and many of them have firsthand knowledge of these collective initiatives. This activist, hands-on commitment gives unique insights into the front-line struggles of producer organizations and infuses the book with a rare immediacy.

Macro-level processes whose interactions describe the "global coffee crisis" are an important element of the story, but these dynamics conceal the many different "worlds" that constitute the networks of coffee production and consumption. This book provides a deeper understanding of these constituent "worlds" by exploring their complex unfolding in Mexico and Central America. More specifically, although the "coffee crisis" is manifest throughout the region as a combined, interrelated livelihood and ecological crisis, the participatory-community-based case studies that are a notable feature of this volume focus attention on the struggles by producers

*outside
influences*

to safeguard their independence as coffee farmers—that is, the collective strategies formulated by producer organizations to mitigate the immediate effects of the crisis, and their efforts to devise more sustainable trajectories of socio-ecological development. These case studies also convey the texture of individual experience as a counterpoint to more aggregative analyses at the local community and regional scales.

This focus on ways to achieve less vulnerable, more sustainable livelihoods and to conserve biodiversity in coffee-producing areas is complemented by explorations of alternative trade practices, of certification processes, and of the growth of niche markets for organic, shade-grown, and Fair Trade coffees. These practices open up new opportunities to attain ecological sustainability and stabilize rural communities by establishing more direct relationships between coffee producer organizations and ethical consumers in the North. Although these alternative commodity networks are market-based, they are distinguished from conventional trading channels by a "logic of relationality," most prominently in the case of Fair Trade coffee, where market access and quality standards are determined by agroecological practices and normative criteria, including social justice, participatory local governance, and a shared moral economy of fairness.

A central research question addressed in this volume concerns how participation by coffee farmers and producer organizations in alternative, relational trade networks has extended livelihood opportunities, reduced producers' vulnerability to the coffee crisis, and stimulated the adoption of more sustainable farm-management practices. This is a complex issue to disentangle. As Silke Mason Westphal observes in chapter 8, small- and medium-scale farmers may adopt traditional shade management in order to reduce purchases of off-farm commercial inputs, notably chemicals and fertilizers, as well as an integral part of "survival strategies." Yet this shift also may be stimulated by the higher premium prices currently paid for specialty organic, shade-grown, and Fair Trade coffees in the niche markets structured and managed by Northern conservation and Fair Trade organizations in their role as gatekeepers and auditors of socio-ecological standards of production.

Although individual authors adopt different methodological strategies to address these complexities, the field-based studies collected in this volume are oriented by a strong shared research agenda—namely, to explore how producers have responded to the crisis, and to evaluate what difference their "projects" have made in socio-economic and ecological terms. Specifically, to what degree do Fair Trade coffee and organic coffee offer a socially and ecologically sustainable alternative for the millions of rural producers who depend on this crop for their livelihood? The chapters by Steve Gliessman, Silke Mason Westphal, and Ernesto Méndez extend this question to explore the effectiveness of crop diversification and different methods of farm management as secondary strategies to resist the debilitating effects of the crisis.

While the findings from this research agenda admittedly do not provide definitive answers to these questions, these field studies offer valuable analytical and empirical insights into the myriad ways in which this process is unfolding.

A more general theoretical point arising from this research agenda is to fully grasp how these "projects" of individual and collective agents articulate or bring into play the social and the ecological. For example, the decision to contain monetary costs by utilizing on-farm, internally generated resources or use-values in the production process is complemented directly by related changes in farm-management practices. A fuller understanding of producer strategies to buffer the coffee crisis therefore calls for research designs and field methods that reject polarities in favor of an integrative framework that combines political-economic and agroecological approaches. The case studies presented below actively embrace this integrative mission. Indeed, this innovative analytical engagement with the interplay between the social and the ecological stands as a major theoretical contribution of this book.

Some Macroeconomic Considerations

By wide consensus, the origins of the present crisis are to be found in the breakdown of the International Coffee Agreement (ICA) in 1989, the ensuing relaxation of supply controls, such as producer export quotas and retention schemes, and the cumulative effect of chronic over-production on the price of "green" coffee in world export markets. The destabilizing effects of global deregulation were exacerbated by Vietnam's rapid emergence as a major producer of coffee in the 1990s, and by technological innovations that have given major roasting companies greater flexibility in using coffees of different quality grades in the production process. On this diagnosis, the current depressed world market conditions are symptomatic of a more deep-seated crisis of governance and institutional structures created by the end of the ICA trade regime, which prevailed in the period 1962–1989.

However, it is important to situate the demise of the ICA regime in the wider geopolitical context of the globalizing world economy, easing Cold War tensions, and the neo-liberal political project epitomized by the "Washington consensus" among Western governments, led by the United States, and multilateral financial agencies. That is, the advocacy of development strategies for the "global South" based on the deregulation of markets, the privatization of state agencies and public services, and the liberalization of international trade. In the post-ICA world, in coffee-producing countries, these neo-liberal "winds of change" have deeply compromised the efficacy of national institutional mechanisms of supply management, farm income maintenance, and price stabilization, including state subsidy programs, extension services, and production quota and export retention programs. This fiscal and ideological withdrawal or "hollowing out" of the state has severely weakened national state and quasi-government organizations, which played important roles in the political negotiation and implementation of the ICA trade regime.

The years following the ICA regime of managed trade have witnessed a profound change in the loci of power in the coffee commodity chain, and correspondingly in the sites or nodes where the appropriation of value added are concentrated. Stefano Ponte (2004, p. 1105) reflects a widely held view when he observes that "from a fairly balanced contest between producers and consumers within the politics of the commodity agreement, market relations shifted to a dominance of consuming country-based operators . . . over farmers, local traders, and producing country governments . . . accompanied by lower and more volatile coffee prices (and) a higher proportion of the income generated in the chain retained in consuming countries."

Estimates by Talbot (1997) reveal the dramatic and disequalizing nature of the income redistributive process underway since the end of the ICA regime. Thus, whereas in the 1980s the shares of producing and consuming countries in total income were 20 percent and 55 percent, respectively, these shares were 13 percent and 78 percent in the early to mid 1990s. The sharp fall in the price of "green" coffee experienced in the late 1990s and the early 2000s probably has reduced the share of producing countries to even lower levels.

In research on North-South flows of primary products and changes in the terms of trade between producers and buyers, analysts in the 1970s and the 1980s developed such meso-level concepts as commodity chains, commodity systems (Friedland 1984), and the French notion of *filière* (Raikes et al. 2000). More recently, the salience of globalization and evidence of increasing inequality in the geographical distribution of gains from global trade have renewed interest in these operational concepts, with the notion of "global value chains" rising to particular prominence (Gereffi and Korzeniewicz 1994; IDS 2001).

Briefly, this perspective, which builds on the Ricardian theory of economic rent, suggests that the actors in the value chain are engaged in a constant struggle to develop new sources of rent or excess profit and to protect these profit streams from the dynamic, erosive forces of competition by erecting barriers to entry. Actors who successfully appropriate value added and secure excess profit also are well placed to exercise governance roles over value-chain activities, including that of supply coordination.

On this approach, therefore, it is important to analyze how and on what terms different actors are inserted into the social division of labor characteristic of specific value chains. Here, Raphael Kaplinsky is worth citing at length:

First, barriers to entry are the determinants of the distribution of rents. That is, they determine who gains and who loses in the chain of production. Those who command rents, and have the ability to create new domains of rent when barriers to entry fall are the beneficiaries. By contrast, those who are stuck in activities with low barriers to entry lose, and in the world of increasing competition, the extent of these losses will increase over time. Secondly, the growing areas of rent are increasingly found in the intangible parts of the value chain. (2000, pp. 14–15)

Brand names and intellectual property exemplify these expanding sources of rent. Extending this approach to the present coffee crisis focuses analytical attention on shifts in the locus of market power in the value chain, the changing balance of power between producers and consumers, the redistribution of value added, and changes in governance that have occurred since the fall of the ICA regime. In primary production, this new institutional framework has resulted in lower barriers to entry, exemplified by the emergence of new producing countries and by increases in new coffee plantings. Similarly, with the abandonment of the institutional fabric of supply management, producers in former ICA member countries have experienced increasing isolation, which has exacerbated the fragmentation or atomization of supply and the loss of market power. Conversely, industry observers have identified diametrically opposite trends in the downstream nodes of the coffee chain: increasing industrial concentration among roasters and international trading companies and correspondingly stronger oligopolistic barriers to entry in these components of the value chain.

Although there is little evidence of increasing merger activity or other means of achieving vertical integration between large-scale roasters and major international traders, concentration has risen in both of these branches. Thus Ponte reports that a wave of mergers and acquisitions has raised concentration ratios in international coffee trading, and that "in 1998, the two largest coffee traders (Neumann and Volcafé) controlled 29 percent of total market share, and the top six companies 50 percent" (2004, p.1107). In the roasted and instant coffee markets, Nestlé and Philip Morris together control 49 percent and, with the addition of Sara Lee, Procter & Gamble, and Tchibo, "the top five groups control 69 percent of the market. Nestlé dominates the soluble market with a share of 56 percent" (ibid., p. 1108). According to Ponte (ibid., p. 1108), "international traders argue that roasters have gained increasing control of the marketing chain in recent years."

The glaring contrast between the primary production and downstream nodes of the value chain in their exposure to competition and ability to create barriers to entry offers analytical insight into the significant shift in income shares observed in the post-ICA coffee industry. In terms of structuralist political economy, the power asymmetries revealed by these distributional changes would suffice to explain the global coffee crisis. Yet this does not go to the heart of the matter, namely, that the crisis is not "global" but rather highly "localized" and one-sided, with the costs

of adjustment—immiseration, out-migration, and income losses—falling disproportionately on coffee-producing countries.

De-Commoditizing Coffee: Specialty Coffees and Socio-Ecological Imaginaries

The relative stability of retail coffee prices, despite the virtual stagnation of global
consumption, vis-à-vis producer prices in precipitous decline on international
markets, is emblematic of this unequal adjustment process. These divergent trajectories between retail and bulk "commodity" coffee prices constitute the "coffee
paradox," according to Ponte and Daviron (2005). To probe this "paradox" further,
consumption and consumers must be brought more directly into the analysis in order
to explore how the social life of coffee and its political economy intersect (Guthman
2002). This analytical move focuses attention on the cultural economy of coffee and
the significance of meaning as a source of durable economic rents. That is, what
are the social meanings of coffee in human interactions, how are these meanings
constructed and disseminated, and how are they transformed and distributed as
value streams of excess profits? An exploration of this terrain again emphasizes that
the *mode of integration* of producers into this cultural economy is an important
determinant of their economic performance.

One entry point into this material and symbolic economy of value is to consider
the efforts of the various actors to elevate their product above generic commodity
status, with its connotations of bulk supply, homogenized attributes, and lack of
refinement. With the limited expansion of coffee consumption in the United States
and in Western Europe, roasters, supermarkets, and café chains have pursued a
variety of product-differentiation strategies to capture premium prices and economic
rents. In the past two decades, and notably since the onset of the present coffee
crisis, producer groups, in alliance with small specialty roasters, have caused further
market segmentation by appealing to the environmental and social ethics of social
activists and reflexive consumers more generally. The rise of the specialty coffee
industry and the increase in out-of-home consumption exemplify this differentiation
process even as the "coffee paradox" is accentuated.

The product-differentiation strategies of large mainstream coffee roasters are
centered on the development of proprietary brands and the deployment of huge
advertising expenditures to gain market share for their homogenized, low-quality
blends. Excess profits arise from the combination of higher prices and scale economies afforded by their significant share of the market for mass-consumption coffee.
Indirect evidence of the success of this strategy is provided by van Dijk et al. (1998),
cited by Ponte (2004), who suggest that the large brand-name roasters have retained
control of the coffee chain despite strong competition from supermarkets and their
own-label coffees.

Brand recognition, with its combined material and semiotic barriers to entry, is one strategy to decommoditize coffee and create long-term economic rents. Nevertheless, mainstream roasters have lost market share since the mid 1980s as a result of the accelerated growth of "quality" coffees. This challenge has come from specialty café chains (epitomized by Starbucks) and from the proliferation of small specialty roasters. Stealing a march on mass-market industrial blends, the specialty coffee industry has chosen to decommoditize coffee by promoting and reinforcing a quality "turn" by consumers toward more distinctive, individualized products.

In this "economy of quality," retail specialty coffees acquire material and symbolic value by distinguishing their type, country, region or estate of origin, flavor, and roast. "Alternative" coffees—Fair Trade, organic, and shade-grown—have created value by building on relational links, fostered by social-movement activists, with Northern consumers who articulate ethics of social justice and ecological sustainability. Using labels and other discursive devices, intensely "local" narratives of coffee-growing communities and their farming practices are transported to distant global markets, building a relational "moral economy" between producers and consumers (Goodman 2004a).

These varied constructions of quality, ranging from the subjectivities of taste to shared socio-ecological imaginaries, arguably mirror a more generalized quality "turn" by movement activists and reflexive, affluent consumers away from mass-market, "placeless" foods toward an interrogation of where foods are produced, how, and by whom. In the coffee sector, the growing interest of consumers in provenance, authenticity, taste, and socio-ecological embeddedness has opened up opportunities for the specialty industry to translate these quality values and ethical norms into economic rents in the form of premium prices. Analytically, this industry has decommoditized coffee by marketing new constructions of "quality" whose conventions differ from those of brand-name roasters, with their focus on product uniformity, efficiency, and price.

The "Starbucks phenomenon" continues to be a major force in the growth of the specialty industry. Bowing to pressure from social activists, Starbucks recently added Fair Trade coffee to its product range.[2] However, the sheer scale of Starbucks' operations and its "McDonaldized" version of European café culture suggest that it is increasingly embracing the brand-recognition strategy of its traditional mainstream rivals.

These new quality conventions offer small-scale coffee growers havens of shelter from the calamitous decline in international bulk coffee prices. Thus, in Central America, Mexico and other growing regions, select producer cooperatives and associations, with financial and organizational support from local and international non-government organizations, have managed to gain access to the Fair Trade, organic, and shade-grown segments of the value chain. The complex and protracted processes of learning, organizational and institutional change, and contractual negotiation involved in reaching and maintaining certification standards are discussed in detail by several authors in the present volume.

As these authors and the wider literature confirm, these segmented, premium markets have provided significant protection from the worst effects of the international export crisis. However, such access is highly selective, and most coffee producers are not formally "qualified" or recognized under these social and ecological norms. Bluntly, in terms of innovation theory, groups that have successfully carried out the necessary farm-level and institutional innovations have enjoyed "first-mover" rents over other, excluded producers. As a corollary, this raises the question of the durability of premium prices in these segments of the value chain. These prices will come under increasing pressure as greater numbers of producers satisfy certification requirements and as other downstream actors implement competing schemes to regain market share from "ethical consumption" organizations.

The pursuit of certification pathways into differentiated premium markets suggests other analytical framings of small producers' responses to depressed international export markets. For individual producers and communities, this market conjuncture represents a crisis of social reproduction—that is, of the subjective identity bound up with the social labor and the ecological practices of coffee growing. Thus, the depressed coffee market serves as a mechanism of social differentiation in coffee-growing communities by threatening small-scale farmers with the loss of their livelihood and enforced proletarianization as wage laborers. The crisis also threatens larger, specialized plantations via indebtedness, insolvency, and foreclosure, bringing lower wages and open unemployment for permanent workers and the loss of part-time income sources for smaller growers and their families. In the Central American coffee sector, it is estimated that permanent employment fell by 54 percent and seasonal employment by 21 percent in the 2000–2001 and 2001–2002 crop seasons, leading to an overall loss of US$140 million in salary income (Varangis et al. 2003).

In the lens of social reproduction and differentiation, the certification pathways offer livelihood strategies to resist proletarianization. At the same time, since access to certification is very restricted, it can be a double-edged sword, protecting select groups but leaving many others to the mercies of *coyotes* (intermediaries) and the harsh forces of the bulk commodity markets.

As the classical political economists emphasized, processes of social differentiation and proletarianization can lead to severe self-exploitation as small producers struggle to retain their land and their livelihoods. Ethnographic research in the following chapters provides vivid evidence of these hardships and their consequences.

Stratagems here include increased reliance on family labor and on-farm inputs in the production process and greater recourse to self-provisioning or subsistence cropping. Van der Ploeg (1994, p. 75) has incorporated such activities into a continuum of "styles of farming" based on farmers' differential relationships "vis-à-vis the markets and dominant technological models." By varying the extent of their insertion in markets for inputs, credit, and commodities, farmers can control their monetary outlays and so retain "room for maneuver" or a relative degree of autonomy in farm management.

As the ethnographic case studies in this book reveal, the move of small producers away from modern, purchased inputs and toward the adoption of traditional shade-grown coffee management practices with their greater reliance on internally generated resources can readily be analyzed in these terms. Concomitantly, however, the resurgence of traditional coffee agroforestry systems is increasing the number of potential participants in the organic and shade-grown coffee markets. The stresses and tensions created by these developments are reviewed briefly in the following section.

Differentiated Markets, Producer-Consumer Politics, and "Mainstreaming"

One critical point of tension, now magnified by depressed export markets, is how to reconcile the "alternative," oppositional ethos of the social movements that gave rise to Fair Trade and organic agriculture with the opportunities for growth offered by closer integration with major mainstream trading companies, roasters, and retailers. A related concern noted by Bacon (2005) is that the recent proliferation of certifications, including Rainforest Alliance and Utz Kapeh, is leading to lower standards of social justice and organic production. Significantly, however, these dilemmas are not equally acute for all the actors engaged in Fair Trade and organic coffee production.

Movement activists fear that "mainstreaming," which has accelerated with the growth of product certifications, involves "selling out" their hopes to build a progressive producer-consumer politics with the capacity to achieve radical structural change in international trade relations and industrial agriculture. On this view, mediation by mainstream retailers in the distribution of Fair Trade products is draining the movement's alternative "project" of political meaning, effectively reducing it to the level of simply another branded marketing niche. Active and explicit engagement with the politics of solidarity and social justice for producers in the global South is being dissipated. The relational politics forged between poor, small-scale farmers and socially responsible consumers are being superseded by an individualistic, subjective consumer politics of choice, taste, and quality (Goodman 2004b).

In effect, Fair Trade is losing its social-movement identity in a bewildering welter of competing labels, brand names, product logos, and other marketing messages.[3] The disempowerment of the Sustainable Agriculture movement in the United States by the "mainstreaming" of production sectors and retailing, which has followed the adoption of the US Department of Agriculture's broadly defined organic standards, lends substance to these concerns (Guthman 2004).

For the corporate downstream actors, "mainstreaming" via partnerships with certification organizations can be seen as opportunistic rent-seeking behavior to circumvent entry barriers that have successfully maintained premium prices for Fair Trade and organic specialty coffees. Such strategic imitation of these specialty products, given the enormous oligopolistic buying power of mainstream roasters and retailers, is likely to reverse recent losses of market share and enhance their ability to capture value. There are also reports that such imitation increasingly involves companies marketing coffees as socially and environmentally "sustainable" without third-party certification (Giovannucci and Koekoek 2003).

Although the Fair Trade guaranteed minimum support price currently underpins producer prices when international markets are depressed, downstream actors will be able to exert greater control over producer profit margins and have more scope to squeeze production costs. This suggests that the economic rents represented by premium prices may prove to be short-lived gains for specialty coffee producers. Such an outcome appears particularly likely in the market for organic coffee in view of the narrowly technical nature of current certification standards and the generic character of this designation.

This discussion emphasizes why Fair Trade activists view "mainstreaming" and re-branding with trepidation. To recapitulate, the profusion of marketing "messages" involved in moving Fair Trade coffee onto supermarkets' shelves and into Starbucks' cafés threatens to obscure the "relational ethics" (Whatmore 1997) and the moral economy that provided the foundation of a distinctive "alternative" politics—that is, a politics that integrates producer empowerment and material livelihood struggles in the global South with ethically engaged consumer practices in the North.

The Fair Trade provisioning networks that are the material expression of these politics undoubtedly have contributed to the survival of thousands of small-scale coffee growers during the present crisis (Rice 2002; Raynolds et al. 2004). Nevertheless, their precarious livelihoods, their subordinate position in the value chain, and the restricted size of Fair Trade markets mean that producers take a pragmatic approach to "mainstreaming" and its alleged drawbacks. Furthermore, it is important to understand that this pragmatism has its origins in producers' long engagement with the activist organizations promoting Fair Trade, organic, and shade-grown coffees. It is the outcome of the long, demanding process of social mobilization and collective institutional learning needed to achieve certification.

The producer politics nurtured by the Fair Trade movement, for example, strongly articulate the principle of individual and collective empowerment through the formation of democratic producer associations and participatory, transparent norms of governance. Access to the Certified Fair Trade market thus is conditional upon the success of parallel processes to build local organizational and administrative capacity for collective decision making. These processes, in turn, strengthen producers' ability to negotiate collectively with buyers, acquire knowledge of specialty markets, and undertake investments in processing facilities to improve the quality of their coffee and add value. Such learning, and the collective empowerment it represents, fosters internal coping mechanisms that buffer livelihood vulnerability against external shocks.

This enhanced resilience is a very beneficial outcome of the processes of social organizational change that "qualify" individual growers and collective actors for specialty certified coffee markets. This quality of resilience now benefits about 800,000 small-scale growers and their export associations in more than 40 countries, according to Fairtrade Labelling Organizations International (known as FLO), which was formed in 1997 to coordinate Fair Trade certification and to monitor processes globally (Bacon 2005). Despite this impressive scope, however, the small size of the Fair Trade market means that the vast majority of certified producer associations must sell a significant part of their members' output in other outlets, including bulk commercial-grade markets. This market size constraint is formidable. Thus, according to FLO estimates cited by Murray et al. (2003), the export capacity of certified Fair Trade growers worldwide is seven times current Fair Trade sales.

In addition to this restrictive market ceiling vis-à-vis potential production capacity, Fair Trade coffee sales in Europe now "are largely stagnant" (Murray et al. 2003, p. 15) and certified "ethical" coffees have so far achieved only limited penetration of the substantial North American specialty market (Bacon 2005). Pragmatism is unavoidable in these circumstances, and the great lure of "mainstreaming" is the promise of market expansion at premium prices.

Nevertheless, this pragmatism is tinged with regret that direct producer-consumer solidarity ties are giving way to an individualistic consumer politics of choice as the Fair Trade labeling system becomes more heavily institutionalized. In their recent survey of selected Fair Trade producer cooperatives in Mexico, Guatemala, and El Salvador, Murray et al. (2003, p. 22) note that leadership cadres lament the decline of direct personal interactions with consumers, which they regard as "part of Fair Trade's shift away from social movement strategies toward a depersonalized niche market plan." This shift is exemplified by reports of increasing competition for Fair Trade buyer contracts, which is "undermining the historical solidarity among Mexican producer organizations" (ibid., p. 16). In a similar vein, Murray et al. (ibid., p. 19) report allegations that "participation in Fair Trade remains dominated by older and better-organized cooperatives." These authors express strong concern that current market ceilings could induce many producers to defect—particularly if prices for bulk commodity coffee reach and exceed the guaranteed minimum Fair Trade level, threatening the long-term commercial viability of Fair Trade.

There is increasing though still fragmentary evidence that Fair Trade standards are at risk from direct marketing arrangements between producer organizations and major downstream corporate actors, including Starbucks, Sara Lee, Carrefour, the Neumann Group, and Phillip Morris. For VanderHoff (2002, cited in Murray et al. 2003, p. 23), such arrangements "with large enterprises, especially with the supermarkets" mark "the third phase" in the development of Fair Trade. In some cases, these agreements match established Fair Trade prices, but "others set a different 'social minimum price,'" demonstrating their potential to erode "Fair Trade certification and sales [and] to confuse and dilute Fair Trade standards" (Murray et al. 2003, p. 23).

According to one leader of a major second-tier cooperative representing 41 producer organizations in the Mexican state of Oaxaca, "the arrival of these giants" will expand the market, but "at risk is that Fair Trade will simply become a market with higher prices and not include the respect for all the policies and ideals that go with it" (Aranda and Morales 2002, cited in Murray et al. 2003, p. 24). Murray et al. (ibid., p. 24) share this skepticism: "Corporate commitment to Fair Trade may well be a temporary strategy. Once corporations have captured the mantle of Fair Trade certification, they may move on to establish their own criteria, labels, and certification processes." Lewin et al. (2004, p. xiv) similarly emphasize that it is important to "improve the clarity of these standards and to support third-party verification as various organizations, including corporations, are developing their own independent sustainability principles and standards." Failure to do so, these authors suggest, could "damage one of the few niches in which small coffee producers have a chance to be competitive in a lucrative global trade" (ibid., p. xiv). This point is well taken. Significant dilution of the transparent and rigorous standards for Fair Trade and certified organic coffees will simply accelerate the current proliferation of competing labels, drawing in new sources of supply and driving prices down. Even with this strong caveat, however, Lewin et al. go on to argue that "differentiated coffees are not a panacea, and industry surveys indicate that two other factors are equally or, perhaps, more important to be competitive in today's coffee markets: quality and consistency" and that "the high value placed on consistency underscores the industry's preference for steady and predictable quality given the costs and risks of sourcing from new suppliers" (ibid., p. xiv). These authors suggest that this preference implies—"particularly for smaller suppliers"—a "need to improve basic business practices, as well as agronomic practices in their cooperatives and organizations" (ibid., p. xiv).

Of course, in the drive for higher quality, "first movers" may benefit from higher prices, but these gains are likely to be short-lived as imitators respond and enter the market in increasing numbers. In other words, producers will be back on the treadmill as first-mover advantages are competed away. The statement by Lewin et al. (ibid., p. xiv) on what is needed to be "competitive in today's coffee markets" raises a more fundamental issue: that the trade strategies of differentiated, certified markets and "quality/consistency" both fail to address the historically subordinate position of producers in the value chain and their consequent vulnerability to cyclical price fluctuations. Bluntly, producers' share of retail value is determined primarily by commodity price movements "because the value-adding costs are independent of the price of green coffee" (ibid., p. xii). These value-adding dynamics are illustrated by the cost structure of the cafés that retail specialty and gourmet coffees. Fitter and Kaplinsky (2005, p. 15) suggest that "the 'product' they are offering is not coffee. It is the ambience. In these markets, the coffee content of the cost of cappuccino is less than four percent."

Since certified, standards-oriented coffees and other "quality" metrics are vulnerable to competitive imitation and market saturation, the economic rents associated with this trajectory are insecure. These trade strategies can afford some temporary respite from supply pressures but they hold little promise of long-term, sustained improvement in the terms of trade of producing countries or in their share of total income. As Fitter and Kaplinsky conclude from their analysis of the global coffee value chain, the gains from differentiation—the capacity to supply products of "higher variety and enhanced quality"—are not "filtering through to producers, either at the farm level or at the national level, and this is a source of serious developmental concern" (2001, p. 16).

In response to this long-term prognosis, value-chain analysis would suggest that producers seek access to more stable sources of value—notably, in the case of coffee, symbolic value. Two embryonic initiatives in Mexico provide some encouraging evidence that producer organizations can create collective institutional networks with the capacity to appropriate value downstream in the consumer segments of the value chain. The Union La Selva cooperative in the state of Chiapas is the exclusive supplier to an autonomous franchise chain of 18 coffee shops, bearing the La Selva logo, located in Mexico, the United States, France, and Spain (Murray et al. 2003, p. 23). The second case involves Comercio Justo, an umbrella organization formed in 1999 by civil-society groups and producer associations, which has established a certification agency and a national Fair Trade label to foster a domestic market for certified Fair Trade products, including coffee (Jaffee et al. 2004). The organizational structure of Comercio Justo thus also challenges the "asymmetries of power at work in deciding what is fair" (Goodman and Goodman 2002, p. 114), which are embedded in the North-South hierarchy of certification and monitoring. Comercio Justo products are distributed through Agromercados, S.A., a marketing company owned by small producer member organizations, which is building a chain of coffee shops, as well as selling Fair Trade coffee and other certified products to supermarkets. As these two promising initiatives develop, their success in creating value and expanding markets for associations of small farmers and the limits they encounter deserve further analysis.[4]

This discussion of the current coffee crisis has now come almost full circle. The critical role of differentiated, certified coffees in sustaining the livelihoods of thousands of small producers should be fully acknowledged. This role is in no way diminished by recognition that thousands of other growers and their families have suffered the consequences of this crisis. Moreover, the limitations of differentiation and quality enhancement as a trade strategy to benefit the majority of producers have been emphasized insistently. Even if cyclical factors raise the international prices of green coffee in the medium term, the barriers to entry to the activities which create product rents for oligopolistic corporate actors will remain formidable.

Leaving aside the possibility of some international re-regulation of supply,[5] most recent analyses of global and regional coffee markets canvass measures of output diversification as a means to sustain rural livelihoods and reduce farmers' dependence on coffee as the principal cash crop (Lewin et al. 2004; Varangis et al. 2003). This alternative demands serious attention in present circumstances, yet it also ignores the parlous, enfeebled state of public-sector institutions after 20 years of fiscal retrenchment and neglect. Moreover, in many rural areas of Mexico and Central America, there are few, if any, alternative crops that can match the attractions that coffee holds for small-scale producers, including its long productive cycle and its characteristics as a marketable asset and a store of value. In addition, the option of traditional shade-grown management, which is open to many small farmers, reduces their vulnerability in hard times, not to mention its other benefits for society, including biodiversity conservation and environmental services, such as watershed protection (Perfecto et al. 1996; Méndez 2004).

Finally, structural change on the scale envisaged in rural diversification programs also must overcome more intangible resistance—that is, the remarkable tenacity with which coffee growers and their families struggle to hold on to their social identity and its collective expression in communities in particular places. Such tenacity, carried to the point of severe self-exploitation in the current crisis, is the other side of the "coffee paradox," and it is at the heart of this book.

Organization of the Book

David Goodman (chapter 1) and Seth Petchers and Shayna Harris (chapter 3) analyze the structural roots of the crisis and reveal the unequal power relations embedded in global coffee networks. These power imbalances work to enrich a few large multinational corporations and compound the devastating consequences of the crisis for small-scale growers and producing countries. Indeed, it is a gross injustice to speak of "the global coffee crisis," since its damaging impacts, in the main, are unknown and unfelt by Northern coffee consumers. These impacts are manifest almost entirely as threatened livelihoods, greater poverty, malnutrition, deforestation, and out-migration, placing family and community relationships under severe strain in the coffee-growing regions, such as those of Mexico and Central America, that are the central focus of this book.

In chapter 2, Stephen Gliessman argues that an agroecological perspective that extends the ecosystem concept to farm systems can elucidate the conditions necessary to achieve sustainable coffee agroecosystems. Sustainability, in turn, is identified as an essential foundation of a strategy to mitigate the impacts of the global coffee crisis on small-scale family producers. However, for this to occur, a transition to sustainable coffee agroecosystems must be accompanied by complementary changes in economic, social, and political processes at wider spatial scales, from local to global, reinforcing one of the main propositions advanced in this volume.

From an agroecological perspective, natural ecosystems provide a template for the co-evolutionary development of locally adapted, diverse agroecosystems, such as the "traditional" coffee systems found in Mexico and Central America. As Gliessman observes, "with their diverse shade tree species and multiple use strategies, traditional coffee systems are actually very sophisticated examples of the application of ecological knowledge, and can serve as the starting point for the conversion to more sustainable agroecosystems in the future." This insight is amplified by chapters 7–9, which furnish rich evidence of the creative interaction between these tacit, vernacular knowledges and the reproduction of small-farm livelihoods embedded in particular places, ecologies, and landscapes.

Against this broader canvas, the chapters in part II explore the crisis through the prism of the individual and collective responses of small-scale growers to the collapse of world market prices for bulk commodity green coffee. These chapters investigate a central question of the book: Have membership in producer organizations and participation in differentiated, specialty markets for Fair Trade, organic, and shade-grown coffees created new livelihood opportunities, reduced growers' vulnerability to external "shocks," and stimulated adoption of more sustainable farm-management practices? Bringing recent scholarship (notably participant-observation and action-research methods) to bear on this question, the authors, in their different ways, bring out the intimate interrelationships between household political economies and agroecological management decisions and practices.

These research methods give "voice" to the many small-scale coffee producers, weaving their stories and their experiences into the political economies and socio-ecological networks in which they and their communities are embedded. These narratives also uncover a vital collective dimension of social agency—encompassing historical processes of peasant struggle, rural social movements, and collective organization—that has shaped producers' innovative responses to the present crisis. A recurrent theme of these chapters, then, is that this powerful legacy of collective struggle is now embodied in organizations of small-scale producers, whose social networks are the sine qua non of their recent emergence as growers of differentiated, sustainable coffees. As many authors (in this volume and elsewhere) emphasize, the Fair Trade movement and international conservation organizations, with their modern institutional capacities and market access, have built on this legacy of collective struggle and social mobilization.

In chapter 4, Laura Trujillo traces the evolution of human-environment relations that are inscribed on coffee as an internationally traded commodity and the contested landscapes of its production in the Huatusco region of central Veracruz, Mexico. The spatio-temporal development of these landscapes is conceptualized as a contingent interactive process involving physiographic characteristics, political economic structures, social relations, and cognitive interpretation. In this perspective, current land-use patterns bear the imprint of the labor-supply and resource-control regimes that have characterized the agrarian social formations of coffee production since the colonial period. Against this historical background of landscape change, Trujillo investigates the role of socio-spatial marginalization and resource access in shaping the complex mosaic of coffee-production strategies and land use in nine counties in contemporary Huatusco. She argues that this causal nexus has produced a "bricolage landscape" of land use and coffee agroecosystem management. This analysis intersects with Carlos Guadarrama-Zugasti's discussion of the environmental impacts of these production decisions in chapter 6.

María Elena Martínez-Torres's assessment of the sustainability of different coffee-production technologies (chapter 5) is based on survey data from members of small-producer cooperatives, mainly Mayan in origin, in the Mexican state of Chiapas. Cooperatives such as these, whose inherited organizational strength reflects the resurgence of peasant movements in the 1980s and the 1990s, account for Mexico's pioneering role in producing and exporting certified organic coffee. Martínez-Torres uses income, yield, and agroecological data to evaluate four different coffee-production systems encountered in Chiapas: passively managed shade-grown or "natural," transitional organic, intensive organic, and technified, chemical-dependent cultivation. Martínez-Torres concludes that in the conditions of rural Chiapas, where small-scale growers are financially poor and family labor often is under-utilized, intensive organic farming of coffee "bests natural technology in economic terms and is superior to chemical technology in ecological terms, providing the best overall combination of productivity today, plus the likely sustainability of that productivity into the future."

In chapter 6, Carlos Guadarrama-Zugasti develops a typology of growers in order to identify the sources of the heterogeneity found in farm production strategies in central Veracruz. This represents a significant advance over previous typologies, since it enables individual components of the farm's "operational structure" to be analyzed separately. A disaggregated approach is particularly important in distinguishing the environmental impacts of different practices, which farmers typically combine in complex "hybrid" systems of agroecosystem management. As Guadarrama-Zugasti demonstrates, this hybridity can juxtapose practices that enhance ecological sustainability with practices that reduce it. Moreover, such hybridity means that the broad sustainability claims frequently made for traditional shade-grown production must be strongly qualified. This finding has significant policy implications, particularly since many shade-grown coffee farms are currently engaged in replanting, which typically involves intensive use of chemicals. Furthermore, Guadarrama-Zugasti suggests that this process of intensification has largely escaped notice, despite its adverse effects on soil quality and Mexico's competitive position in world coffee markets.

In chapter 7, Christopher Bacon formulates a livelihood vulnerability framework to evaluate the significance of access to Fair Trade, organic, and specialty coffee markets for members of producer cooperatives in northern Nicaragua. His survey results clearly demonstrate that households in cooperatives participating in certified markets are significantly less vulnerable to low coffee prices than members of cooperatives whose sales are directed exclusively into conventional marketing channels. Bacon's analysis also suggests that a livelihood vulnerability approach is helpful in articulating more integrated and diversified responses to the current crisis in rural Nicaragua.

A focus on how household livelihood strategies and agroecological practices are interwoven also distinguishes chapter 8, in which Silke Mason Westphal analyzes coffee agroforestry in the populous Meseta region of western Nicaragua. Drawing on rich household survey data, Westphal investigates the recent convergence on shade-grown coffee systems by two groups of small growers from radically different social backgrounds. The first group, originally landless wage workers and now known as *parceleros,* emerged in the 1990s as beneficiaries of the re-distribution of land on large plantations previously expropriated and collectivized under the Sandinistas. The second group comprises growers who acquired land through inheritance or purchase before the Sandinista Revolution. The common catalyst of change for these groups was the collapse of agricultural modernization policies after 1990, which led to a wholesale withdrawal of subsidies on rural credit and purchased inputs (notably chemicals, agricultural price supports, and rural development services). Westphal's principal finding is that the convergence on shade-grown management is not an unreflexive, passive return to an idealized "traditional" system, but rather an original, adaptive diversification of these practices to meet changing household needs. Her analysis exemplifies how poor rural households integrate livelihood strategies and agroecological decisions in responding to individual circumstances and the manner of their insertion in much wider political economies.

In chapter 9, V. Ernesto Méndez brings socio-ecological and participatory action-research methods to bear to investigate the rationale of shaded coffee management and its potential for biodiversity conservation in El Salvador, a country whose "resource base has reached a critical level of degradation." His household and farm survey research of cooperative members in the municipality of Tacuba in western El Salvador documents that shade trees, whether grown for fruit, timber, or firewood, are significant sources of alternative income, and suggests that species richness and abundance result directly from their integration in family livelihood strategies. Although native tree richness and abundance vary between members of the three different types of cooperative analyzed in this study, these households, taken together, currently grow at least 88 tree species. However, Méndez cautions that tree composition differs very significantly between the cooperatives and the nearby forest. He also notes that tree species valued by coffee farmers may not be important for conservation purposes. Reflecting on his experience in Tacuba, Méndez suggests that action research can help to better inform the aims of producer organizations and provide access to new networks.

The chapters in part III examine the institutional structures, regulatory procedures, and management systems established to certify differentiated specialty coffees and the complex processes involved in delivering certified coffees into the cups and mugs of Northern consumers.

As David Bray and his colleagues lucidly relate in their case study of certified organic coffee in Chiapas (chapter 10), La Selva Ejido Union, like many other strong peasant groups, benefited during the 1980s from "a shifting political and social matrix that led to the relative empowerment of small producers organizations," which subsequently propelled Mexico to the forefront of organic coffee production. Situating the production decisions of La Selva producers within public-and-private-goods framework, Bray et al. present a careful analysis of the potential and limitations of certified organic coffee as "a vehicle for both social justice and biodiversity conservation." In this regard, they conclude that "strategies that go far beyond certification and consumer consciousness campaigns are necessary." They also make the salutary point that shade-tree organic coffee is only one component of peasant agroecosystems, emphasizing that the continuing and unsustainable clearance of forest for corn fields (*milpa*) is "a major priority for habitat conservation in the region."

In chapter 11, Tad Mutersbaugh analyzes the paradoxes and contradictions that arise in organic coffee certification processes, primarily as a result of the tense juxtaposition of two disparate cultural milieus—that is, the transnational certification agencies based in the northern hemisphere and the peasant-inspectors appointed to unpaid (*cargo*) labor service positions in indigenous villages in Oaxaca, who must reconcile transnational certification standards with local "agri-cultural" practices. Mutersbaugh's fine-grained participant-observation methods get "inside" the certification process, detailing the complex management technologies and practices of the "audit chain" that runs from international certification agencies down to village-level and farm-level inspections. Mutersbaugh draws on recent labor-process theory to situate his case study, and he concludes by suggesting that product certification can be regarded as a form of workplace surveillance.

In chapter 12, Sasha Courville focuses on recent developments in the global regulation of organic and social certification standards. Tracing the institutional trajectory of such "pioneer" certification organizations as the International Federation of Organic Agriculture Movements (IFOAM) and Fairtrade Labelling Organizations International, she examines the implications of "a new generation of sustainability initiatives [that] has recently emerged, particularly in the coffee sector." The responses of the "pioneers" to the challenges represented by standards proliferation, the growing involvement of public-sector regulatory agencies, and the costs of certification for producer groups are considered in detail, with emphasis on the formation in 1999 of the International Social and Environmental Accreditation and Labeling Alliance (ISEAL) to participate in multi-stakeholder, multilateral negotiations to harmonize certification processes. Courville welcomes these developments but concludes that "whether the certification pioneers of Fairtrade, organic, and eco-friendly coffee systems [will] move fast enough to address these challenges remains an open question."

In chapter 13, Roberta Jaffe and Christopher Bacon examine the action-research agenda of the Community Action Network (CAN), which was created in 2001 to link communities of small-scale coffee farmers in Mexico and Central America with reflexive consumers and communities of praxis in the United States. This innovative response to the coffee crisis grew out of dissatisfaction with the top-down certification processes that were dominating the governance structures of value chains in Fair Trade and organic coffee networks and their increasing integration in mainstream marketing channels. As Jaffe and Bacon emphasize, CAN represents an explicit effort to move beyond certification markets by "taking certification as a starting point instead of a final goal." CAN espouses a multi-dimensional agenda that integrates several mutually reinforcing mechanisms to build inter-community relationships. In addition to direct coffee marketing, these mechanisms include inter-community modes of knowledge production and transmission operating along both vertical (North-South) and horizontal (South-South) axes to develop more secure livelihoods and ecological sustainability. Jaffe and Bacon vividly describe the organizational and bureaucratic challenges encountered in the genesis of CAN and the imaginative ways in which these were overcome. This hopeful narrative of the role that action researchers can play in embedding market relations in a wider structure of community-to-community networks also examines the possibilities of scaling up as these networks "thicken" and grow. The authors conclude by suggesting that CAN compares creditably with alternative agro-food networks on several dimensions, including empowerment, enhanced livelihoods, and biodiversity conservation.

In chapter 14, Christopher Bacon, Ernesto Méndez, and Jonathan Fox draw together the main empirical findings, thematic foci, and analytical perspectives presented in the preceding chapters. In this synthesis, articulations of the relationship among smallholder livelihoods, coffee communities, and rural landscapes, which is central to the book, are traced across different spatial scales and through diverse networks, notably the sustainability initiatives of certified Fair Trade, organic, and shade-grown coffee production. Bacon et al. briefly consider recent trends and institutional developments that may threaten the capacity of these initiatives to contribute significantly to household livelihoods, biodiversity conservation, and the socio-cultural survival of vibrant coffee-growing communities. Future research agendas, they suggest, should pay careful attention to the implications of further "mainstreaming," to the erosion of niche market premiums when world markets recover, and to the effects of out-migration on land-use patterns, including the eradication of coffee trees and the concentration of farm ownership. The volume lays the groundwork for further explorations of socio-ecological change in Mesoamerica.

Notes

1. As an illustration of the dramatic consequences of this decline, Varangis et al. (2003) report that in the 2000–2001 crop season permanent employment in the Central American coffee sector fell by 350,000 and seasonal employment by 1.7 million.

2. Starbucks purchases 1–2 percent of its coffee from certified Fair Trade growers, equivalent to 2.1 million pounds in 2003. In addition to bags of beans for sale, Fair Trade coffee is featured once a month as Starbucks' "coffee of the day" (Rogers 2004). Starbucks recently committed to purchasing 10 million pounds of certified Fair Trade coffee.

3. Giovannucci and Koekoek (2003) surveyed specialty coffees in twelve Western European countries and found that consumers were confused by the plethora of certification schemes, labels, and terms.

4. These initiatives recall a marketing campaign of the 1960s that featured the fictitious peasant producer Juan Valdez. In an attempt to compete more effectively in the specialty coffee industry, the Colombian Coffee Federation recently opened a chain of Juan Valdez coffee shops in the United States.

5. Ponte (2002) notes that discussions are underway in the International Coffee Organization concerning the adoption of a minimum quality export standard by coffee-producing countries as a means of reducing the global supply.

References

Aranda, J., and C. Morales 2002. Poverty alleviation through participation in Fair Trade coffee: The case of CEPCO, Oaxaca, Mexico. http://www.colostate.edu.

Bacon, C. 2005. Confronting the coffee crisis: Can Fair Trade, organic and specialty coffees reduce small-scale farmer vulnerability in northern Nicaragua? *World Development* 33, no. 3: 497–511.

Fitter, R., and R. Kaplinsky. 2001. Who gains from product rents as the coffee market becomes more differentiated? A value chain analysis. *IDS Bulletin* 32, no. 3: 69–82.

Friedland, W. 1984. Commodity systems analysis: An approach to the sociology of agriculture. *Research in Rural Sociology and Development* 1: 221–236.

Gereffi, G., and M. Korzeniewicz, eds. 1994. *Commodity Chains and Global Capitalism.* Praeger.

Giovannucci, D., and F. Jan Koekoek. 2003. *The State of Sustainable Coffee: A Study of Twelve Major Markets.* IISD, UNCTAD, and ICO.

Goodman, D., and M. Goodman. 2001. Sustaining foods: Organic consumption and the socio-ecological imaginary. In *Exploring Sustainable Consumption*, ed. M. Cohen and J. Murphy. Elsevier Science.

Goodman, M. K. 2004a. Reading fair trade: Political ecology imaginary and the moral economy of fair trade foods. *Political Geography* 230: 891–915.

Goodman, M. K. 2004b. Muddling transparency: The changing consumer politics of place in Fair Trade commodity cultures. Unpublished manuscript, Issues in Environment and Globalization Seminar Series. King's College, London.

Guthman, J. 2002. Commodified meanings, meaningful commodities: Re-thinking production-consumption links through the organic system of provision. *Sociologia Ruralis* 42, no. 4: 295–311.

Guthman, J. 2004. *Agrarian Dreams: The Paradox of Organic Farming in California.* University of California Press.

Institute of Development Studies. 2001. The Value of Value Chains. Special issue of *IDS Bulletin* (32, no. 3).

Jaffee, D., J. Kloppenburg Jr., and M. Monroy. 2004. Bringing the "moral charge" home: Fair Trade within the North and within the South. *Rural Sociology* 69, no. 2: 169–196.

Kaplinsky, R. 2000. Spreading the Gains from Globalization: What Can Be Learned from Value Chain Analysis? IDS Working Paper 110.

Lewin, B., D. Giovannucci, and P. Varangis. 2004. Coffee Markets: New Paradigms in Global Supply and Demand. World Bank Agriculture and Rural Development Discussion Paper 3.

Méndez, V. E. 2004. Traditional Shade, Rural Livelihoods, and Conservation in Small Coffee Farms and Cooperatives of Western El Salvador. Ph.D. thesis, University of California, Santa Cruz.

Murray, D., L. Raynolds, and P. Taylor. 2003. One Cup at a Time: Poverty Alleviation and Fair Trade in Latin America. Fair Trade Research Group, Colorado State University. http://www.colostate.edu.

Perfecto, E., A. Rice, R. Greenberg, and M. Van Der Voort. 1996. Shade coffee: A disappearing refuge for biodiversity. *BioScience* 46: 598–608.

Ponte, S. 2002. The "Latte Revolution"? Regulation, markets and consumption in the global coffee chain. *World Development* 30, no. 7: 1099–1122.

Ponte, S., and B. Daviron. 2005. *The Coffee Paradox: Commodity Trade and the Elusive Promise of Development.* Zed Books.

Raikes, P., M.F. Jensen, and S. Ponte. 2000. Global commodity chain analysis and the French filière approach. *Economy and Society* 29, no. 3: 390–417.

Raynolds, L., D. Murray and P. Taylor. 2004. Fair Trade coffee: Building producer capacity via global networks. *Journal of International Development* 16: 1109–1121.

Rice, P. 2002. Fair Trade: A more accurate assessment. Chazen Web Journal of International Business (www.gsb.colombia.edu).

Rogers, T. 2004. Small coffee brewers try to redefine fair trade. *Christian Science Monitor*, April 13.

Talbot, J. M. 1997. Where does your coffee dollar go? The division of income and surplus along the coffee commodity chain. *Studies in Comparative International Development* 32, no. 1: 56–91.

VanderHoff Boersma, F. 2002. Poverty alleviation through participation in Fair Trade coffee networks: The case of UCIRI, Oaxaca, Mexico. http://www.colostate.edu.

van der Ploeg, J. D. 1994. Agricultural production and employment: Differential practices and perspectives. In *The Functioning of Economy and Labour Market in a Peripheral Region: The Case of Friesland*, ed. C. Verhaar and P. de Klaver. Fryske Academy.

van Dijk, J. B., D. van Doesburg, A. Heijbroek, M. Wazir, and G. de Wolff. 1998. *The World Coffee Market*. Rabobank International.

Varangis, P., P. Siegel, D. Giavannucci, and B. Lewin. 2003. Dealing with the Coffee Crisis in Central America: Impacts and Strategies. Policy Research Working Paper 2993, Development Research Group, World Bank.

Whatmore, S. 1997. Dissecting the autonomous self: Hybrid cartographies for a relational ethics. *Environment and Planning D* 15: 21–37.

2

Agroecological Foundations for Designing Sustainable Coffee Agroecosystems

Stephen R. Gliessman

Achieving sustainability in coffee farming systems is more than just focusing on the economic activities designed to produce a crop or making as large a profit as possible on the coffee that is produced. A coffee farmer can no longer only pay attention to the economic objectives and goals for his or her coffee farm and expect to adequately deal with the concerns of long-term sustainability. As in other sectors of the food system, our knowledge about the design and management of sustainable agriculture must go far beyond what happens within the fences of any individual coffee farm. A sustainable coffee agroecosystem must be viewed as part of a much larger system with many interacting parts, including environmental, economic, and social components (Gliessman 2001; Flora 2001; Francis et al. 2003). It is the complex interactions and balance among all of these parts that have brought us together in this book to discuss coffee agroecosystem sustainability, to determine how we move toward this broader goal, and to demonstrate how an agroecological perspective focused on sustainable agroecosystems is an essential foundation for confronting the recent crisis affecting coffee-growing regions of the world.

Much of modern agriculture has lost the balance needed for long-term sustainability (Kimbrell 2002). With their excessive dependence on fossil fuels and external inputs, most industrialized agroecosystems are overusing and degrading the soil, water, genetic, and cultural resources upon which agriculture has always relied. Problems in sustaining agriculture's natural resource foundation can only be masked for so long by modern practices and high input technologies. In a sense, as we borrow ever-increasing amounts of water and fossil fuel resources from future generations, the negative impacts on farms and farming communities will continue to become more evident. Thus, the conversion to sustainable agroecosystems must become our goal (Gliessman 2001).

In an attempt to clarify our thinking about coffee agroecosystems, we can think of coffee farms located on a landscape through which small streams run. These small streams come together to eventually form rivers that drain a region. In coffee land-scapes, coffee farms form part of a mosaic of land use and conservation activities that can impact, as well as be impacted by, what happens at different points along any one of the streams. If we expand this analogy, and think of an individual coffee farm as a "pool" in a calm eddy at some bend in a stream's flow, we can imagine how many things "flow" into the farm, and we also expect that many things will "flow" out of it as well. Each coffee farmer works hard to keep their own pool in the stream (their farm) clean and productive. They try to be as careful as possible in how they care for the soil, the shade trees, the level of diversity maintained, how pests and diseases are controlled, and how to market the harvest. Back in the days when coffee was grown mostly on small farms, when there was less coffee on the market and less concentration of the coffee commodity chain, farmers could keep their farms in pretty good shape. They could keep their "pools" in the stream pretty clean, and they did not have to worry very much about what was going on "down-stream" from their farms.

But such a strategy has become much more difficult. Each coffee farmer finds that he has less and less control over what comes into his "pool." Farmers face a variety of "upstream impacts" that in combination can threaten the sustainability of the individual farm. This includes the inputs into the farm that are purchased or that arrive from the surrounding area. These could include pesticides or fertilizers that are applied, the quality of the water that reaches the farm, soil cultivation practices in the surrounding region, the cost and availability of labor, market access for the coffee and other products that are produced, legislated policies that determine if trees are planted, harvested, or simply cut down, not to mention the vagaries of the weather! An individual "pool" in the stream can become quickly muddied.

Each coffee farmer must also increasingly consider that the way he cares for his "pool" can have "downstream effects" in the watershed below. Soil erosion and groundwater depletion can negatively affect other farms. Inappropriate or inefficient use of pesticides and fertilizers can contaminate the water and the air and can leave potentially harmful residues on the coffee or other products that the farm family and others will consume. How well each individual farm is managed is reflected regionally in the viability of the rural farm economy, the health of the local commu-nity, and general cultural reproducibility. Among the indicators of these conditions are the loss of coffee land to other activities, such as pasture or annual crops, and the loss of people to farming in general as they migrate to other locations for work or abandon the farm completely. Upstream and downstream factors are linked in complex ways, often beyond a farmer's control, and they weigh heavily on the sus-tainability of each farm.

The Agroecology Perspective

The Coffee Agroecosystem

Any definition of sustainable agriculture must include how we examine the production system as an agroecosystem. We need to look at the entire system, or the entire "stream" using the analogy introduced above. This definition must move beyond the narrow view of agriculture that focuses primarily on the development of practices or technologies designed to increase yields and improve profit margins. These practices and technologies must be evaluated on their contributions to the overall sustainability of the farm system. The new technologies have little hope of contributing to sustainability unless the longer-term, more complex impacts of the entire agricultural system are included in the evaluation. The agricultural system is an important component of the larger food system (Francis et al. 2003).

A primary foundation of agroecology is the concept of the ecosystem, defined as a functional system of complementary relations between living organisms and their environment, delimited by arbitrarily chosen boundaries, which in space and time appears to maintain a steady yet dynamic equilibrium (Gliessman 1998; Odum and Barrett 2005). Such an equilibrium can be considered to be sustainable in a definitive sense. A well-developed, mature natural ecosystem is relatively stable, is self-sustaining, recovers from disturbance, adapts to change, and is able to maintain productivity using energy inputs of solar radiation alone. When we expand the ecosystem concept to agriculture, and consider farm systems as agroecosystems, we have a basis for looking beyond a primary focus on the common measure of system outputs (yield or economic return). We can instead look at the complex set of biological, physical, chemical, ecological, and cultural interactions determining the processes that permit us to achieve and sustain yields. This is especially the case for coffee agroecosystems that have traditionally retained more of their natural ecosystem structure.

Agroecosystems are often more difficult to study than natural ecosystems because they are further complicated by human management that alters normal ecosystem structures and functions. There is no disputing the fact that for any agroecosystem to be fully sustainable a broad series of interacting ecological, economic, and social factors and processes must be taken into account. Still, ecological sustainability is the foundation upon which other elements of sustainability stand.

A coffee agroecosystem is created when human manipulation and alteration of the pre-existing ecosystem takes place for the purpose of establishing coffee production. This introduces several changes in the structure and function of the natural ecosystem (Gliessman 2004b), and as a result it changes some important system-level qualities. These qualities are often referred to as the emergent qualities or properties of systems—aspects that manifest themselves once all of the component parts of the system are organized. These same qualities can also serve as indicators of agroecosystem sustainability (Gliessman 2001). Some of the emergent qualities of ecosystems, and how they are altered as they are converted to coffee agroecosystems, are as follows:

Energy Flow   Energy flows through a natural ecosystem as a result of complex sets of trophic interactions and relationships, with certain amounts being dissipated as metabolic heat at different stages along the food chain, and with the greatest amount of energy within the system ultimately moving along the detritus pathway (Odum and Barrett 2005). Annual production of any system can be calculated in terms of net primary productivity or biomass, each component with its corresponding energy content. Energy flow in agroecosystems is altered greatly by human interference (Pimentel and Pimentel 1997). Although solar radiation is obviously the major source of energy, many inputs are derived from human-manufactured industrial sources and are most often not self-sustaining. The herbicides used to control weeds in a coffee system without shade is an example, where the shade from trees associated with coffee provided the same service previously. Coffee agroecosystems too often become "through-flow" systems, with a high level of fossil fuel input and considerable energy directed out of the system at the time of each coffee harvest. Biomass is not allowed to otherwise accumulate within the system or contribute to driving important internal ecosystem processes (e.g. organic detritus from the leaf litter of shade trees returned to the soil serving as an energy source for microorganisms that are essential for efficient nutrient cycling). For sustainability to be attained, non-renewable energy must be replaced by renewable sources of energy, and solar energy must be maximized to fuel the essential internal trophic interactions needed to maintain other ecosystem functions.

Nutrient Cycling   Small amounts of nutrients continually enter an ecosystem through several hydrogeochemical processes. Through complex sets of interconnected cycles, these nutrients then circulate within the ecosystem, where they are most often bound in organic matter (Odum and Barrett 2005). Biological components (the biodiversity) of each system become very important in determining how efficiently nutrients move, ensuring that minimal amounts are lost from the system. In a mature ecosystem, these small losses are replaced by local inputs, maintaining a nutrient balance. Biomass productivity in natural ecosystems is linked very closely to the annual rates at which nutrients are able to be recycled. In a simplified sungrown coffee agroecosystem, recycling of nutrients can be minimal, and considerable quantities are lost from the system with each harvest or as a result of leaching or erosion due to a great reduction in permanent biomass levels held within the system (Hamel et al. 2004). The frequent exposure of bare soil between coffee plants creates "leaks" of nutrients from the system. Modern sun-grown coffee has come to rely heavily upon nutrient inputs derived or obtained from petroleum-based sources to replace these losses. Sustainability requires that these "leaks" be reduced to a minimum and recycling mechanisms be reintroduced and strengthened.

Population-Regulating Mechanisms   Through a complex combination of biotic interactions and limits set by the availability of physical resources, population levels of the various organisms are controlled, and thus eventually link to and determine the productivity of the ecosystem. Selection through time tends toward the establishment of the most complex structure biologically possible within the limits set by the environment, permitting the establishment of diverse trophic interactions and niche diversification. Owing to human-directed genetic selection, as well as the overall simplification of many coffee agroecosystems (i.e. the loss of species diversity and a reduction in trophic interactions), populations of coffee plants are rarely self-reproducing or self-regulating. Human inputs in the form of replacement coffee plants or pest-control agents, often dependent on large energy subsidies, determine population sizes. Biological diversity is reduced, natural pest-control systems are disrupted, and many niches or microhabitats are left unoccupied. The danger of catastrophic pest or disease outbreak is high, often despite the availability of intensive human interference and inputs. A focus on sustainability requires the reintroduction of the diverse structures and species relationships that permit the functioning of natural control and regulation mechanisms. We must learn to work with and profit from diversity, rather than focus on coffee agroecosystem simplification.

Dynamic Equilibrium   The species richness or diversity of mature ecosystems permits a degree of resistance to all but very damaging perturbations. In many cases, periodic disturbances ensure the highest diversity, and even highest productivity (Connell 1978). System stability is not a steady state, but rather a dynamic and highly fluctuating one that permits ecosystem recovery after disturbance. This promotes the establishment of an ecological equilibrium that functions on the basis of sustained resource use which the ecosystem can maintain indefinitely, or can even shift if the environment changes. At the same time, rarely do we witness what might be considered large-scale disease outbreaks in healthy, balanced ecosystems. But owing to the reduction of natural structural and functional diversity, much of the resilience of a traditional shade-grown coffee system is lost, and constant human-derived external inputs must be maintained. An over-emphasis on maximizing harvest outputs upsets the former equilibrium, and can only be maintained if such outside interference continues. To reintegrate sustainability, the emergent qualities of system resistance and resiliency must once again play a determining role in coffee agroecosystem design and management.

We need to be able to analyze both the immediate and the future effects of coffee agroecosystem design and management so that we can identify the key points in such systems on which to focus the search for alternatives or solutions to problems. We must learn to be more competent in our agroecological analysis in order to prevent problems or negative changes, rather than struggle to reverse them after they have been created. The agroecological approach provides us with such a foundation (Altieri 1995; Gliessman 1998, 2004a).

Applying Agroecology

The process of understanding the basis for coffee agroecosystem sustainability has its foundations in two kinds of ecosystems: natural ecosystems where coffee is grown and traditional (also known as local or indigenous) coffee agroecosystems. Both provide ample evidence of having passed the test of time in terms of long-term productive ability, but each offers a different knowledge base from which to understand this ability. Natural ecosystems are a reference system for understanding the ecological basis for sustainability in a particular location. Traditional agroecosystems provide many examples of how a local culture and its local environment have co-evolved over time through processes that balance the needs of people, expressed as ecological, technological, and socio-economic factors. Agroecology, defined as the application of ecological concepts and principles to the design and management of sustainable agroecosystems (Gliessman 1998), draws on both to become a research approach that can be applied to converting unsustainable and conventional coffee agroecosystems to sustainable ones.

Natural ecosystems reflect a long period of evolution of local resources and adaptation to local ecological conditions. They have become complex sets of plants and animals that co-inhabit in a given environment, and as a result, provide extremely useful information for the design of more locally adapted agroecosystems. "The greater the structural and functional similarity of an agroecosystem to the natural ecosystems in its biogeographical region, the greater the likelihood that the agroecosystem will be sustainable." (Gliessman 1998) If this suggestion holds true, natural ecosystem structures and functions can be used as benchmarks or threshold values for measuring the sustainability of coffee agroecosystems. By better understanding the forested ecosystems most commonly found naturally in most coffee-growing regions, we can build an important foundation for sustainable coffee agroecosystems that conserve and protect the environment and natural resources. In Mexico, for example, Moguel and Toledo (1999) found that coffee is predominantly grown in the biogeographically and ecologically important altitudinal belt between 600 and 1,200 meters elevation. Tropical and temperate vegetation types overlap, with rain forest merging into cloud forest on Atlantic slopes and tropical dry forests merging into pine-oak forests on Pacific slopes. Many of the coffee-growing regions of Mexico are considered biodiversity "hot spots" for conservation, and because of the mountainous landscape and the rainfall patterns they are also important for maintaining important watershed processes (Méndez 2004). This knowledge would imply that coffee agroecosystems that maintain such diversity and ecosystem processes should be of considerable value (Méndez et al. 2002; Rosa et al. 2003).

Traditional coffee agroecosystems are different from modern-day conventional systems in that they developed in times, in places, or under production conditions where inputs other than human labor and local resources were generally not available or desirable to the local farmers. Production takes place in ways that respond to household and community needs to consider long-term sustainability, rather than solely maximizing output of a single crop and/or net profit. Multiple criteria are used to design and manage the complexity of diverse systems that include multiple crops, rotations, trees, and animals. Two traditional coffee systems are encountered in Mexico and Central America, as classified by Moguel and Toledo (1999). One is called "rustic" coffee, where mostly indigenous and local communities substitute the forest understory with coffee, without any major change in the forest overstory. Within these systems natural forest watershed processes continue, and much of the forest's original biodiversity is maintained and sometimes enhanced (Méndez 2004). The management is usually much less intensive (although it can be ecologically intensive if the farmers are actively managing soil fertility with organic amendments or using biological pest control), no agrochemicals are applied, and coffee yields are markedly lower.

The other traditional system found in Mexico and elsewhere is called "traditional polyculture." This results in a shaded coffee plantation that was probably initially introduced under the native forest canopy, but with time and considerable manipulation, removal, and replacement of the native tree species, the agroecosystem takes on the appearance of a "coffee garden." Wild and domestic species are mixed together, with coffee the predominant shrub layer, but with many adjustments to the tree and herb layers that create considerable vegetational and architectural complexity, as well as the greatest "useful diversity." Use of agrochemicals is negligible in most cases. Coffee may remain the most important economic component of this traditional system, but a considerable array of other products are grown for the market and for local subsistence. Traditional coffee agroecosystems continue to be important as alternative or complementary producers of food for a large part of the populations of many coffee-growing countries, especially in times of crisis like those that are addressed in this book. With so much coffee still being grown by small-scale farmers and their families in diverse polyculture "agroforests," it is important to understand the ecological foundations upon which these systems are designed (Altieri 1990). With their diverse shade-tree species and with multiple-use strategies, traditional coffee systems are actually very sophisticated examples of the application of ecological knowledge, and can serve as the starting point for the conversion to more sustainable agroecosystems in the future.

Traditional coffee agroecosystems dependent on the maintenance of native tree cover for shade have been replaced in many parts of Latin America by what Moguel and Toledo (1999) refer to as "commercial polyculture." The original forest canopy species are removed and replaced with tree species more specifically adapted to use with coffee production. Tree species are selected for their particular shade-producing abilities, adaptability to pruning, for being leguminous nitrogen contributors to the soil, or having some commercial use or marketable product. Much of the original forest structure is possible with this heavily managed system, with both native and non-native trees being employed (Méndez, this volume), and multiple shrub-like or herbaceous species (such as citrus, bananas, and other cash crops) add understory diversity. The contributions of such heavily managed systems to maintaining biodiversity and watershed services, especially when they are regularly and heavily pruned, can be limited, and even more so when intensive agrochemical use is included (Méndez et al. 2002). However, the diverse tree canopy can still provide considerable value as habitat for plant and animal species (Perfecto et al. 1996).

From an agroecological perspective, when sustainability is a criterion in the design and management of coffee agroecosystems, trees growing over coffee must provide much more than shade. Monoculture coffee with one or a few shade trees subject to intensive pruning loses most of the ecological advantages of diverse shade, either from native species or planted species, and must almost obligatorily apply agrochemicals. This calls for a re-classification of what is considered for coffee to be shade-grown, taking into account contributions to biodiversity and watershed process.

How can agroecology link our understanding of natural ecosystem structure and function with the knowledge inherent in traditional coffee agroecosystems? On the one hand, the knowledge of place that comes from understanding local ecology is an essential foundation. Another is the local experience with coffee growing that has its roots in many generations' living and working within the limits of a particular landscape. We put both of these approaches together when we work with coffee farmers going through the transition process to more environmentally sound management practices, and thus realize the potential for contributing to long-term sustainability. This transition is already occurring. Many farmers have abandoned environmentally degrading and agrochemical dependent sun-grown coffee without shade trees. Despite the heavy economic and social pressure on coffee-growing regions, growers are converting their coffee to more sustainable design and management (Guadarrama-Zugasti 2000). Shade is being reintroduced, and diversity is beginning to reappear. In Latin America the dramatic increase in organic acreage for coffee has been based largely on farmer innovation (Bray et al. 2002). It is incumbent on agroecologists to take a more active role in this conversion process.

Converting an unsustainable coffee agroecosystem to a more sustainable design is a complex process. It is not just the adoption of a new practice or a new technology, and there is no single solution. Instead, this conversion uses the agroecological approach described above and the principles listed in table 2.1. The coffee farm is perceived as part of a larger system of interacting parts—an agroecosystem within a landscape. We must focus on redesigning that system in order to promote the

Table 2.1
Guiding principles for the process of conversion to sustainable coffee agroecosystems design and management. Modified from Gliessman 2004a.

---

1. Shift from through-flow nutrient management to recycling of nutrients, with increased dependence on natural processes such as biological nitrogen fixation from legume shade trees and mycorrhizal relationships in the soil.

2. Use renewable sources of energy instead of non-renewable sources, such as firewood from shade tree pruning for coffee drying.

3. Eliminate the use of non renewable off-farm human inputs that have the potential to harm the environment or the health of farmers, farm workers, or consumers.

4. When materials must be added to the coffee plantings, use naturally occurring materials instead of synthetic, manufactured inputs.

5. Manage pests, diseases, and weeds in the coffee system instead of "controlling" them.

6. Reestablish the biological relationships that can occur naturally in the coffee system instead of reducing and simplifying them.

7. Make more appropriate matches between the ecology of the coffee plant and the productive potential and physical limitations of the farm landscape.

8. Use a strategy of adapting the biological and genetic potential of the coffee plant to the ecological conditions of the farm rather than modifying the farm to meet the needs of the coffee crop.

9. Value most highly the overall health of the agroecosystem rather than the outcome of the individual coffee crop.

10. Emphasize conservation of soil, water, energy, and biological resources.

11. Incorporate the idea of long-term sustainability into overall agroecosystem design and management.

---

functioning of an entire range of different ecological processes (Gliessman 1998). In studies of the conversion of conventional coffee growing to organic management, several changes have been observed. When farmers first make the transition from conventional management practices that are generally dependent on agrochemicals to certified organic production, yields often fall dramatically (Bacon 2005; Damiani 2002). Several research experiences point to a recuperation of yields over time (Nigh 1997; Lyngbaek et al. 2001; Bray et al. 2002; Grossman 2003). In El Salvador, a cooperative that had made a transition to organic 13 years earlier reported an initial yield decline of about 30 percent, a recuperation to conventional levels of production at about 6–7 years into the conversion process, and higher yields after 11–12 years of organic management (Méndez et al. 2006).

As the use of synthetic chemical inputs was reduced or eliminated, and as recycling was emphasized, agroecosystem structure and function changed. A range of processes and relationships began to transform, beginning with improvement in basic soil structure, an increase in soil organic matter content, and greater diversity and activity of beneficial soil biota. In Chiapas, Mexico, Grossman (2003) found that organic coffee farmers had an excellent understanding of some soil processes and not of others. For example, they had a good understanding of leaf litter decomposition and mineralization. However, their understanding of the role of moisture influence, nutrient uptake, and soil organisms was not as great. The results of this study indicate that training in organic production has not provided farmers with a complete understanding of soil processes, and there are gaps in knowledge regarding any phenomena that they cannot directly observe. Major changes begin to occur in the activity and relationships among weed, insect, and pathogen populations, and in the functioning of natural control mechanisms (Staver et al. 2001), requiring more complex understanding of the entire agroecosystem. Coffee pest and disease management without agrochemicals requires a deeper understanding of natural population-regulating mechanisms and management practices that ensure that they function (Hillocks et al. 1999).

Ultimately, nutrient dynamics and cycling, energy use efficiency, and overall agroecosystem productivity are affected. Changes may be required in day-to-day management of the farm, which include planning, marketing, and even philosophy. The specific needs of each farm will vary, but the principles for conversion listed in table 2.1 can serve as general guidelines for working through the transition. An agroecological focus can help the coffee farmer evaluate and monitor these changes during the conversion period in order to guide, adjust, and evaluate the conversion process. Such an approach provides an essential framework for determining the requirements for and the indicators of the design and management of a sustainable coffee agroecosystem.

Comparing Ecosystems and Coffee Agroecosystems

The key to moving toward more sustainable coffee farms is building a s[ ]
logical foundation under the agroecosystem, by using the ecosystem knowledge of
agroecology described above. This foundation then serves as the framework for
producing the sustainable harvests needed by farmers, their families, and their communities. In order to maintain sustainable harvests, though, human management is
a requirement. Agroecosystems are not self-sustaining, but rely on natural processes
for maintenance of their productivity. The coffee agroecosystem's resemblance to
natural ecosystems can allow the system to be sustained, in spite of the long-term
human removal of biomass, without large subsidies of non-renewable energy and
without detrimental effects on the surrounding environment.

Table 2.2 compares natural ecosystems with three types of coffee agroecosystems in terms of several ecological criteria. Traditional coffee systems most closely
resemble natural ecosystems, since they most often are focused on the use of locally
available and renewable resources, maintenance of forest structure, local use of
agricultural products, and the return of biomass to the farming system. Sustainable
agroecosystems are very similar in many properties, but due to the more intense
focus on export of harvested coffee to distant markets, the need to purchase a significant part of their nutrients externally, and the much stronger impact of those
market systems on agroecosystem diversity and management, they are more dissimi-
lar in others. Compared to conventional sun-grown monoculture coffee systems,
sustainable coffee agroecosystems have somewhat lower and more variable yields
due to the weather variation that occurs from year to year. Such reductions in yields
can be more than offset, from the perspective of sustainability, through the advantages gained in reduced dependence on external inputs, more reliance on natural
controls of pests, and reduced negative impacts of farming activities off the farm.

Looking toward the Future

Current problems in agriculture create the pressures for the changes which will
bring about a sustainable agriculture. But it is one thing to express the need for
sustainability and yet another to actually quantify it and bring about the changes
that are required. Designing and managing sustainable coffee agroecosystems, as an
approach, is in its formative stages. It builds initially upon the fields of ecology and
agricultural science, and emerges as the science of agroecology. This combination
can play an important role in developing the understanding necessary for a transition to sustainable coffee-production systems.

Table 2.2
Emergent properties of natural ecosystems, as well as traditional, conventional, and sustainable coffee agroecosystems. The agroecosystem properties listed below are most applicable to the farm scale and for the short- to medium-term time frame. Modified from Gliessman 1998.

Emergent ecological property	Natural ecosystem	Coffee agroecosystem type		
		Traditional	Conventional	Sustainable
Productivity (process)	Medium	Medium	Low/medium	Medium/high
Production (yield)	Medium	Low	High	Medium
Species diversity	High	Medium/high	Low	Medium/high
Structural diversity	High	Medium/high	Low	Medium
Functional diversity	High	Medium/high	Low	Medium/high
Output stability	Medium	High	Low/medium	High
Biomass accumulation	High	High	Low	Medium/high
Nutrient recycling	High	High	Low	High
Trophic relationships	High	High	Low	Medium/high
Natural population regulation	High	High	Low	Medium/high
Resistance	High	High	Low	Medium/high
Resilience	High	High	Low	Medium
Dependence on external human inputs	Low	Low	High	Medium
Autonomy	High	High	Low	High
Human displacement of ecological processes	Low	Low	High	Low/medium
Sustainability	High	Medium/high	Low	High

But sustainable coffee is more. It takes on a cultural perspective as the concept expands to include humans and their impact on the environments in which coffee is grown. Agricultural systems are a result of the co-evolution that occurs between culture and environment, and sustainable coffee values the human as well as the ecological components. Our small farm "pool" in the "stream" becomes the focal point for changing how we grow coffee, but that change must occur in the context of the human societies within which coffee is grown, sold, and consumed, the whole stream in this analogy.

Coffee systems can no longer be viewed as strictly production systems driven primarily by economic pressures. We need to reestablish an awareness of the strong ecological foundation upon which coffee production originally developed and ultimately depends. Too little importance has been given to the "downstream" effects that are manifest off the farm, either by surrounding natural ecosystems or by human communities. In this book we present the interdisciplinary frameworks necessary to evaluate these impacts.

In the broader context of sustainability, we must study the environmental background of the agroecosystem, as well as the complex of processes involved in the maintenance of long-term productivity. We must first establish the ecological basis of sustainability in terms of resource use and conservation, including soil, water, genetic resources, and air quality. Then we must examine the interactions among the many organisms of the coffee agroecosystem, beginning with interactions at the individual species level, and culminating at the ecosystem level as our understanding of the dynamics of the entire system is revealed.

Our understanding of ecosystem level processes should then integrate the multiple aspects of the social, economic and political systems within which coffee-production systems function, making them even more complex systems. Not only will such an integration of ecosystem and social system knowledge about coffee lead to a reduction in synthetic inputs used for maintaining productivity; it will also permit the evaluation of such qualities of coffee agroecosystems as the long-term effects of different input/output strategies, the importance of the environmental services provided by coffee landscapes, and the relationship between economic and ecological components of sustainable coffee management. By properly selecting and understanding the "upstream" inputs into coffee systems, we can be ensured that what we send "downstream" will promote a sustainable future.

Acknowledgments

I am extremely grateful for support provided by the Alfred E. Heller Endowed Chair for Agroecology. An earlier version of this chapter focusing on the general applications of agroecology to sustainable agriculture was published in Agronomy Society Monograph 43, edited by Diane Rickerl and Chuck Francis, under the title Agroecosystem Analysis. I greatly appreciate the detailed edits from the editorial team of this book that gave it a clear focus on coffee.

References

Altieri, M. A. 1990. Why study traditional agriculture? In *Agroecology*, ed. C. Carroll, J. Vandermeer, and P. Rosset. McGraw-Hill.

Altieri, M. A. 1995. *Agroecology: The Scientific Basis of Alternative Agriculture*, second edition. Westview.

Bacon, C. 2005. Confronting the Coffee Crisis: Nicaraguan Smallholders Use of Cooperative, Fair Trade and Agroecological Networks to Negotiate Livelihoods and Sustainability. Ph.D. thesis, University of California, Santa Cruz.

Bray, D. B., J. L. Plaza-Sanchez, and E. Contreras-Murphy. 2002. Social dimensions of organic coffee production in Mexico: lessons for eco-labeling initiatives. *Society and Natural Resources* 15, no. 6: 429–446.

Connell, J. H. 1978. Diversity in tropical rain forests and coral reefs. *Science* 199: 1302–1310.

Damiani, O. 2002. Organic agriculture in Guatemala: A study of coffee producer associations in the Cuchumatanes Highlands. Office of Evaluation and Studies, IFAD.

Flora, C., ed. 2001. *Interactions between Agroecosystems and Rural Communities*. CRC Press.

Francis, C., G. Lieblein, S. Gliessman, T. Breland, N. Creamer, R. Harwood, L. Salomonsson, J. Helenius, D. Rickerl, R. Salvador, M. Wiendehoeft, S. Simmons, P. Allen, M. Altieri, J. Porter, C. Flora, and R. Poincelot. 2003. Agroecology: The ecology of food systems. *Journal of Sustainable Agriculture* 22: 99–118.

Gliessman, S. R. 1998. *Agroecology: Ecological Processes in Sustainable Agriculture*. Lewis/CRC Press.

Gliessman, S. R., ed. 2001. *Agroecosystem Sustainability: Toward Practical Strategies*. CRC Press.

Gliessman, S. R. 2004a. Integrating agroecological processes into cropping systems research. *Journal of Crop Improvement* 12: 61–80.

Gliessman, S. R. 2004b. Agroecology and agroecosystems. In *Agroecosystems Analysis*, ed. D. Rickerl and C. Francis. Agronomy Monograph 43.

Grossman, J. M. 2003. Exploring farmer knowledge of soil processes in organic coffee systems of Chiapas, Mexico. *Geoderma* 111: 267–287.

Guadarrama-Zugasti, C. 2000. The Transformation of Coffee Farming in Central Veracruz, Mexico: Sustainable Strategies? Ph.D. thesis, University of California, Santa Cruz.

Hamel, C., C. Landry, A. Elmi, A. Liu, and T. Spedding. 2005. Nutrient dynamics: Utilizing biotic-abiotic interactions for improved management of agricultural soils. In *New Dimensions in Agroecology*, ed. D. Clements and A. Shrethsa. Harworth.

Hillocks, R. J., N. A. Phiri, and D. Overfield. 1999. Coffee pest and disease management options for smallholders in Malawi. *Crop Protection* 18: 199–206.

Kimbrell, A, ed. 2002. *Fatal Harvest: The Tragedy of Industrial Agriculture*. Island.

Lyngbaek, A. E., R. G. Muschler, and F. L. Sinclair. 2001. Productivity and profitability of multistrata organic versus conventional coffee farms in Costa Rica. *Agroforestry Systems* 53: 205–213.

Méndez, V. E. 2004. Traditional Shade, Rural Livelihoods, and Conservation in Small Coffee Farms and Cooperatives of Western El Salvador. Ph.D. thesis, University of California, Santa Cruz.

Méndez, V. E. Farmer livelihoods and biodiversity conservation in a coffee landscape of El Salvador. In this volume.

Méndez, V. E., C. Bacon, S. Petchers, D. Herrador, C. Carranza, L. Trujillo, C. Guadarrama-Zugasti, A. Cordón, and A. Mendoza. 2006. Sustainable Coffee from the Bottom Up: Impacts of Certification Initiatives on Small-Scale Farmer and Estate Worker Households and Communities in Central America and Mexico. Research Report, Oxfam America.

Méndez, V. E., D. Herrador, L. Dimas, M. Escalante, O. Diaz, and M. Garcia. 2002. Cafe con sombra y pago por servicios ambientales: riesgos y oportunidades para impulsar mecanismos con pequeños agricultores de El Salvador. Estudio de Caso para el Proyecto "Pago por Servicios Ambientales en las Américas." PRISMA/Fundacion FORD, San Salvador.

Moguel, P., and V. M. Toledo. 1999. Biodiversity conservation in traditional coffee systems of Mexico. *Conservation Biology* 13: 11–21.

Nigh, R. 1997. Organic agriculture and globalization: A Maya associative corporation in Chiapas, Mexico. *Human Organization* 56, no. 4: 427–436.

Odum, E. P., and G. W. Barrett. 2005. *Fundamentals of Ecology*, fifth edition. Thomson Brooks/Cole.

Perfecto, I. R., R. A. Greenberg, and R. Van der Voort. 1996. Shade coffee: A disappearing refuge for biodiversity. *BioScience* 46, no. 8: 598–609.

Pimentel, D., and M. Pimentel, eds. 1997. *Food, Energy, and Society*, second edition. University of Colorado Press.

Rickerl, D., and C. Francis. 2004. *Agroecosystem Analysis*. American Society of Agronomy Monograph 43.

Rosa, H., S. Kandel, and L. Dimas, with N. Cuellar and E. Méndez. 2003. Compensation for Environmental Services and Rural Communities: Lessons from the Americas and Critical Themes to Strengthen Community Strategies. PRISMA, San Salvador. www.prisma.org.sv.

3

The Roots of the Coffee Crisis

Seth Petchers and Shayna Harris

Though recent improvements in the international price of coffee provide some relief for small-scale, family farmers and coffee farm workers, the dynamics of the coffee market have not shifted in ways that guarantee long-term stability for those at the bottom of the supply chain. At the macro level, a shift from a managed market to a free market, along with increasingly unbalanced supply-chain power that favors multinational buyers and roasters, has created an environment in which small-scale, family farmers cannot compete. At the farm level, lack of access to farm credit, over-dependence on a single source of income, inability to access markets directly, and farmer organization weaknesses exacerbate supply-chain power imbalances and leave farmers vulnerable to the coffee market's cyclical price fluctuation. Oxfam urges the governments in coffee-producing and coffee-consuming countries, along with non-government organizations and the coffee industry, to focus their attention on the coffee sector on three priority areas: sustainability and sustainable pricing, development assistance for small-scale, family farmers, and farmer and farm worker representation in international debate.

The Coffee Crisis: Sustained Disequilibrium in Supply and Demand

Market Restructuring: From Managed to Flooded
In December 2001, the international coffee price crashed to 30-year lows, in real terms the lowest they have been in 100 years (ICO 2003), creating hardships and reducing incomes for millions of coffee farmers around the world. The collapse in the price of coffee was largely a function of oversupply. Production in 2001–02 was estimated at 115 million bags,[1] compared with consumption of 105–106 million bags (FO Licht 2002). Though a predicted shortfall during the 2005 Brazilian harvest could bring supply and demand into closer alignment, it is likely that resulting price increases will trigger increased production that will once again send prices falling.

Three reasons explain how supply and demand can became so misaligned in today's coffee market: the end of the managed market in 1989, major new entrants into the market, and lagging demand in traditional Western markets.

The Breakdown of the Managed Market

The coffee market changed radically in the period 1990–2005. Until 1989, coffee—like most commodities—was traded in a managed market, regulated by the International Coffee Agreement (ICA). Governments in both producing and consuming nations agreed on pre-determined supply levels by setting export quotas for producing countries. The aim was to keep the price of coffee relatively high and relatively stable, within a price band or "corset" ranging from $1.20/pound to $1.40/pound. To prevent oversupply, countries had to agree not to exceed their "fair" share of coffee exports. If, however, prices rose above the ceiling level, producers were permitted to exceed their quotas to meet the surge in demand.

Disagreement between members led to the effective breakdown of the ICA in 1989. Opposition from the United States, which subsequently left as a member, was a major factor. The ICA survives, administered by the International Coffee Organization (ICO), but it has lost its power to regulate the supply of coffee through quotas and the price corset. Prices for coffee are determined on the two big futures markets based in London and New York, with each market trading particular varieties and grades of coffee. The London market is the benchmark for robusta coffee, the New York market for arabica.[2] The price of coffee is influenced by the huge number of contracts for coffee that are traded, which far exceeds the physical amount of coffee that changes hands.

From the perspective of producer countries, the ICA brought a golden era of good and stable prices, compared with the present development disaster. As figure 3.1 shows, though prices fluctuated significantly, from 1965 to 1989, they remained relatively high and rarely fell below the ICA price floor of $1.20/pound. In sharp comparison, once the ICA broke down and the price corset ended in 1989, prices dropped dramatically. Apart from two sharp price spikes in 1995 and 1997 caused by frost ruining the Brazilian crop, prices remained very low, even below the average cost of production. The recent rise in coffee prices by no means signals the end to the coffee crisis, as prices remain unstable because of market booms and busts.

Critics point to many reasons for the ICA's breakdown. There was a struggle to capture larger quotas, and it was difficult for new producers trying to enter the market. Despite agreed quotas, additional volumes leaked out to countries outside the ICA, undermining its intended prices and undermining trust. Some in the industry believe that the price corset laid the ground for overproduction because the

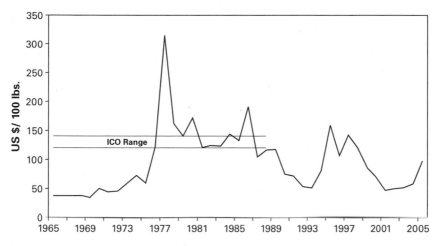

Figure 3.1
Monthly averages of New York Coffee Futures prices for 100-pound sacks, 1965–2005.
Source: ICO 2005.

coffee price was artificially set too high; others argue that the current glut probably owes its origins more to the price hikes of 1994–95 and 1997 than to the high coffee prices of the 1980s.

Proposals to revive the ICA are impeded by the apparent lack of political will to make it work. Consumer nations show no willingness to participate at present, and producer nations may not be willing or able to abide by their own rules. In the absence of consumer country support, producer countries did attempt to limit their own exports, but the initiative collapsed in 2001. The lack of will to revive this approach to managing markets through quotas does not mean that other approaches could not work, especially those that would operate through market mechanisms. In recent years, the ICO proposed one such approach: a scheme to reduce the amount of coffee traded on grounds of quality.[3] However, owing to objections from consumer countries over so-called market intervention, this initiative is no longer on the table.

Enter the Giants: Brazil and Vietnam
Brazil and Vietnam have reshaped the world's coffee supply. Before 1990, Vietnam was barely a statistical blip in the coffee world, producing just 1.5 million bags. Its agricultural economy was opened to the world market during the 1990s, with the government providing irrigated land and subsidies to encourage resettlement by farmers into coffee production. By 2000, it had become the second-largest producer

in the world, with 15 million bags to its name, largely produced on small farm holdings. Brazil, on the other hand, is not a newcomer: it has long been the world's largest producer, but production has recently been boosted by changes in how and where coffee is grown. Increased mechanization, intense production methods, and a geographical shift away from the traditional, frost-prone growing areas have all increased yields.

Though production levels in Brazil will rise and fall over time, the country's ability to produce profitably even at low world prices renders growers in many other regions powerless in the market. As a result of Brazil's unprecedented production capacity and efficiency, the impact for traditional coffee-producing countries is serious. Patrick Installe, the Managing Director of Efico, a green coffee trader, elaborates: "To give you an idea of the difference, in some areas of Guatemala, it could take over 1,000 people working one day each to fill the equivalent of one container of 275 bags, each bag weighing 69 [kilograms]. In the Brazilian *cerrado*, you need five people and a mechanical harvester for two or three days to fill a container. One drives, and the others pick. How can Central American family farms compete against that?" (Installe 2002)

What were the triggers for the jump in world coffee production and the resulting oversupply? Freak price hikes in 1994, 1995, and 1997, due to frosts in Brazil, certainly encouraged countries, and their farmers, into the market. But other factors were also at play in producer countries. National policies, new technologies, and currency movements were also important influences.

Lagging Demand
The United States, Germany, France, and Japan collectively consume over half of world coffee exports (ICO 2005). While coffee production has grown rapidly, demand for coffee in the developed world has seen sluggish growth. The big coffee companies spend millions of dollars on advertising each year, but they have failed to stop rich consumers turning to alternative drinks. Figure 3.2 compares coffee consumption with the growth of soft drink intake in the United States, the world's largest consumer market. This is not a worldwide picture, however. Nestlé, whose share of the US market is relatively small, states that it boosted consumption of Nescafé by 40 percent in the 1990s.

The combination of oversupply, increased production, and lagging demand has created a severely imbalanced market which cannot simply be left to its own devices if supply and demand are to be brought back into line in the long term. The human toll of such an approach is unacceptable: the market makes no suggestions as to what farming families are supposed to live on while waiting through cyclical price lows for the market to "clear."

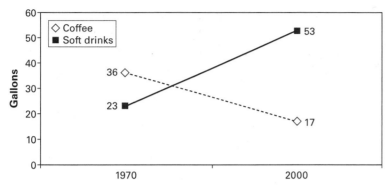

Figure 3.2
Gallons of coffee and soft drinks consumed annually per capita in the United States, 1970–2001. Sources: US Department of Agriculture, Davenport and Company. Lines represent trends, not data points, between 1970 and 2001.

The Coffee Crisis Today: David vs. Goliath

All around the world, small-scale farmers face stiff competition from large agribusiness. In the case of coffee and other tropical commodities, an increasing portion of crop value is being captured by transnational firms, many of which not only control collection and processing infrastructure in the countries where crops are grown but also control brands and distribution networks in countries where end products are sold to consumers. As the global trading system consolidates, increased pressure is felt at the farm level. The "bigger, faster, cheaper" mentality has created a supply-chain dynamic that exploits the most vulnerable at the bottom of the supply chain, including small scale coffee farmers and farm workers. Regardless of price swings and market stagnation, transnational companies are increasing their control of the supply chain, squeezing out profits from producing countries and passing on costs to consumers. As a result, family farmers and farm workers are losing out as the gains from globalization shift to the top. If this persists, trade will continue to be a negative force in poverty reduction for export-oriented economies.

Manifestation of Power Imbalances in the Market: Penniless Farmers, Profiting Roasters

The fundamental changes in the market structure, along with the challenges that face small-scale farmers and farm workers, have led to a situation in which the profits from the coffee trade are disproportionately reaped by the coffee industry in the global North. While the coffee crisis has been devastating communities in coffee-producing countries, coffee has been a bonanza market for the transnational roaster companies. Far from getting a fair share of its profitability, producer countries have collectively been receiving a smaller and smaller share of the market's value:

• In the 1990s, producer countries went from earning $10–12 billion from a coffee market worth around $30 billion, to receive less than $5.5 billion of export earnings from a market that is worth more than $70 billion (ICO 2002). That represents a drop in their share from over 30 percent of the market to under 10 percent (Ponte 2004).

• Today coffee farmers receive 2 percent or less of the price of a cup of coffee sold in a coffee bar. They receive roughly 6 percent of the value of a standard pack of ground coffee sold in a grocery store.

The following numbers illustrate how marginal the actual coffee beans have become to the whole business of selling the beverage to consumers. In 1984, green bean costs constituted 64 percent of the US retail price. When prices crashed in 2001, the raw material price as a proportion of the final retail value had fallen to 18 percent. In 2004, growers received on average 17 percent of the final retail value coffee—and as low as 9 percent for South American producers of robustas.[4] Some markets may be giving consumers a better deal than others, but in all of them the importance of coffee beans to the final retail price has fallen.

Increasing Presence of Transnationals in Producing Countries

Farmers face a whole series of challenges, starting with the volatile nature of the international price for coffee. However, some farmers interviewed for an Oxfam survey also complained of having to accept the price and terms of trade offered by the trader and of having very little or any power to negotiate. Often, local buyers behave dishonestly. Farmers in Peru, for example, reported being short-changed: "We see that the coffee is dry, but the buyers say 'give us a discount.' . . . I don't know what grade it is, but I think they are taking advantage of us because they know we have to sell to them," says Carmela Rodriguez of Sauce, Peru (2002).

International traders are increasingly active in producer countries. Roaster companies rely on these traders to supply very large volumes of coffee from diverse origins at short notice. This has changed the way traders work, according to Stefano Ponte (2001): "The need to guarantee the constant supply of a variety of origins and coffee types has prompted international traders to get even more involved in producing countries than they would have done simply as a result of market liberalization."

With fewer restrictions on foreign investment, international traders have either established local subsidiaries or now deal directly with producers—in rare cases they own the coffee farms themselves. Some of these traders have very close links with the major roasters. This shift is confirmed by Lorenzo Castillo (2002) of Peru's

Junta Nacional de Café: "The transnational companies want to reduce costs. To do so they are seeking to reduce intermediaries between themselves and the producer."

Local millers, middlemen, and even larger domestic trading companies are struggling to stay afloat as they do not have the financial resources of the international trading companies. In Uganda, for example, the number of exporters has shrunk since the mid 1990s from 150 to 20, according to one European trader, and many of those who have lost out have been local entrepreneurs. This has left a gap that the better capitalized and stronger international traders have stepped into. In Tanzania, transnationals now "control more than half of the export market through direct subsidiaries, and another substantial proportion through finance agreements with local companies" (Ponte 2001). The concern is that the crisis is undermining an important local entrepreneurial base, while the profits generated for the international trading companies flow back to industrialized countries.

Most Value Is Added in Consuming Countries

Even though traders squeeze extra margins for themselves out of farmers, the real margins in the market are made after export by coffee roasting companies and marketers. In sharp contrast with losses, or at best tiny margins, made by farmers and exporters in developing countries, the roaster companies in the United States and Europe are making extraordinary profits on their retail coffee business.

As the price of coffee reached its low point in 2001, Oxfam interviewed many players in the supply chain in Uganda to trace the rising price of coffee beans as they made their journey from the farmer's trees to the cans sitting on supermarket shelves in the United Kingdom. Oxfam found that in this case the farmer got just 2.5 percent of the retail price of the coffee. In the United States, the figure would be 4.5 percent of the retail price. Beyond the story in Uganda, Oxfam commissioned a consultant to construct an indicative value chain to try to assess what percentage of the end value farmers were getting in different countries around the world. The research found that farmers of the cheapest type of coffee, doing no processing to their coffee cherries, are getting an average of just 6.5 percent of the final retail value of that same coffee. This value chain uses official price data, where available, weighted to take into account different market shares.[5] Even this figure is probably an overestimate, since official data on prices to producers may overstate what farmers actually get.

In 1997, an analyst report on Nestlé's soluble business concluded: "Martin Luther used to wonder what people actually do in heaven. For most participants in the intensely competitive food manufacturing industry, contemplation of Nestlé's soluble coffee business must seem like the commercial equivalent of Luther's spiritual meditation." (Deutsche Bank 2000) Referring to Nestlé's market share, size of sales, and operating profit margins, the same author said: "Nothing else in food and beverages is remotely as good. The report estimates that, on average, Nestlé makes 26 cents of profit for every $1 of instant coffee sold. (ibid.) Another analyst believes that margins[6] for Nestlé's soluble businesses worldwide are higher, closer to 30 percent. For Nestlé, the rich markets of the United Kingdom and Japan are particularly profitable.

Roast and ground coffee is less profitable than soluble, but the profits are still enviable. In 2002, Sara Lee, buffeted by competition in the US market, still managed a more than decent operating profit margin of nearly 17 percent for its beverages unit,[7] which deals mostly in coffee.

A quick glance at profits made in other food and drink markets reveals just how mouth-watering these profit levels are. The Heineken beer group, for instance, managed a margin of around 12 percent in 2001. Sara Lee's margins on its deli meats and sausage business were under 10 percent in 2002[8]; its profits on breads and bakery were even lower, at 5.5 percent. Dannon's dairy and yogurt business managed around 11 percent in 2001. Coffee—and especially soluble coffee—is a cash cow by comparison.

Obstacles at the Farm Level

While the dynamics of the market at the macro level have shifted against the interests of small-scale farmers and farm workers in recent years, many of the problems that directly confront these farmers, farm workers, and their communities are not new. Small-scale coffee farmers face obstacles such as lack of farm credit, direct access to markets, and alternative economic opportunities to help ensure sustainable livelihoods. Coffee farm workers are subject to labor abuses and substandard wages often below the legal minimum, and remain constantly at risk of losing their jobs. The coffee price crisis of 1989–2005 has brought this reality to light. Even as prices rise, farmers and farm workers are not able to capitalize on these market gains because of the aforementioned circumstances. To the contrary, small-scale farmers and farm workers assume the most risk and are the most vulnerable to the price shocks of the volatile coffee market.

Small-Scale Family Farmers: Fighting to Stay Afloat

Small-scale, family farmers represent over 70 percent of the world's coffee producers. These producers generally depend on family members for farm labor and farm on very small plots of land that they own.[9] Lack of access to farm credit, over-dependence on a single source of income, inability to access markets directly and underdeveloped organizational management stack the cards against small-scale farmers. These circumstances make it next to impossible for them to weather the boom and bust cycles of the commodity market.

Caught in the Middle with No Access to Finance   Much attention has been paid to credit and financing for micro-enterprises in the developing world. However, coffee farmer marketing cooperatives are typically too large to qualify for micro-credit and are overlooked by commercial lenders. They are stuck in a financial "no-man's land." Marketing cooperatives and associations play an integral role in providing small-scale coffee farmers with direct access to international markets. Nonetheless, a lack of access to capital has limited the potential of the cooperative business model, and thus its ability to function as a means to increase direct market access and the higher incomes associated with it. "The principal challenge of the coffee producers at [our cooperative] is access to financing," says Dagoberto Suazo of the La Central Cooperative in Honduras. "This includes short-term credit for farm maintenance, fertilization, and harvest; medium and long term financing for investment in productive and commercial infrastructure, diversification projects, and land purchases." (Oxfam America interview, 2004)

The recent rise in the international price of coffee has increased marketing cooperatives' need for working capital so that they can buy coffee from their members. Independent of price fluctuation, marketing cooperatives need financing to invest in capital improvements essential for quality production and to make pre-harvest farm credit available to farmers for investment that increases yields and improves quality.

Small-scale farmer organizations are in constant need of low-interest working capital to finance the purchase, processing, and sale of coffee. This financing allows the cooperative to maintain a positive cash flow, maintain high quality, and meet contract terms. Additionally, financing for on-farm infrastructure is used for construction of small mills and drying patios, and can come from informal lenders, development project budgets or cooperative credit funds. Financing for centralized infrastructure is used for cooperative-level mills, transport vehicles, quality equipment, or for drying patios.

At the individual level, farmers who are encouraged to diversify cannot do so successfully without transitional financing for investment in the new activity or crop, income substitution for basic needs, debt payments prior to generation of new income sources, and support for required training. There is a general shortage of this type of funding. Nonetheless, many farmers struggling to maintain financial stability through the coffee crisis have increased their debt load. In El Salvador, this translated to widespread farm seizures when farmers were forced to default. René Rivera Magaña (2004) of Fundación Nacional para el Desarrollo (FUNDE), in El Salvador, comments: "This is an extremely severe problem to have this debt, because it isn't only a problem of credit. Because of the crisis, profitability isn't possible so there aren't resources to continue investing in their farms. Practically, the farmers are just waiting for this period to pass. There's nothing they can do, except hope their land isn't taken."

In El Salvador, the Foro de Café (Coffee Forum) has successfully negotiated with the government to reduce the total amount owed by coffee farmers and the allowance of a grace period on repayment. This has somewhat stemmed the tide of land seizures. However, debt refinancing and forgiveness remain priorities for small-scale farmers (Rivera 2003). Often, smallholders owe less in terms of total debt, but they have far more to lose than medium and large farmers. In response to the most recent fall in prices, Costa Rica, El Salvador, and Honduras instituted emergency funds for farmers when the price dropped. The repayment of these funds depends on the recovery of coffee prices. If the price does not recover over a sustained period of time, farmers will fall further into debt and may lose their land as a result. Several countries are also restructuring the debt of farmers, although these usually reach only the medium and large-scale farmers who receive formal credit (Varangis 2003).

Getting the Beans to Market   The coffee market is extremely consolidated. As a result, marketing and market access are two considerable roadblocks for small-scale farmers. One of the central challenges of small-scale farmer organizations is competition with exporters who have financing, infrastructure, risk-management tools, extensive market knowledge, and existing contracts. This combination of capital, infrastructure, and experience allows exporters to reduce their cost of goods and achieve economies of scale. Add to this dishonest, unaccountable traders and the result is a highly competitive and unfavorable environment for smallholder organizations, especially emerging ones.

Achieving economies of scale can be virtually impossible for small-scale farmer organizations when they are competing with large exporters with consolidated processing and established relationships with international buyers. However, farmer organizations want to export their products directly and are increasingly making direct links with buyers in consuming countries. This requires a certain level of organizational development in order to meet the legal, quality, and volume requirements of exporting (Lopez 2004). There is also a concern that direct coffee buying by transnational companies is increasing in producing countries. Rather than bringing increased benefits to producers, roasting companies are using this as a way to sidestep intermediaries and cut their own costs by paying the same prices as the local buyers. This is not the type of "market access" that is beneficial to smallholders.

Farmers require consistent and reliable information on the coffee market and worldwide production trends to plan accordingly. Currently, no dependable system exists, and much of the information comes from national coffee institutes and ad hoc information gleaned from market contacts. Trade fairs and conferences are opportunities for information gathering, but they can present a skewed version of the market depending on how representative the trade fair is of actual commerce. Much of the best information is held by large traders and thus inaccessible to farmer cooperatives (Lewin 2004).

Along with low prices, small-scale farmers point to high transaction costs for themselves, and their organizations, as a barrier to markets. These costs include taxes and technical assistance fees to various national agencies. The Salvadoran Foro del Café has successfully campaigned with the government to reduce these fees by approximately 60 percent. Transaction costs also refer to the costs of entry to specialty markets (certification, administrative, and quality-control costs) and export-related fees.

Brewing Up Opportunities to Increase Coffee's Quality and Value   While not a solution for all small-scale coffee farmers, the specialty (i.e., gourmet) segment of the coffee market can provide farmers with price premiums that can make the difference between profit or loss. Farmers recognize the advantages and opportunities available with a higher-quality coffee.

Small-scale farmers, especially those in very remote areas, need more investment for on-farm infrastructure in order to maximize their quality and minimize bean defects. Farmer organizations also need centralized infrastructure to process the coffee that comes from their many small-scale members. Lucas Silvestre, Manager of Asociación Guaya'b, a farmer-marketing cooperative in Guatemala, notes: "A priority for us is to improve the quality of our coffee because the market for high quality coffee is growing. Without the infrastructure to process coffee it is difficult for a cooperative to achieve this goal."

Centralized processing allows many organizations to maintain the standard and consistent quality necessary for sales to the specialty market. But there are some farmers who live too far from a centralized buying/processing station and thus need affordable, efficient on-farm processing methods along with training on quality control. According to Miguel Paz (2004) of the Peruvian farmer cooperative CECOVASA, "centralization of processing is not really an option because of the poor roads and difficulties in transporting coffee to a central location. Rain can come in the middle of the harvest and stay for a month, which can ruin the coffee. It needs to be processed on farm." Some experts suggest that on-farm processing also encourages farm investment by small-scale farmers, and promotes individual empowerment. However, farmers need financial support and training to ensure that standards for quality and consistency can be met.

CRECER, a Guatemalan business development association, is considering a micro-mill project to confront this challenge of farm-level processing. It presents considerable organizational and training challenges, but also considerable potential in knowledge dissemination and long-term farm-level sustainability. Another need is farmer accessible "cupping" (i.e. tasting) labs. Many farmers have never tasted their coffee and do not understand the ways to control taste and reduce practices that damage flavor. Finding a way of disseminating basic cupping technology is a need for those organizations, which are looking to sell to the specialty market. In Nicaragua, Thanksgiving Coffee Company worked with the leading Fair Trade-certified cooperatives to successfully build nine professional cupping labs. These smallholder cooperatives have used these labs to improve their coffee quality, and sell more coffee at better prices into the specialty markets (Bacon 2001). CLSUSA also worked to install "mini-labs" Nicaraguan in farmers' cooperatives (Kenney 2002).

Knowledge Is Power: Strengthening Farmers' Organizations Many small-scale producers are still unorganized or affiliated with small cooperatives that have relatively little economic or political power. These organizations need more long-term financial and institutional resources in order to solidify basic organizational structure. The farmers who belong to these organizations also face the most intractable obstacles due to the remoteness of their location. They lack transport and communication infrastructure, education, and experience in business practices. Some may have rich community traditions of cooperation that can be strong foundations of collective action, but organizational strengthening and alliance building is still direly needed.

Farmer organizations also need financial and organizational management capacity to operate as successful businesses in the competitive and marginally profitable world of green coffee exporting. These organizations, especially the emerging ones, need to acquire the ability to reduce high operational costs, maximize efficiency, and achieve economies of scale in post-harvest processing. They need to develop dependable internal control systems (ICS) for information management, which are in turn many times required by certification programs oriented to quality or sustainability (e.g. Fair Trade and organic).

Organizations in poor rural areas are hard pressed to find and retain skilled management and technical staff. They often need, but do not have, at least one person proficient with English, in order to be able to communicate with buyers, development workers or roasters. Antonio Cordón (2004) of the Asociación CRECER in Guatemala commented on this topic: "The principal challenge to improving the situations in these cooperatives is to find skilled human resources. Many cooperatives have been recipients of assistance from aid agencies and the government over a long period of time, but they haven't been able to capitalize on this. There is a high turnover rate of personnel in these cooperatives because first, it is difficult to find qualified people and second, the ability of small cooperatives to pay attractive salaries is also a problem. When a skilled person finds a better employment opportunity, they leave. We also have the problem that many technicians leave for the US looking for better work."

Traders regularly use tools to manage the inherent risk in commodity trading. These tools have not been generally available to smallholder organizations due to limited access to information and training. Farmer organizations that are aware of these tools are interested in using them, after receiving the necessary training and credit they require for implementation (World Bank 2004). The World Bank is supporting pilot projects to train smallholder organizations to use tools such as hedging and price and weather insurance. There has been considerable learning in this area over the last few years, and several organizations are successfully using these risk-management tools. The emphasis now is on disseminating the information in order to educate both the smallholders and national bank sectors. Many smallholder organizations have heard about risk management, but they express a lack of understanding of how these tools can be used. This represents a key obstacle for them to gain access to them. Thus, access to organizational capacity building and training is essential in order to create opportunities for small-scale farmers to compete.

Lack of Alternatives to Coffee as a Cash Crop   Despite calls for several decades, for countries to diversify away from commodity dependence, this has rarely happened. Sub-Saharan Africa, for instance, is now more dependent on commodities than it was in the early 1980s (World Bank 2002). Livelihood diversification is often cited as the answer to low commodity prices, but finding alternatives that can deliver equal benefits as cash crops like coffee is exceedingly difficult. Farmer leaders mention diversification as a priority, but emphasize the difficulty of finding legal, profitable alternatives to coffee. In the words of Dagoberto Suazo of the La Central cooperative in Honduras (2004):

Our members have had some success in diversification programs, but on a very small scale. Examples are honey, pigs, cattle, vegetables, basic grains, roasted and ground coffee, and plantain. To expand and promote rural diversification, the following services and/or programs are needed:

Assistance in feasibility studies
Long-term lines of credit
Market studies and access to national and international markets

Brand design and development
Design and development of marketing strategies
Technical assistance (quality control, knowledge of the final consumer, etc)
Exchange of experience and information

For some farmers, diversification means expanding into new roles in the supply chain, such as coffee processing and marketing, in order to increase value. For others, diversification means access to viable sources of income to complement coffee production, such as intercropping with other cash crops. Without other options, some small-scale farmers are forced to diversify by cultivating illegal drugs or migrating.

Capturing More Value up the Supply Chain: Vertical Diversification   Farmers can increase their incomes by diversifying their coffee-related activities. This is achieved through quality differentiation and adding value to capture upstream margins. Varangis et al. (2003) refer to this:

Producers' share of total value has declined considerably: from approximately 30 percent to less than 10 percent in the last two decades. To increase their share of the total value and to add value, producers need to simultaneously develop downstream supply chain linkages and pursue promotion strategies that feature their coffee's comparative advantages.

The reasons for the decrease in producer value include improvements in roasting technology, improvements in processing efficiency, and the combined leverage of cash and knowledge of roasters to add value and reduce costs of raw materials. Thus, roasters' profit margins have increased as producers' profits have fallen.

New Ways to Earn a Living: Horizontal Livelihood Diversification   While many small-scale farmers have found other sources of income to augment coffee sales, few have the training, the technical assistance, the financing, or the market access to take these alternative income sources to the scale needed to generate significant economic improvement. This situation does not call for programs that switch coffee farmers to monocrop cultivation of other commodities with volatile prices. A better alternative would be for initiatives that help farmers create diverse income portfolios that best match their individual situations. Jerónimo Bollen (2004) of Manos Campesinas, a second-level cooperative in Guatemala, put it this way:

Our second priority [after increasing price through quality] is diversification of production. We understand the concept of diversification to be the amplification of the range of products grown in association with coffee. It is not the change of one monoculture (coffee) to another. Through diversification, producers continue working their coffee and at the same time cultivate other products, such as avocado, citrus, and local fruits. Diversification in association with coffee is able to partially alleviate the crisis caused by low prices, improve the family diet and provide possibilities of generating new and better sources of income for rural families.

Diversification in addition to or out of coffee into alternative crops is particularly important for those farmers who are unable to produce coffee for specialty markets due to the environmental conditions of their farms or other factors.[10] Coffee farmers have used strategies to diversify for some time, but there exists the challenge of finding markets for new products. As expressed by Merling Preza, PRODECOOP, Nicaragua (Wisconsin Coordinating Council on Nicaragua 2004): "Diversification is a theme that I don't like very much. And not because I don't think it's necessary. The small producers have diversification on their farms -beans, bananas, oranges, and a little coffee- but their farms are so small that they basically produce for their own consumption, with just a little for sale internally. Coffee is what generates the income to buy clothes, for school, for health."

When Coffee Is Not an Option For some farmers, coffee production is not a profitable option in the long term. Yet the costs of switching out of coffee are substantial, and farmers lack feasible alternatives. This is partly due to the failure of international aid donors and national governments to promote efficient rural development and diversification. In addition, it is due to the protectionist policies of the European Union and the United States, which have effectively prevented farmers in developing countries from benefiting from other crops.

It may seem economically irrational for farmers to continue to sell coffee at a price that does not allow them to cover their basic needs, but in fact the decision is entirely rational. First, the costs of replacing their coffee trees with an alternative crop are high. Even if their land is suitable for, say cocoa, they may lack the skills or training to grow it, and most farming families have no savings to live off while waiting for the new crops to bear fruit. Second, there is a severe lack of compelling alternatives. Coffee farmers know better than most how dangerous it is to rely wholly on this fickle crop for their income, so most smallholders intercrop their coffee with subsistence and other cash crops, or raise chicken and cattle. Domestic markets for their produce tend to be too small and too low-priced to replace the income that coffee used to generate (i.e. the returns on many alternative crops are as bad as coffee, or worse). Abarya Abadura (2002), a coffee farmer from Jimma in Ethiopia, reported: "Three years ago I received $105 a year from my corn. Last year, I received $35." The price paid for corn—a staple—is estimated to have fallen over 60 percent since 1997. Abarya explains the connection: "When the coffee price falls, people don't have enough money to buy corn."

A leading diversification option for coffee farmers is the production of illegal drug crops like coca and poppies in South America or chat in Ethiopia, which can deliver 3–10 times the price of coffee (Oxfam 2003). Other farmers have replaced coffee with environmentally damaging activities such as cattle grazing. Farms are being subdivided and sold, sometimes through land seizures due to debt default. There are many projects dedicated to diversification, but they have had little success relative to the overwhelming need.

Migration is another common diversification option for struggling farmers. In some Mexican coffee-producing communities, over 70 percent of the population has left in search of better incomes in the north (Hernandez 2004). Luis Hernández Navarro (2004) comments: "The relationship between coffee and migration has undergone a fundamental change since the crisis in international coffee prices in 1989. Today there is a new coffee migration. This time it isn't related to the productive cycle of the plant. The new migration is fleeing the coffee fields and the enforced poverty of low prices. This migration heads north, to the United States, and stays there."

In an Oxfam field report, John Crabtree (2002) notes that Peruvian farmers are diversifying their activities through changing the gendered division of labor. Women and children are increasingly providing the majority of the farm labor, and make up for labor shortfalls as men migrate for work or when there is no income to hire occasional farm labor. Children working on farms generally attend school less.

Gaining a Foothold and a Voice: Participation in International Debate

While international forums to address the coffee crisis exist (e.g. ICO), legitimate small-scale farmer and farm worker organizations typically have limited participation in these debates. Participation of delegations from coffee-producing countries, does not necessarily mean that small-scale farmers and/or farm workers are represented. Organizations like the Global Alliance on Commodities and Coffee (GLACC) do exist, but there are still many obstacles to the participation of these marginalized groups in international debates. These barriers include language, ability to travel to international meetings, and skilled human resources who can monitor situational developments at the global scale. Stefano Ponte (2004) describes the situation: "The post-International Coffee Agreement regime exhibits many of the characteristics of a 'buyer-driven chain' [in which] the institutional framework is moving away from a formal and relatively stable system where producers had an established "voice" toward one that is more informal, inherently unstable and buyer-dominated."

The fact that small-scale farmers produce most of the world's coffee could be leveraged for significant political power if farmers had effective international representation. While farmer representatives realize the difficulty in this, it remains one of their high priorities. Regardless of the obstacles, without effective representation, small-scale producers will continue to draw the short straw in international negotiations. Oxfam helped establish the Global Alliance on Coffee and Commodities

(GLACC), which plays a representative role for farmers. There is a need for sustained institutional and financial support for GLACC's long-term development and sustainability as an effective alliance and as a representative body. Coordinadora de Pequeños Productores de Comercio Justo (Coordinator for Small-scale Fair Trade Producers) also serves this role, although only for organizations on the register of Fairtrade Labelling Organizations International.

The Most Vulnerable Actors in the Coffee Market: Farm Workers

Hired coffee farm workers, or short-term contract farm laborers, are perhaps the population hardest hit by the coffee crisis. The World Bank estimated that for two crop cycles, seasonal employment in Central American coffee cultivation fell by more than 20 percent and permanent employment by more than 50 percent. Lost salaries for coffee farm workers across the region are estimated at $140 million (Varangis et al. 2003).

The livelihoods of these farm workers are affected by the economic viability of large estates, by job loss, and by a lack of political voice and control of their situation. As cutting labor is an easy way to trim costs, farm workers assume much of the market risk. Workers are constantly vulnerable to farm layoffs or foreclosures and have no legal recourse in the instance of unfair loss of employment or labor abuses. These workers do not benefit as prices rise, either. Historically, even when coffee prices are high, many farm workers suffer from poor working conditions, low wages, and contractual abuses. The coffee crisis has exacerbated these problems through widespread layoffs or cost cutting by plantation owners. Coffee prices are currently rising and there is the real threat that the need for action may be seen as less urgent. While a small number of estates provide decent working conditions and pay for their workers, concerns for the majority of farm workers will continue to be ignored without strong action from coffee buyers and national governments. The US/Guatemala Labor Education Project (1996) put it this way: "For years, US companies have sold us coffee picked by workers who are paid such miserable wages that they can't afford an adequate diet, basic health care, or an extra change of clothes for their children."

Workers require fair wages to cover their basic costs of living, and achieve more sustainable livelihoods. Some coffee-producing countries have minimum wage laws, but few are enforced in the agricultural sector. The US/Guatemala Labor Education Project's case study of Guatemalan coffee workers states that Guatemala's minimum wage is approximately $5.40 below the cost of basic living expenses. There is research verifying that between 60 and 80 percent of coffee plantations pay below the minimum wage (US/GLEP 1996).

Advocacy is needed to improve the conditions and benefits for both temporary and permanent workers, to ensure fair treatment in labor disputes, and above all to foster economic development for those unemployed by the coffee crisis. It is an immediate and pressing need to increase and improve the assistance to these people.

Given the central place of coffee in many countries' economies, the amount of research done on working conditions on farms that hire laborers is proportionately very low. This is due in part to the difficulty of finding farm workers in coffee countries who feel secure enough to talk openly about their situations, as well as researchers who are willing to investigate pertinent information. As a result, it has proven difficult to find conclusive information on the conditions and situation of coffee farm workers.

Beyond Coffee: A Damaged Global Trading System

Coffee is not alone in its crisis. Many commodities, such as sugar, rice, and cotton, face long-term price declines as increased productivity and greater competition boosts their supply beyond demand. Like coffee, many of these other commodities are subject to the boom and bust of volatile prices. This long-term decline is not only closing down alternatives for farmers, it is also devastating national economies. The greater a country's reliance on primary commodities, the more devastating the impact of falling prices on government budgets. Commodity dependency has worsened in sub-Saharan Africa, with 17 countries depending on non-oil commodity exports for 75 percent or more of their export earnings (World Bank 1999). Many of these countries still face heavy debt burdens, while their capacity to repay has been severely undermined.

The World Bank and the International Monetary Fund (IMF) have exacerbated this problem with the "one size fits all" approach that they have pushed onto all low-income countries, using structural adjustment lending. This approach has focused on the need to generate export-led growth and to facilitate foreign investment through liberalizing trade barriers, devaluing exchange rates, and privatizing state enterprises—essentially moving to a free market in which each country supposedly develops its own "comparative advantage." However, little attention has been paid to the direct impact of this approach on poor people. Historically, the lowest-income countries have depended on producing primary commodities, and in many cases the focus on liberalization and comparative advantage has increased that dependence. At the same time, the removal of tariffs or support to domestic industry in the interests of a fully free market has made it increasingly difficult for countries to diversify "upstream" or into more value-added industrial enterprises. Attempts to try to protect new industries have met strong opposition from the World Bank and the IMF. For example, Ugandan programs to protect infant industry by promoting strategic export areas for activities like fish processing have "been met with derision by World Bank and IMF officials" (Greenhill 2002).

Beyond declining commodities, producers that should have a comparative advantage in global trade, are effectively suppressed into a position of subservience at the whims of a rigged and unfair global trading system that works to benefit the rich and further weaken the poor. Global trade is undermining the effects of development aid, since for every dollar given to poor countries in aid two dollars are lost due to unfair trade (Watkins 2002). It is not possible to address single commodity issues without acknowledging the detrimental effects of a larger, damaged system. The roots of the coffee crisis are deeply embedded in this defective system.

Response to the Crisis: Recommendations for Public Policy, Non-Government Organizations, and the Coffee Industry

Despite increases in the international price of coffee since 2001, the dynamics of the coffee market have not shifted in ways that guarantee long-term stability for those at the bottom of the supply chain. Moreover, there is great risk that as the price of coffee rises, the sense of urgency for dealing with the crisis will wane.

As a result of disproportionate power in the supply chain, family farmers and farm workers are relegated to destitution in a boom and bust market. While there is no easy solution to the crisis, Oxfam urges the governments in coffee-producing and coffee-consuming countries, along with non-government organizations and the coffee industry, to focus attention on the following:

price stability and sustainability
access to farm credit
technical assistance in quality improvement, marketing and management
livelihood diversification
market information
participation in international debate on the working of the market

Development programs and supply-chain reforms that address these needs can help to level the playing field and leave the poor less vulnerable to the market's boom and bust cycle in the future. Moving forward, global trade policy and corporate coffee sourcing practices must be adjusted to ensure that the profits realized at the top of the supply chain are spread in a way that truly rewards all its participants.

Acknowledgments

This chapter draws heavily on two previous reports: Mugged: Poverty in Your Cup, written by Charis Gresser and Sophia Tickell, and The Crisis Continues, written by Stephanie Daniels and Seth Petchers. Oxfam gratefully acknowledges the contribution of the authors of these reports.

Notes

1. The data given are for crop years.

2. There are two main coffee varieties: robusta and arabica..Robusta, traded on London's LIFFE exchange, is used widely for soluble coffee and in the stronger roasts. The better-quality arabica, traded on New York's NYBOT Exchange, is typically grown at higher altitudes. It is harder to grow and more susceptible to disease, but it commands a higher price. It is sold in specialty coffee markets, as well as being used in soluble coffee blends for its flavor.

3. ICO Resolution 407.

4. This was calculated with available data from the International Coffee Organization, including prices paid to growers in exporting member countries in constant terms for 2004, and the average retail price paid of roasted coffee in the United States.

5. This research was done by Karen St Jean Kufuor, a commodities economist and consultant.

6. "Margins" refers to gross profits once such operating costs as marketing, salaries, and processing are deducted.

7. Financial results for the 9 months to March 2002 of fiscal year 2002.

8. Financial results for the 9 months to March 2002 of fiscal year 2002.

9. Small-scale farmers are defined as those who farm on average 1–3 hectares in Latin America, and even less in Africa, where farms are sometimes so small that they are counted by the number of trees on a property.

10. Some low-altitude coffee can be sold for decent prices in the mainstream market, Central American coffee being an example.

References

Abadura, A. 2002. Interview by Oxfam. April.

Bacon, C. 2001. Cupping What You Grow: The Story of Nicaragua's Coffee Quality Improvement Project. Unpublished project evaluation.

Castillo, L. 2002. Interview by Oxfam. February.

Chemonics. 2003. CADR QCP Quarterly Activity Report, October 1–December 31. www.dec.org.

CLUSA. Nicaragua: Small-Scale Farmer Income and Employment Project. http://www.ncba. coop/clusa_work_nicaragua.cfm.

Coelho, J. 2004. Coffee Prices Perking Up 14 percent. www.Reuters.com.

Commission for the Verification of Codes of Conduct (COVERCO). 2000. Coffee Workers in Guatemala: A Survey of Working and Living Conditions on Coffee Farms. www.usleap.org.

Common Code for the Coffee Community. 2004. www.sustainable-coffee.net.

Common Fund for Commodities. 2004. Partners in Sustainable Development: Basic Facts. www.common-fund.org.

COVERCO. 2002. Women and Children: The Precarious Lives behind the Grains of Coffee. www.usleap.org.

Crabtree, J. 2002. Visit to Peru. Oxfam Coffee Campaign. www.maketradefair.com.

Crabtree, J. 2003. Coffee Collapse Is Leading to Global Drug Boom. Press Release, Oxfam, Oxford.

CTD. 2004. "Declining Commodity Prices in the Spotlight." *Bridges* 8, no 7. www.ictsd.org.

Deutsche Bank. 2000. Soluble Coffee: A Pot of Gold?

Economic Research Service, US Department of Agriculture. 1999. Agricultural Outlook.

Elton, C. 2004. Guatemalan labor disputes fueling violent stand-offs. *Miami Herald*, December 1.

FO Licht. 2002. Estimate of World Coffee Production.

Freedman, P. 2004. Designing Loan Guarantees to Spur Growth in Developing Countries. http://www.usaid.gov.

Giovannucci, D., and F. Koekoek. 2003. The State of Sustainable Markets: A Study of Twelve Major Markets. www.iisd.org.

Green, D. 2004. Tropical Commodities and the WTO. *Bridges* 8, no. 10. www.ictsd.org.

Harris, B. 2004. Coffee market perking up, but growers still smarting and cautious. *Miami Herald*, November 22.

Hernández Navarro, L. 2004. *To Die a Little: Migration and Coffee in Mexico and Central America*. Interhemispheric Resource Center.

Installe, P. 2002. Interview by Oxfam.

Institute for Agriculture and Trade Policy. 2004. 10 Ways to Fix Agricultural Trade. www .tradeobservatory.org.

International Coffee Organization. 2002a. The Global Coffee Crisis, a Threat to Sustainable Development. www.ico.org.

International Coffee Organization. 2002b. Diversification in coffee exporting countries. Report to ICO Executive Board. May. www.ico.org.

International Coffee Organization. 2004. Colombia launches a campaign to increase domestic consumption of coffee. www.ico.org.

International Coffee Organization. 2005. "Trade Statistics—Monthly Imports." www.ico .org.

International Coffee Organization and World Bank. 2003. Address on the Coffee Crisis. http://web.worldbank.org.

International Union of Food, Agricultural, Hotel, Restaurant, Catering, Tobacco and Allied Workers' Associations (IUF). 2004. International Coffee Code Must Bring Real Changes for Workers and Farmers. Press release, September 12. www.iuf.org.

Kenney, J. 2002. Nicaraguan Co-ops Seek Growing Markets. *Cooperative Business Journal*, November: 10–11.

Koning, N., et al. 2004. Fair Trade in tropical crops is possible: International commodity agreements revisited. Wageningen University North-South Discussion Paper 3. www.north-south.nl.

Kraft Foods. Growing a Better Future for Coffee. www.kraftfoods.co.uk.

Lewin, B., Giovannucci, D., and P. Varangis. 2004. *Coffee Markets: New Paradigms in Global Supply and Demand*. World Bank.

López, O. 2004. Informe de Reunión del Comité Ejecutivo del Código Común de la Comunidad del Café: 4C o CCCC. Frente Solidario.

Maizels, A. 2000. Economic Dependence on Commodities. UNCTAD X High-Level Round Table on Trade and Development: Directions for the Twenty-First Century. United Nations Conference on Trade and Development

May, P., et al. 2004. Sustainable Coffee Trade: The Role of Coffee Contracts. www.iisd.org.

Murphy, S. 2004. UNCTAD XI: Challenging the Commodity Crisis. Institute Agriculture and Trade Policy.

Oxfam. 2003. Coffee Collapse Is Leading to Global Drug Boom. Press release. www.oxfam.org.

Oxfam International Briefing Paper 44. 2003. Walk the talk: A call to action to restore coffee farmers' livelihoods. In The Commodities Challenge: Towards an EU Action Plan. www.oxfam.org.

Petchers, S., and C. Bacon. 2004. Assessing the Impacts of Fair Trade at the Farmgate: Does Access to the Fair Trade Market in Nicaragua Impact the Income of Coffee Farmers? Unpublished.

Pomareda, C., and C. Murillo. 2003. The relationship between trade and sustainable development of agriculture in Central America. http://ase.tufts.edu.

Ponte, S. 2001. Coffee Markets in East Africa: Local Responses to Global Challenges or Global Responses to Local Challenges. Working Paper 01.5, Center for Development Research, Copenhagen.

Ponte, S. 2004. *Standards and Sustainability in the Coffee Sector: A Global Value Chain Approach*. IISD.

Potts, J. 2004. Multi-Stakeholder Collaboration for a Sustainable Coffee Sector: Meeting the Challenge of Anti-Trust Law. www.iisd.org.

Rivera, R., et al. 2003. El Impacto de la Crisis de Café en El Salvador. Fundación Nacional para el Desarrollo, San Salvador.

Rodriguez, C. 2002. Interview by Oxfam.

Schomer, M. 2003. Nicaragua Faith-based Specialty Coffee Alliance: Appraisal of Proposed Project and Options. www.dec.org.

Segal, M. 2004. Multinationals, producers agree on Common Code industry standards. *Café Europa* No. 19, November. www.scae.com.

Starbucks C.A.F.E. (Coffee and Farmer Equity) Practices Program. www.scscertified.com.

Stecklow, S., and E. White. 2004. How fair is Fair Trade?" *Wall Street Journal*. June 8.

Sustainable Agriculture Platform. 2003. Common Understanding of Mainstream Sustainable Coffee. www.saiplatform.org.

Sweet Maria's Coffee Cupping Reviews. Central America: Honduras. www.sweetmarias .com.

T&G endorses new Charter of Rights for migrant workers in agriculture. 2004. www.tgwu .org.uk.

Technoserve in Nicaragua: An Overview. www.technoserve.org.

UK Department for International Development/Overseas Development Institute. 2004. Rethinking Tropical Agricultural Commodities. http://dfid-agriculture-consultation.nri.org.

USAID Press Office. 2004. USAID Supports Coffee Growers around the Globe. Press Release. www.usaid.gov.

US/GLEP. 1996. Justice for Coffee Workers Campaign. www.usleap.org.

Varangis, P., and B. Lewin. 2002. *The Coffee Crisis in Perspective*. World Bank.

Varangis, P., Siegel, P., Giovannucci, D., and Lewin B. 2003. 2003. *Dealing with the Coffee Crisis in Central America: Impacts and Strategies*. World Bank.

Watkins, Kevin. 2002. *Rigged Rules and Double Standards*. Oxfam International.

Wisconsin Coordinating Council on Nicaragua. PRODECOOP: Quality coffee and women's leadership. www.wccnica.org.

World Bank. 2002. Presentation for Coffee Association.

World Bank. ITF pilot project reports. http://www.itf-commrisk.org.

World Food Programme, UNICEF and the Salvadoran Ministry of Public Health. 2003. Estudio de campo sobre caracterización socioeconómica de las familias afectadas por la crisis del café en el occidente del país. San Salvador. (Quoted in R. Rivera et al., El Impacto de la Crisis de Café en El Salvador, FUNDE, 2003).

List of Interviewees

In place of references to specific statements made during some interviews, citations have been substituted with the following list of interviewees:

Walter Rodríguez Vargas, Asociación de Pequeños Productores de Talamanca, Costa Rica

Lucas Silvestre García, Asociación Guaya'b, Guatemala

Miguel Paz, CECOVASA, Peru

Preston Motes, Chemonics, United States

Frank Hicks, Consultant, Costa Rica

Kari Hamerschlag, Consultant, United States

Bill Harris, Cooperative Coffees, United States

Antonio Cordón, CRECER, Guatemala

Doug Hellinger, Development Group for Alternative Policies, United States

Robert Mack, EARTH University, Costa Rica

William Foote, Ecologic Finance, United States

Eric Poncon, ECOM Trading, Mexico

Karen Cebreros, Elan Organic Coffees, United States

Guillermo Denaux, Sn., Fairtrade Labelling Organizations, El Salvador

Gerardo de León, FEDECOCAGUA, Guatemala

Sigfredo Benitez, FUNDACAFE, El Salvador

René Rivera Magaña, FUNDE, El Salvador

Rodney Nikkels, Green Development Foundation, The Netherlands

Jason Potts, International Institute for Sustainable Development, Canada

Jonathan Rosenthal, Just Works Consulting, United States

Ann Vaughan, Legislative Assistant, Office of Congressman Sam Farr, United States

Dagoberto Suazo, La Central (CCCC), Honduras

Jerónimo Bollen, Manos Campesinas, Guatemala

Eli Landa, Oxfam America, El Salvador

Gawain Kripke, Oxfam America, United States

Katherine Daniela, Oxfam America, United States

Stephanie Weinberg, Oxfam America, United States

Sonia Cano, Oxfam International/Global Alliance on Coffee and Commodities, Honduras

Chris Wille, Rainforest Alliance, Costa Rica

Ernest Van Punhuys, Technoserve, Nicaragua

Kimberly Easson, TransFair USA

Yamile Slebi, TransFair USA

Muriel Calo, Tufts Global Development and Environment Institute, United States

Tim Wise, Tufts Global Development and Environment Institute, United States

Daniele Giovannucci, Senior Consultant to International Organizations (UN and World Bank), United States

David Griswold, Sustainable Harvest, United States

Ben Corey-Moran, Thanksgiving Coffee, United States

II

Ecological and Social Dimensions of Producers' Responses

4

Coffee-Production Strategies in a Changing Rural Landscape: A Case Study in Central Veracruz, Mexico

Laura Trujillo

This chapter analyzes how particular land management regimes become dominant over time and the landscape through the confluence of socio-economic, cultural, and ecological drivers. I do this by studying the coffee-production strategies playing out in Central Veracruz, which embody a complex set of social, political, and ecological interactions that come into play within the coffee commodity agro-chain. In order to understand the development of coffee farming in this sector of the global coffee agro-chain, I suggest that the current land-use model and the production strategies performed by the growers conform what I call a "bricolage landscape," as an outcome of the interactions between marginalization, landform heterogeneity, and farming choices. I contend also that these relationships are driven by new ways of access to the means of production, locally based perceptions of environmental risk, and processes of small-scale farmer empowerment. Differently than in former periods where landscape homogeneity and high-tech were encouraged by international market and regulation policies, a bricolage landscape expresses how marginalization and land heterogeneity factors are performing particular roles in specific coffee-production regions.

In the first section, I discuss the use of the landscape concept as an integrative part of my approach. This view is then applied explicitly to the examination of coffee production in nine counties within the Veracruz region, focusing on an in-depth analysis of coffee-production strategies. In the final section, and for a more complete comprehension on the genesis of this varied landscape, a qualitative analysis on its farming strategies is integrated to a historical review of coffee landscape evolution in the state of Veracruz, and other coffee-producing countries.

Conceptualizing Landscape

Landscape as an outcome of human action and perception is materially and perceptually constructed. Landscape is a way of seeing and embracing, rather than just a scene or image, since it is experienced through cognitive processes where discursive and ideological struggles, over place and power, are played out in concrete physical settings. The physical landscape is materially constructed as it is perceived and imagined by any social group or individual through their everyday life. Their interpretations and applications on the land are done through inter-personal strategies, varying in time and space, and resulting in a multifunctional landscape (Kristensen et al. 2004; Brandt et al. 2001; Wilson 2001).

Following Terkenli (2001), landscape can be theorized as form, meaning and function. Landscape form involves representation, perceptibility, and imageability of the physical environment. Landscape meaning engages cultural identity, symbolism, values, mythological archetypes or any social characterization. Landscape function implies both bio-ecological processes and humanistic and biophysical relationships within the context of socio-economic distinctiveness. In other words, the distinctiveness of landscape function is integrated with a particular socio-economic system by embracing and expressing its own uniqueness on geographical, historical and environmental properties, as associated to cultural processes. Landscape as cultural process interprets the physical, economic, and political systems within a social order, and develops an unparallel identity.

The Rural Coffee Landscape

The concept of rural landscape is used here to include the overall outcomes that are played out in social and environmental change, and which reflect the global, institutional, social, and livelihood practices associated with land use and agricultural commerce. Such practices include the ways in which different groups define and enclose the biophysical features, and how they work to distinguish the appropriateness of places for certain kinds of production activities. In turn, these activities rely on the cultural interpretations of the physical, economic, and political systems within a similar space and ecological realm which lead to the adoption of agricultural technologies and insertion into social, political, and economic processes of a specific agricultural commodity market. A rural landscape is constituted when specific land-management systems emerge and are dominant over time in specific places, and through their promotion of economic relationships that shape cultural process and sustain ecological resilience. Rural landscape is the outcome of historical land practice (Nightingale 2003; Mansfield 2003). In the case of raw commodities like coffee beans, these relational human components, along with environmental and agricultural production, constitute a rural landscape. In addition, form, meaning and function are instilled into agricultural production processes.

A rural coffee landscape is not just a geographical area with symbolic expressions of human-environment relationships. The rural coffee landscape may be viewed as a cultural image, the symbolic expression of human environment relationships, which are formed through a distinctive political and economic history. It has been constructed through a range of globalization projects related to the coffee commodity that involve institutional relationships, social networks, labor organization, technology and market development, as well as rural livelihood transitions. Rural coffee landscapes and their livelihoods have never been independent of historical processes of internationalization, transnationalization and/or globalization, since production and reproduction occurs at all scales ranging from the local to the global.

Perhaps the most striking particularity linking the physical coffee landscape form to its respective human component is not the degree of interaction between these two realms but its relation with global market and their production technologies and social networks. Rural people have tried to address their survival through a constant engagement in more globalized sets of social and economic relationships, either throughout institutional linkages, social relationships, or labor (Bebbington 2001). That is why the coffee landscape's distinctiveness stems from its environmental-global commodity production relationship.

Production Strategies and Evolving Coffee Landscapes

Coffee landscape changes did not occur in absolute terms, but developed as a multifunctional landscape with two important components: (1) empowerment as vulnerability and risk as cultural perception and (2) access to means of production.

In this section, I elaborate these two components on the grounds of theoretical contributions sketched in the first section, which are linked to a historical sequence of coffee-production models in the study area. Such synthesis seeks for an explanation of the reasons why the regional coffee landscape evolved and crystallized into the current variety of farming strategies that we observe today.

The physical landscape is transformed as it is perceived and imagined by any social group or individual in everyday life. Their perception about themselves and the risks they face are materially constructed on the land through inter-personal strategies, varying in time and space, and resulting in a bricolage landscape.

The perceptions of social groups or individuals about risk and vulnerability deeply affect the choices they make about production and social involvement.

The establishment of any particular technology is strongly based on if the return will guarantee social reproduction no matter what changes occur in the weather, prices, or other external variables. Therefore, risk aversion is the foundation for household livelihood strategies. Consequently, peasants' perception of risk may dominate their production strategies. So, risk perceptions shape their structural production in time and space. Options that reduce risk include crop diversification such as inter-cropping, plot scattering, and subsistence production.

Besides environmental (drought, frost, etc.) and agricultural (pest, diseases) risks, there are some production elements that can also alter risk perception. Credit is an economic risk element that is usually available only for large capitalist producers and it is often biased against small-scale farmers and women because of their lack of economic reliability. Historically, most agricultural credit programs have been tied to the purchase of specific "Green Revolution" technologies which are highly risky to people perceiving themselves as vulnerable.

Labor can also be a significant constraint in the formation of risk perception. Coffee production has a seasonal peak in labor requirements, which occurs during harvest season, since planting, weeding and pruning have a wider seasonal period. High-yielding coffee plantations not only add to total labor requirements, and exacerbate seasonal peaks in harvest labor requirements, but they also increase the costs of hiring labor, while lowering profitability in a very risky market. There is also the risk of harvest loss because of absence of available labor at a critical time. Property rights and farm size are usually mentioned as an important factor that shapes the use of different technologies (Knox 1998; Bromley 1991).

The power perception that each social group or individual believes they have to cope with those risks, make them consider themselves more or less vulnerable. Following Blaikie et al. (1994, p. 9), vulnerability is "the characteristics of a person or group in terms of their capacity to anticipate, cope with, resist and recover from any change." Any changes on environment, market, and/or political organization inevitably alter the ability of different actors to earn a livelihood. Thus, the capability that actors recognize they have to overcome new situations, will either empower or make them vulnerable. The response to any change will depend on the reasons farmers use to explain to themselves their own ability and power, or their vulnerability.

However, everyday life embodies a decision process about future risks. The imaginative future scenario includes the way to cope with eventual difficulties. Coping is "the manner in which people act within existing resources and range of expectations of a situation to achieve various ends" (Blaikie et al. 1994, p. 62). Resources include production means (land, environment, skills, technology, capital, market access), and social means (households members, kinship patrons, political or/and business organization, religion and other social values that give them rights). There are different coping strategies: mitigation (e.g. secure needs through lowering them; minimum level of food and shelter rather than an increment in income); production maintenance (e.g. labor safeguarding through large number of children rather than acquiring land); diversification of production strategies; diversification of income resources (pluri-activity); and development of social support networks (Bryant and Bailey 1997; Blaikie et al. 1994)

Production Strategies and the Central Veracruz Rural Landscape

The rural coffee landscapes of Mexico have been built through a complex set of environmental and social processes, as well as international commodity market relationships. A coffee landscape can be seen as a good example of how global trade processes can shape it, and how local people constantly struggling with economic survival, mold the places they inhabit. Coffee landscapes have been developing as complex, contingent and elaborate schemata of environmental and natural resource perceptions, evaluations and management embracing a unique nature-culture image.

Nowadays there are 350,000 small scale coffee growers in Mexico, mainly in the states of Chiapas, Veracruz, and Oaxaca. These growers produce 40 percent of the total of national coffee production, while the remaining 60 percent of production is done by large landholdings (Renard 1999; Nolasco 1985).

Veracruz—the number-two coffee-producing state, after Chiapas—is the number-four state in terms of indigenous population and the first state with more *ejido* land tenure type. However, it also holds the smallest farm size average at the national level. In addition, it possesses the highest number of small towns in the nation, which have populations of less than 500 inhabitants. Veracruz is perceived as one of the richest states of Mexico, because of its abundance of natural resources including oil, water and energy, and 19 different climates. In reality, however, it is one of the poorest states, with high to very high marginalization indexes, including one of the four most marginalized counties of Mexico (INEGI 2005).

The form of the Veracruz coffee landscapes can be described as mosaics of intermingled multi-crops (mainly coffee, sugarcane, and pasture) over an abrupt mixture of physiographic land forms. Its humid, cloudy climate produces plenty of water as runoff (25 percent of the Mexico's water).

There are three main coffee regions in Veracruz: La Huasteca in the northern region of the state divided by the big mountain system called "Eje Neovolcanico" which involves Atzalán, Misantla, Chicontepec, and Papantla subregions; the gold triangle in Central Veracruz, conformed mainly by the Coatepec-Xalapa, Córdoba-Zongolica, and Huatusco subregions, with over 70 percent of the state's coffee production, including the main economic and political cities; and thirdly, the Los Tuxtlas region in southern Veracruz, which is mostly indigenous (figure 4.1).

Central Veracruz is conformed by five subregions: Coatepec, Córdoba, Huatusco, and Tezonapa y Zongolica. This study was conducted in the Huatusco subregion

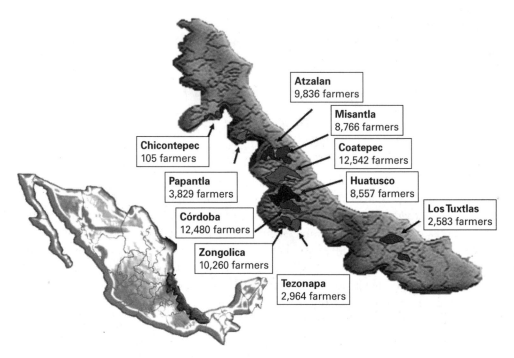

Figure 4.1
Coffee regions in Veracruz state.

which includes nine counties: Comapa, Huatusco, Ixhuatlán del Café, Sochipa, Tenampa, Tepatlaxco, Tlacotepec, and Totutla y Zentla.

The Landscapes of the Huatusco Subregion

The Huatusco subregion, where the present study was conducted, is situated in a complex set of different physiographical elements. Twenty-nine percent of its territory is covered with mountains, 28 percent of the region with hills, 17 percent with valleys, and 8 percent with ravines and cliffs. Most of its 85,000 inhabitants are living under medium- to high-marginalization conditions (the marginalization index varies from 0 to 2) within the rural landscape (table 4.1).

A general view of the regional production in hectares for all nine counties shows a diversified landscape, where coffee, sugarcane plantations, pasture, and corn plots are intermingled in a mosaic (table 4.2). The amount of land devoted to coffee production tells how each county is integrated to the coffee agro food chain. Sochiapa arises as the more specialized county, followed by Ixhuatlán del Café and Totutla.

Table 4.1
Physiographic and socio-economic characteristics of the nine counties within the Huatusco subregion (INEGI 2005).

	Land area (km²)	Dominant land form (%)	Urban cities	Urban residents	Rural villages	Rural population	No. ejidos	Ejido area (ha)	Rural roads (km)	Marg. index
Comapa	320	Hill-valley (83.7)	1	4,340	71	12,754	14	5,968	3.5	0.78
Huatusco	212.2	Mountain-valley (83.4)	1	26,848	70	19,629	11	1,462	51	0.79
Ixhuatlán	134	Valley (68.1)	1	6,427	30	13,518	9	2,525	38	0.74
Sochiapa	21.4	Valley (54.7)	0	0		3,105	0		5.5	0.74
Tenampa	69.9	Sierra/gully (58.9)	0	0	15	5,900	6	1,567	12	0.74
Tepatlaxco	99.5	Sierra/gully (64.2)	0	0	12	7,844	5		4.1	0.99
Tlacotepec	90.5	Hill (69.1)	0	0	7	3,624	2	949	12.7	0.33
Totutla	80.6	Sierra/gully (88.5)	1	3,390	32	11,562	6	0	60	1.19
Zentla	241	Valley (44.3)	0	0	54	12,339	3	4,256	64	1.08

Table 4.2
Crop production areas by county in Veracruz, Mexico (hectares).

	Coffee	Sugarcane	Livestock	Corn	Other
Comapa	2,723	473	15,246	4,306	489
Huatusco	5,159	432	5,785	4,306	245
Ixhuatlán	4,836	174	1,227	472	40
Sochiapa	804	0	0	0	0
Tenampa	2,059	180	730	266	357
Tepatlaxco	2,488	29	0	435	161
Tlacotepec	1,033	279	678	212	22
Totutla	4,167	721	204	0	1,001
Zentla	3,947	1,712	15	874	9

However, counties such as Totutla, Tenampa, Tlacotepec, and Tepatlaxco explicitly reported their land as multi-crop (the two last ones reported 70 percent and 50 percent respectively) (INEGI 2005). Furthermore, through the interviews, most of the coffee producers explained that they have developed mainly a multi-crop strategy. A regional small-scale coffee farmer generally owns up to 2 hectares and usually cultivates more than one crop on the farm, depending on his or her articulation to a particular socio-economic system, and the uniqueness of his or her geographical, historical and environmental land properties. Thus, through time and space they develop a multi-crop strategy. If a crop is a community formed by a complex mix of different populations of cultivated plants, weeds, insects, etc. (Gliessman 2002), production management matters when it is classified as monocrop or multi-crop.

It is important to point out that the only crop formally typified regionally as multi-crop is coffee. Scholars have described the following five different coffee agroecosystems based on shade canopy structure and composition (Toledo and Moguel 1996): Traditional rustic coffee, traditional multi-crop, commercial polyculture, shaded monoculture, and full-sun coffee. However, choosing coffee agroecosystem management only represents one of the many responses that peasants may generate in response to their physical environment and their marginalization condition along time and space. Coffee growers, depending on their marginalization condition, their market connectivity, and their environmental context, will employ more than one production strategy to ensure their social reproduction.

Therefore, characterizations of coffee agroecosystems based only on their shade structure are insufficient for a meaningful understanding of coffee production and management. The character and rate of expansion and adaptation of different coffee agroecosystems varies accordingly with landform, marginalization, and economic processes. This variation is expressed through spatial structure of the production strategies, since the spatial dispersion of fields in a mosaic landscape can only indirectly be brought under the guise of strategy.

With coffee production looked at as a strategy, coffee landscape can be described as different mosaics of intermingled multi-crop and/or coffee monocrops (either under shade or at full sun), with other subsistence or cash crops. Each strategy has incorporated a time-space structuralization, in which distance, and yearly and daily activities are calculated based on landform, weather, economic and social conditions, and risks. This time-space organization is expressed in a structured routine, where the spatial distribution of the farm elements and labor shape the production strategy. Therefore, each of the production strategies described below has its particularities in space and time, which are associated with a specific set of geographic and social conditions.

I found eight production strategies in the study region after clustering the results of household interviews and census data. The first four strategies have these characteristics: they are labor intensive, the demand for capital and technical assistance is high, producers are highly integrated with trade and local political networks, and all income comes from the sale of their products at local and regional markets. The first four of these strategies are as follows.

Cash Mosaic: A patchy farm devoted to cash crops, where shade coffee monocrop is cultivated with the same level of importance as sugarcane, banana or chayote monocrop, and, exceptionally, livestock.

Coffee Mosaic: It differs from the Cash Mosaic strategy, since it is based on coffee production, which can be either in a monocrop or multi-crop system, with a small plot of sugarcane and/or chayote monocrop or any other cash crop.

Coffee Monocrop: The only source of cash income is coffee. The coffee agroecosystem model used incorporates two or three high-yielding varieties of coffee such as *Caturra*s, *Catuaí*, and some *Bourbón*. Its shade trees are mostly with legumes, especially from the genus *Inga* or *Erythrina*.

Livestock Mosaic: Cattle raising for dairy and meat production is the main activity. Most farmers also have small patches of shade coffee monocrop and sugarcane or another full-sun monocrop intermingled in their pastures.

The subsequent four strategies have the following characteristics: family labor is used, capital requirements are low, technology is developed through experience and farmer-to-farmer exchanges, they are partially integrated with regional trade networks, they rely on community social networks, they are partially integrated with local commerce, and barter and consumption of what they produce are important to their survival strategies.

Coffee Multi-Crop: The main monetary source of family income is coffee and secondarily banana leaves.[1] Non-monetary income is obtained through bartering fruits harvested from shade trees. The coffee agroecosystem model used incorporates two or three varieties of coffee (*Típica*, *Mundo Novo*, *Bourbón*, and/or *Caturra*), intermingled mostly with trees, including orange, banana, Inga, pepper, avocado, guava, and some native species.

Coffee Subsistence: The main crop within the farm is corn for consumption. Intercropped coffee is grown as a cash crop. These farmers have no other monetary income.

Livestock Subsistence: The main crop is corn for consumption. Excess production is sold for additional income. Farmers with this strategy also raise a few cattle for dairy and meat.

Subsistence: Corn is cultivated for self-consumption, and cash income is obtained from selling family labor for work in coffee farms or through non-agricultural labor.

In accordance with the production strategies I found, I analyzed the number of growers with farms that were producing coffee, and documented the area devoted to other crops in the same household, and at the community level, using census data (INEGI 1995, 1996, 1997), empirical information from surveys, and direct observations. The results show that most of the counties have developed a diversified production strategy (table 4.3).

The diagrams in figure 4.2 visually display the great differences that can occur between the production strategies in each county. Totutla and Sochiapa show a strong orientation to one or two strategies of production, focused mainly on coffee production. Tlacotepec and Ixhuatlán adhere to the same pattern oriented toward coffee monocrop and coffee subsistence. Tepatlaxco, Tenampa, and Zentla manage diversified strategies. Huatusco is strongly oriented toward cash production with Cash Mosaic, Coffee Monocrop, and Livestock as main strategies.

As local space, the Huatusco landscape's form and function are linked by the relationship between the biophysical environment and its human interpretation and management. Human interpretation of abruptness and environment of each particular geographical place can be translated as a particular management of land over space and time. Also, this geographical interpretation will be influenced by

Figure 4.2
Production strategies of counties in Huatusco region. Source: Trujillo 2000.

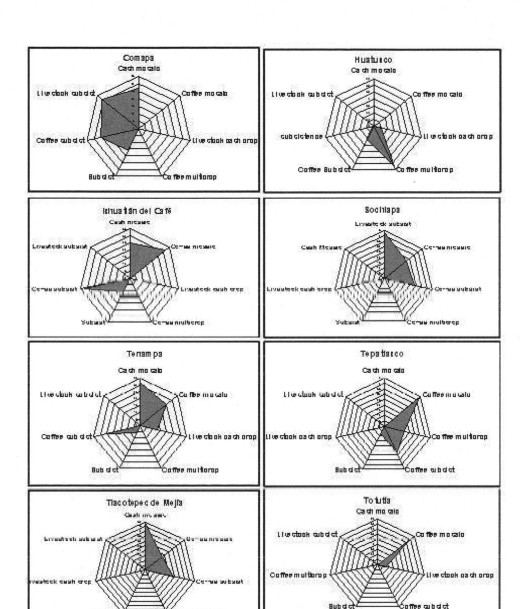
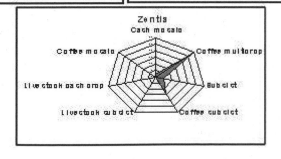

Table 4.3
Percentage of households using different production strategies in the Huatusco subregion of Central Veracruz, México, by county. (Dash indicates data not available.)

	Cash mosaic	Coffee mosaic	Livestock mosaic	Coffee multicrop	Subsist	Coffee subsist	Livestock subsist
Comapa	26.9	0	0	0	16	25.6	31
Huatusco	0	0	10	65	3.4	21.6	—
Ixhuatlán del Café	25	32	0	0	11	32	0
Sochiapa	0	26.6	0	0	0	28.4	43
Tenampa	32	25	12.4	0	5.5	25	0
Tepatlaxco	—	43	0	20	7.6	28	0
Tlacotepec de Mejía	50	21.4	0	0	0	28.5	0
Totutla	0	83	14	0	—	2.6	0
Zentla	4	0	0	58	10	28	0

producers' perception of vulnerability, which is related with their social conditions of marginalization. In order to explain how rural coffee landscapes have produced a multitude of local responses—as figure 4.2 shows—depending on land form geography (abruptness), and social conditions (marginalization), I first calculated the land form percentage of each community, and summarized it in two groups depending on form and slope: abrupt (sierra, mountain, ravine, and gully), and none (valley, hills, and soft mountains). Then I integrated those physiographical elements into a landscape heterogeneity index[2] by calculating the heterogeneity of landforms for each community within the nine counties in the Huatusco subregion (Sklenicka and Lhota 2002). Second, I elaborated a marginalization index[3] that took into account social conditions such as energy and water access, educational level, income, etc.

Abruptness is an important characteristic, which was correlated with the marginalization index. Results showed that there is a significant correlation between the social marginalization index and physiographical features. Where there are more valleys there is less marginalization, and vice versa (figure 4.3).

Small-scale coffee growers with high marginalization were situated not only in the most abrupt, but also in the most heterogeneous landscapes (the ones that include all physiographic forms). This means that these small-scale growers are spatially marginalized; contrarily, the ones with medium marginalization were located in more homogeneous landscapes, just one or two land forms, mainly valleys and mountains, so, they are not spatially marginalized.

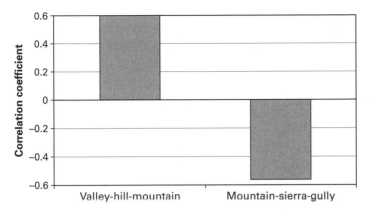

Figure 4.3
Correlation coefficients for percentages of land forms and marginalization indexes in Central Veracruz coffee region.

In order to confirm that multi-crop strategy characteristics are influenced by marginalization and physiographic features, I developed a spatial marginalization index by correlating the coefficient for the heterogeneity index and the social marginalization index by county (figure 4.4). After that, I correlated the index with production strategies (figure 4.5).

Results show that counties which correlated positively, have a strong segregation between people living and producing in better geographical and social conditions, and the ones with high marginalization and heterogeneous landscapes. For example, Ixhuatlán del Café with the highest positive correlation, has a group of highly marginalized communities at the most abrupt region, and less marginalized communities in valleys and hills. In contrast, Huatusco, Totutla, Zentla, Tepatlaxco, and Comapa counties have marginalized communities all along its territories, with no influence from land form heterogeneity. The remaining counties with negative correlation, such as Sochiapa county, where the more important economic activity is coffee, have marginalized areas with less abrupt land forms and with subsistence livestock covering 43 percent of their territory.

Spatial marginalization and production strategies showed no correlation, except for the Subsistence strategy (figure 4.5). This may signify that only those that are completely marginalized will adopt this particular strategy.

Since spatial marginalization had no significant correlation with production strategies, I decided to correlate marginalization index and heterogeneity index independently.

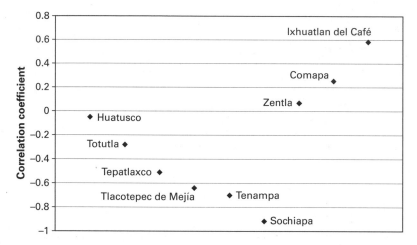

Figure 4.4
Spatial marginalization index (expressed as correlation coefficients of heterogeneity and marginalization) in Central Veracruz.

The correlation between marginalization and production strategies (figure 4.6) indicates that counties with Coffee Subsistence strategies are mostly those with high marginalization indices ($r = 0.694$), showing a trend where the more marginalization results in higher areas devoted to Coffee Subsistence. Inversely, an increase in the Coffee Mosaic strategy results in a lower marginalization index ($r = -0.645$). However, cash mosaic (0.56) and coffee monocrop (0.5) are correlated directly with marginalization. It seems that Cash Mosaic is a way to scatter risk, and coffee monocrop is a social process where people linked to political and economic policies of coffee national production resist changing their strategy and their way of living.

These results point out that the two main strategies, Coffee Mosaic and Coffee Subsistence, use coffee as their main cash crop. The difference between them is the way they are related to social marginalization (figure 4.4). Coffee subsistence is directly related, whereas mosaic coffee has an inverse relationship to social marginalization. It is interesting to notice that the counties that received the most support from government during the cartelization period—Ixhuatlán del Café, Huatusco, Zentla, and Totutla—are the ones with either none or a few examples of the Cash Mosaic strategy. This fact highlights the importance of the cartel period, which is further described below.

So far, I have shown that, depending on their marginalization condition, coffee growers will use more than one strategy to ensure their social reproduction. To establish a linkage with landscape form, I correlated the heterogeneity index to pro-

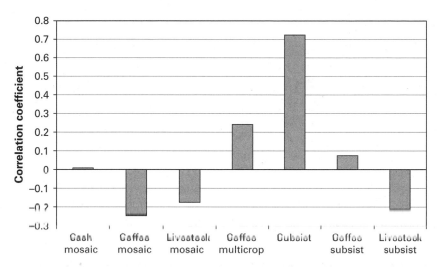

Figure 4.5
Correlation between spatial marginalization and production strategies in Central Veracruz.

duction strategies (figure 4.7). The Subsistence strategy, for example, is not related to landform heterogeneity, but is strongly related to marginalization. This may mean that regardless of geographical or environmental conditions farmers will cultivate maize. On the other hand, the Coffee Mosaic strategy is strongly correlated to both marginalization and landform, and since cash crops are highly labor and capital demanding, they are better adapted to areas with lower environmental and economic risks.

On the other hand, the more heterogeneous the landscape, the greater the area devoted to Coffee Multi-Crop, and less livestock and sugarcane monocrop. Coffee Multi-Crop appears as a strategy to avoid environmental and economic risk, since it does no matter where land is geographically situated and what the farmers social conditions are, and it assures at least some cash gain if not profit. Because livestock raising requires low slopes, it is correlated to landscape heterogeneity.

I decided to separate sugarcane from Cash Mosaic because national policy on the industrial processing of sugarcane establishes that the sugar industry itself will do the harvesting. In order for the crop to be purchased, it is required that it is grown in valleys. The reason why the correlation coefficient is not equal to one is that there are small sugarcane processors that use hand crafted equipment, and they will purchase from marginal sugarcane areas. These results demonstrate that landscape function is articulated independently from geographical-environmental characteristics, and particular social conditions, however, it comprehends both as historically complex cultural processes expressed as spatial landscape configuration.

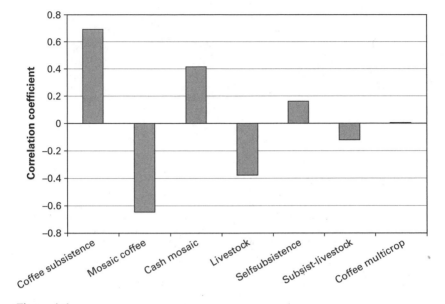

Figure 4.6
Correlation coefficients for marginalization index and production strategies in the Central Veracruz coffee region.

The preceding discussion reveals that coffee producers developed specific production strategies depending on their social conditions of marginalization and landscape heterogeneity. Thus, each production strategy and its relation with its own social conditions of marginalization must be viewed in the context of landscapes of place. In addition, there are struggles over power, since commodity production and land use are a combination of state policies, control over infrastructure, and access to markets and technology. The strength of the link between local landscape and global strategies is such that the latter shapes local landscapes as contested sites of institutionalized, political and economic power, which in turn shape the spatial configuration of rural landscapes

Briefly, the environmental constraints that topography and microclimate represent, the social conditions of marginalization, and the socio-economical restrictions that limit profit and access to market, have an effect on each of the actor's risk and vulnerability perception, which in turn will shape decisions about farm structure and production strategies.

The cash crop strategies such as Coffee Mosaic, Coffee Monocrop, etc., require a perception of low production risk and high ability to overcome deteriorating conditions. Since coffee production is an international commodity, many of its risks are

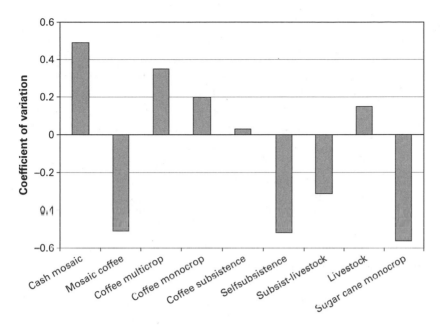

Figure 4.7
Correlation coefficient for landscape heterogeneity and production strategies in the Central Veracruz coffee region.

shaped by political forces. Therefore, coffee producers must be highly integrated with technical assistance, trading, and regional/national political networks, in order to adopt these strategies.

Inversely, socially and spatially marginalized coffee producers, isolated from and forgotten by mainstream development and trade agents, will experience different risks as impossible to overcome. Thus, they will use different coping strategies, and the type and frequency with which they use them will depend on their own vulnerability perception.

The Coffee Monocrop strategy is a good example of how farmers cope. It was promoted in the 1970s and the 1980s by the government as a way to obtain foreign exchange. It organized coffee farmers and local markets, subsidized credit and agricultural inputs, and provided technological assistance. Once government withdrew from controlling coffee production, most of the coffee producers that had adopted this strategy switched to a multi-crop strategy if they were in marginalized communities, or into Coffee Mosaic production if they were less marginalized. In fact, it is now difficult to find coffee monocrop among small-scale coffee producers, but it is still present in larger landholdings.

Access to Means of Production

Social and spatial marginalization and the environmental constraints that land form imposes are insufficient elements to explain Huatusco's bricolage landscape. Each production strategy is not only the result of unequal relations between human-environment interaction and social development, but they are also the outcome of power interplays within commodity production. Struggles over power result in the historical social differentiation manifested in social and spatial marginalization, and the consequent segregation of the coffee market and its surpluses.

Access to production means, in agricultural terms, has a long history of struggles over land property, labor control, and access to market. Each historical period developed commodity production models where different actors contested, adapted or resisted each dominant power. Coffee has been an international commodity since its origin, and it has been playing an important role on kingdoms, nations, and the wealth and power of multinational corporations.

In order to bring this factor into the analysis of the Huatusco landscape, I identified six historical models of the coffee commodity in Mexico. In each period, land tenure, labor control, agro-industrial process property, access to trade and market, and surplus appropriation have varied, and, as a result, have shaped the rural coffee landscape. Below I present these six historical periods and integrate them to my initial discussion on production strategies as drivers of the Huatusco coffee landscape.

Colonial Plantation Model

Colonialism played an active role in the physical transformation of the landscape by promoting extensive cash-monocrop plantations. Extensive plantations as a production strategy in new territories not only secure high incomes, but also avoid bio-environmental risk. Coffee expansion through tropical colonies during the seventeenth and eighteenth centuries went together with massive forest clearing as landscape domination, and slavery as a social way of labor control and power continuity. Landlords not only were part of an aristocracy, but also owned any coffee surplus from land to market. Coffee was a high-value commodity that used to be traded through protectionism or free colonial commerce for luxury markets.

The Spanish crown also tried to introduce coffee plantations into its colonies during the eighteenth century in order to benefit some important resident families in Mexico, mainly in Veracruz. However, they were unsuccessful, owing to the high profitability of previously established tobacco cultivation as the colonial plantation model. This was mainly in the Huatusco region, including the current Huatusco, Sochiapa, Tenampa, Tlacotepec de Mejía, and Totutla counties.

The Huatusco region has been traditionally hacienda territory. In 1740 it reported more than 40 large landholdings with livestock intermingled with tobacco and sugarcane, and indigenous populations as slave-servants (Archivo Huatusco 1915). The 1882 population census reported 33 haciendas with 5 million coffee plants that produced 60,000 sacks of coffee (Archivo Huatusco 1915). The switch from tobacco to coffee was based on the higher profitability of the latter.

Modern Industrial Plantation

This type of production was characterized by the use of migrant or servitude labor, rather than slaves, with either a small payment or the rights to land for subsistence cultivation. This production strategy attempted to maximize the use of labor over a short period of time, as well as to avoid responsibility for providing labor subsistence (Duncan 2002). There is strong evidence that modern coffee plantations that originated as colonial plantations, transformed into full-sun monocultures, which covered thousands of hectares. However, these modern prototypes were now surrounded by subsistence crops in a broad mosaic, in ways that enabled landlord relations of labor control (Jones 1929). The modern coffee plantation model fit perfectly with Latin American colonies because of a social system which was largely based on servitude.

These changes resulted in a radical appropriation of the landscape by the "new migrant/servant farmers." On the one hand, they shaped landscapes through the meaning they assigned to them, which was based on their own region's background. On the other hand, it was shaped through their social culture, as Duncan (2002, p. 318) has pointed out: "modern plantations can be understood as a site of heterogeneous networks, radically reconstituted into a new socio-material configuration."

Modern colonial industrial plantations not only allowed landscape changes to occur as a result of workers activities in their subsistence plots, but also developed a new form of social control and continuity through their power over access to land and market. In this period, the post-colonial modern industrial landlords became the new state elite that not only dominated peasants, but the different classes were bound to each other through personalized patron-client relationships, especially in rural areas where peasants relied on the propertied upper class for access to the land they farmed.

The first "modern" coffee plantation in Huatusco county was introduced by Carl C. Sartorius (a German naturalist) in 1849 (Café Veracruzano 2004). Sartorius cultivated 25 hectares of coffee on the 4,500 hectares he planted with sugarcane, tobacco, and pineapple. By 1872, the whole hacienda was devoted to coffee. Land was loaned to indigenous inhabitants with the commitment "to farm any kind of crop they want, if they take care of any coffee work required by the owner," as well as to ensure the production of at least half of the hacienda's coffee exports (Von Mentz 1990). In post-independence Mexico, policy for rural colonization by foreigners was developed with the specific objective of introducing labor. Along with that process, Antonio Santa Ana, the erstwhile Mexican president and Veracruz governor, facilitated the first large modern colonization movement by foreigners into Veracruz in 1861. They were provided with land, credit, and economic support (Zilli 1981). They planted "sugar cane, coffee and tobacco as main crops, along with grapes, morera, sorghum" (Zilli 1981, p. 267). It would be the first coffee mosaic farmed by small-scale family farmers.

Post-Colonial Coffee Plantations

The first significant change in the transformation of the landscape (in the second half of the nineteenth century for Latin America, in the twentieth century for Asia) was to open up governance processes by empowering the new independent nations' governments, and in this way to change relationships of local resource control. The main driver for the establishment of modern plantations shifted as the newly independent nations focused on developing foreign exchange through coffee commerce (Groppo 1996).

The former patron-client relationships also tied the peasants into a new political system, where agrarian oligarchs, especially coffee planters in Brazil, Colombia, and Central America, dominated state structures and controlled labor to guarantee coffee production. For example, they promoted coercive labor by outlawing vagrancy, debts, and by enforcing public labor on poor people (Barickman 1996). However, as the coffee market expanded to include the lower social classes in the United States and Canada, the international demand for coffee also increased. In order to meet this new demand, small-scale farmers were encouraged to move into coffee production through the initiation of a system of buying unprocessed coffee cherries by local landlords.

The transformation in the control of coffee production introduced changes in the way coffee was cultivated. The need for small-scale farmers to articulate their coffee production with their reproduction strategies created elaborate procedures to project and manage their resources. Shade-grown coffee could have been the technological response to this change. Some recent papers refer to the innovation by Mesoamerican cultures (Perfecto 1996; Rice 1993), where coffee was readily incorporated into existing, indigenous agroforestry systems, such as shaded cacao cultivated beneath the forest canopy. Perhaps shaded coffee started making inroads into the "natural" forest because of labor insufficiency. Shade trees currently are used as a strategy to reduce labor inputs in the plantation, since the closed shade of the trees retards the growth of weeds and grasses, prolongs the harvest period, and provides non-agricultural resources such as firewood. This combination of characteristics allowed laborers the time they needed to cultivate their subsistence crops.

Small-scale coffee producers' landscape meaning and function were important for production design, in the absence of strong and active external institutions. For this reason, smallholders were given the power to decide on their own production strategies and associational activities. However, modern plantations, using temporary labor and buying coffee from smallholders, continued selling their products to exporters and/or directly to roasters. Thus, most of the value of coffee remained in the hands of the landlords.

Regionally, the integration of the immigrant small-scale farmers into coffee production was a natural step. However, immigrant small-scale farmers became coffee producers through two processes: (1) land acquisition from a civil association supported by Carl Sartorius, and integrated to a movement by indigenous tenants lead by Agustín Chicuellar, who bought more than 26,000 hectares, including Totutla, Axocuapan, and Sochiapa and Tenampa counties, and (2) land leasing from inhabitants of Tlacotepec de Mejía and Zentla counties. Regarding the first, Sartorius pointed out in 1865 that "the land was fragmented, and the land use switched to coffee orchards." He continued: "They planted coffee trees, bananas and fruit trees between the coffee furrows. The first two or three years they crop corn, beans and tobacco, after that they are able to harvest coffee." (Sartorius 1995) These farmers were the first ones to use a Coffee Multi-Crop strategy; It was possible due to both a privileged policy for coffee plantations, and exceptional international commercial agreements between the Mexican government and Canada, the United States, and France in favor of Veracruz's coffee (Córdova 2003).

Regional rural landscapes were transformed into a coffee-production subregion (Huatusco, Sochiapa, Tenampa, Zentla, and Tlacotepec de Mejía counties) surrounded by crop subsistence production regions (Comapa and Ixhuatlán del Café counties).

Coffee Cash Mosaic is a symbol of what remains of the modern plantations: medium to large landholdings with their own coffee industrial processing and trade linkages, using temporary labor from smallholders. Huatusco and Sochiapa counties correspond mostly to this model.

The Coffee Multi-Crop and Coffee Subsistence strategies have their origin in this period. This can explain the high coffee production specialization with different strategies in Tlacotepec de Mejía and Zentla (100 percent and 86 percent respectively).

The Peasant-Modern Plantation

The new economic and social role that resulted from becoming commercial farmers changed the meaning and the function of peasant landscape. The disruption of spatial and social barriers led to fragmentation, differentiation, and complexity in both the physical geographical and the human components of place. Inside human geographical components of place, heterogeneous networks were drastically reconstituted into a new socio-material configuration (Roseberry 1995). A new sense of property and rights for land in the social conscience of the peasantry was developed.

This process, along with the economic depression between 1892 and 1933, the consequent bankruptcy of large landholdings devoted to coffee exporting, and the communist threat during the second half of the twentieth century, pushed a second "modernization" of the coffee commodity landscapes. This new phase was based on land tenure changes and trade controls, both under the new global institutions. While the aims, pace, and scale of agrarian reforms in this period differed among countries, their success was related closely to the strength of the political commitments of each government and the new global institutions. The United Nations declared: "Land reform may be necessary to remove impediments to economic and social progress resulting from inadequate system of land tenure." (FAO 2003, p. 18)

The transformation of coffee-producing tenants into small-scale family landowners represented a new change in the coffee landscape, where land tenure shifted from hacienda-dominated landscapes to those of *campesino* (peasant) agriculture. Since then until the present time, smallholder-based commercial crops acquired national importance in most developing countries, allowing for a turning point in the spatial arrangement of coffee farming choices. Landscapes of heterogeneous networks were developed. Modern coffee plantations with processing facilities (coffee mills), small-scale coffee farms, and communal corn cultivation regions were linked together through a new social configuration where industrial infrastructure (transport, mills, and credit) and market linkages remained in the hands of the local oligarchs.

Land reform started in Mexico around the 1930s as a result of civil war. The *ejido* land tenure system, which carried with it the requirement that to keep land it had to be cultivated. Non-cultivated land would be given to other farmers. Most of the current *ejidos* in Huatusco, Tenampa, Tlacotepec de Mejía and Zentla were founded in the late 1930s and the 1940s within indigenous settlements with subsistence strategies.

Further, and as a result of the postwar events, Mexican agricultural policy changed dramatically. The state's support for growing subsistence crops to maintain food security, as privileged agrarian policy, shifted toward export crops to provide foreign exchange. Coffee became the main agricultural commodity in order to promote industrial development. Most peasants with *ejido* land tenure abandoned subsistence crop cultivation and replaced it with Coffee Multi-Crop or Coffee Subsistence strategies.

The Cartel Model

The strong dependency of national governments on export crops to provide foreign exchange made them vulnerable to prices fluctuations. Commodity agreements through cartels have usually been formulated in terms of the stabilization of the trend of prices or earnings at higher levels than would be produced in a free market. From 1963 until 1989, coffee was under a system of export quotas in order to stabilize prices, under the direction of the International Coffee Organization (ICO). Producing countries agreed to stay under the assigned export quotas by sponsoring marketing boards, which were in charge of purchasing coffee from producers and selling it for export. This meant that coffee acreage, prices, producers' incomes, and associational willpower were controlled or influenced through subsidies and/or taxes handed out by those national marketing boards.

The cartel production strategy changed the distribution of coffee value, redirecting it from local oligarchies to government elites through high export taxes. It also modified coffee-production systems. Yield increment was perceived as a need, since coffee consumption increased threefold in Canada, Western Europe and the United States from 1947 to 1968 (Kravis 1968). It was assumed that an increment on production would be achieved through coffee farming intensification. This had to be accomplished through the green revolution production paradigm, which included new high-yielding varieties, full-sun plantations, high-density plantations (increased from 1,500 to 10,000 coffee plants per hectare), synthetic fertilizers, pesticide use, and monoculture farming.

Coffee-production intensification implied a farming system that uses any appropriate technology that maintains production, and where capital and market opportunity are secured. Lacking appropriate physical and economic infrastructure or ability to market their output would lead to a rejection of that production system. However, these specific strategies arose from institutions that were part of a specific power structure. Consequently, the political actions expected to promote its adoption, such as subsidies, marketing opportunities, and so on, promoted that model. Small-scale coffee producers' landscape meaning and function were not important in this intensive farming design because the power was controlled by active external institutions.

In Mexico, the Instituto Mexicano del Café (INMECAFE) was created as a marketing board. It was in the state of Veracruz that INMECAFE had its most powerful intervention in economic, technical, and socio-political terms during the 1970s and the 1980s. Almost 75 percent of INMECAFE's national infrastructure was built in the Veracruz coffee regions through the purchasing of *beneficios* (coffee mills). It was until the second land reform in the 1970s when Ixhuatlán del Café's small-scale farmers became not only *ejidatarios* but also coffee producers, with monocrop coffee strategies strongly linked to national institutions. Ixhuatlán del Café county obtained not only INMECAFE support but national recognition as the best *modern* small-scale coffee producer county, with the highest national coffee yield. Ixhuatlán was not randomly chosen. It was selected because of its high valley landform rate, which favored the implementation of the monocrop model.

INMECAFE controlled production through regional coffee price differentiation, and by targeting certain organized coffee grower groups (Unión Estatal de Productores de Café, UEPC) with credit, technical assistance, and commercialization support. This quota system rapidly developed differences among Mexico's small-scale coffee producer regions and coffee producer types. This led, in the 1980s, to social organizing that demanded better coffee prices, a bigger share of coffee quotas, credit, and the opportunity to sell coffee directly.

In the Huatusco subregion, a strong UEPC was formed, which was linked to the Central Independiente de Obreros Agrícolas y Campesinos (CIOAC) Section 25 as part of a national political social movement, which incorporated all small-scale coffee growers that were left without quotas.

By the 1980s, the expansion of multinationals that used to buy in the "coffee black market (out of quota)," the collapse of the Soviet Union, and a renewed enthusiasm for free markets, swept away the cartel period and any interest in social development. Most of coffee institutions were closed in Latin American countries, along with the support systems for coffee production and marketing strategy.

The Postmodern Model

Changes in coffee consumption, trade and value led to a new arrangement of rural coffee landscapes from the 1990s to the present. Land, labor, and commodity production became a secondary way to produce value, and financial trade, social organization and bio-environmental management became the prominent forces.

The cartel production model, with its mass production, modern distribution, marketing techniques, and homogenized consumption, currently challenges the few entrepreneurs that created brands based on quality and differentiated consumption. Whereas multinationals had been concerned only with retail coffee price and consistency, this new coffee industry considers origin, quality, processing, cultivation methods, and associational ways of production as relevant qualities of coffee production. The almost extinct small-scale roasters became influential international actors in the evolving specialty coffee industry. They were able to bring into the market differentiation process, the growing consumer concern about biodiversity conservation, environmental change, and social justice and equity. This led to the development of the "conscious consumer" niche market, with two primary outcomes: Organic coffee and Fair Trade coffee.

The innovations of satellite and digital communications changed patterns of trade markets through the unprecedented development of a commodity exchange market, where the price for coffee is fixed. This became a knowledge-intensive activity in which financial value frames trade rather than volume and commodity value. Traders and exporters that were able to manage fluctuations not only on coffee commodity price and foreign exchange, and instead invest the financial surpluses they generated, became the powerful coffee traders. Furthermore, they sought to control homogenized mass consumption, and developed a differentiated market for new consumers through special liquid coffee beverages. These are the outcome of high and sophisticated investment in industrial and commercial technology.

Traders and exporters have used this commodity exchange system to reduce their exposure to price fluctuation or any other risk. Since 1990, however, 80 percent of the coffee commodity trade deals have been made on paper (i.e. futures markets), with only 20 percent based on sales of real coffee (Dicum 1999). This fact is putting national coffee exporters (from small to large) out of business, because they do not have access to hedge markets. This sector tends to only make their decisions on volatile market prices, based on largely unrelated supply and demand. Coffee exporters as market players are poorly informed, and they frequently make incorrect financial decisions.

Regionally, local oligarchies that owned private coffee processors, had large cash reserves, managed Coffee Mosaic landholdings, and who had the financial support of national banks, were favored by the end of the cartel period. This translated into their having plenty of cheap coffee from unorganized and confused small-scale coffee growers. In addition, they had direct linkages with American and European roaster markets. However, the collapse of the Mexican economy in 1995, the subsequent monetary devaluation, the rise of interest rates to more than 100 percent, and their ignorance of financial management resulted in the bankruptcies of small-scale processors and coffee exporters. Currently, many local coffee processors are under financial contract with transnational exporters such as AMSA (Ecom), Expogranos (Ed & F Man), Cafés California (Newman Grouppe), and Becafisa (Volcafé).

On the other hand, the most important element in the conscious consumer niche market is its independence from commodity exchange markets, since traders and roasters can set a price based on quality and the conscious consumers' willingness to pay. This led to a redistribution of surplus in a more equitable way.

The first coffee producers to go into a conscious niche market were small-scale farmer groups, which became cooperatives that were able to sell coffee outside of the cartel production model. Those small-scale farmers were organized and supported by non-governmental organizations and/or religious or political groups. Consequently, they were able to become the first organic and fair trade coffee providers because of both their independent organizations, and the international support networks associated with alternative social movements in coffee consumer countries.

In the Huatusco subregion there was only one large coffee grower organization that remained as a political force and was big enough to be able to buy some of INMECAFE mills, the Unión de Pequeños Cafeticultores de la Región de Huatusco, located in Huatusco County (formerly CIOAC Sección 25). Their leader, Professor Manuel Sedas, was able to organize almost 4,500 members. However, when it was necessary for members to donate some, or all, of their coffee harvests to buy INMECAFE's second-largest coffee mill, the only ones that remained were those that did not have a monocrop strategy. This amounted to over 2,500 small-scale coffee farmers, and the purchase took place in 1991, with the support of alternative coffee organizations within social coffee markets. Currently, they sell 30,000 sacks of coffee, mostly to Fair Trade markets. Membership, however, has been reduced to an estimated 1,500. Of these, only 200 coffee farmers are certified organic, represented by those with additional economic sources and the ability to take the risk of new technology adoption.

Most of the coffee farmers holding *ejido* tenure, or small-scale individual farms in Huatusco that were once organized into UEPCs, have left this mode of organization. Additionally, in 1989, when INMECAFE withdrew, most of the coffee harvest was lost because of frost. Migration out of the state was the outcome. Every week a bus left from Huatusco and Ixhuatlán del Café to the *maquiladoras* (industrial factories) along the US-Mexican border. The most affected were those with monocrop coffee and Coffee Multi-Crop strategies. (Ixhuatlán del Café had 32 percent of its inhabitants with monocrop, and Huatusco county with 60 percent with Coffee Multi-Crop.)

There was a ray of hope when coffee prices rose again in 1997 to historical highs. However, coffee producers were hit again by the lowest coffee price in history during the period 1999–2004. Now migration is not just to other parts of Mexico, but for the first time it is also to the United States. Another outcome is a speculative land market, where small-scale farmers sell their land in order to migrate or just to survive a family emergency. Presently, the main economic activity is out-of-farm labor, which represents a form of pluri-activity as the main coping strategy.

Discussion

The current production strategies that determine Huatusco's rural landscape are the result of historical relationships between different social groups, their physical landscape, and the production means and political issues that shape them. In this way, the Coffee Monocrop strategy, which practically disappeared among small farmers, still persists among medium to large landholdings. This is partly due to the fact that any Cash Mosaic strategy is related to technologies that arose from institutions that were part of a specific power structure. Consequently, the positive political outcomes expected to accompany its adoption, such as subsidies, marketing opportunities, and others, promoted its adoption. Therefore, peasants who perceive themselves as political default actors will not adopt the model, while peasants who identify themselves as part of the political system will adopt it, even though their physical environmental conditions are not the most suitable. The existence of a political co-optation system may explain why in marginalized areas Cash Mosaic and Coffee Monocrop continue to exist as production strategies.

The Coffee Mosaic strategy is found from small to large farms that have a history of cash trade. Currently, most of them represent a form of pluri-activity farmers that have either non-agricultural income or socio-political power that guarantees access to capital and appropriate technology. It is present mainly in Huatusco, Ixhuatlán del Café, Tepatlaxco, and Tlacotepec de Mejía counties. Coffee multi-crop strategy currently denotes two possible responses depending on a community's history. If it is an old settlement it most likely represents the privileged modern plantation tenants that acquired land independently of land tenure reforms. If it is a marginalized group, they were the ones that remained independent of the cartel system. Sochiapa, Ixhuatlán del Café, Tepatlaxco y Tenampa counties represents the counties with this strategy. The Coffee Subsistence strategy epitomizes the second modern plantation model, which is made up of smallholders with property rights on land and who sell their labor and coffee to the local oligarchs. Most of them could not be inserted into another production model because of their spatial marginalization. Tenampa, Tepatlaxo, Tlacotepec y Zentla counties hold this strategy.

Subsistence and livestock subsistence strategies characterize the spatially marginalized region, where coffee production is risky, and there is high social marginalization. Most of those regions became crop subsistence suppliers and self-consumers. Comapa and Sochiapa counties are the main examples.

Production strategies of the Central Veracruz coffee region can be seen as a result of not only a historical process of human-environment relationships with the production processes of a commodity, but also as the struggles over place and power that are played out around a specific commodity. In this way, the complexity of different production strategies composes landscapes that express the cultural representation of place of each social group within a rural commodity network as occurred in the nine counties studied.

Notes

1. Leaves from banana trees are widely used in Mexico to wrap a traditional meal called *tamal*.

2. The heterogeneity index was calculated applying Shannon-Wiener diversity index $H' = (\sum pi \log pi)$, where pi = each physiographic land form percentage per community area from each county, on data obtained from aerial photographs and topographic maps for each community within the nine counties in the Huatusco subregion of Central Veracruz.

3. The marginalization index was obtained by following the methodology of the Consejo Nacional de la Población (the government agency for population statistics). It involves the following variables: number of houses per community with access to energy, potable water, and sewage; number of illiterate inhabitants over 15 years old; number of landless inhabitants; number inhabitants with less than US$100 as monthly income; and number of communities with fewer than 1,000 inhabitants per county. Each of them was ranked by using a categorized coefficient based on national statistics. Next I applied the formula $M = (\sum 8 * 0.2527) - 3.3736$. In this formula, I multiplied the negative addition of the eight variables by a coefficient assigned to include conditions specific to state of Veracruz, and I then subtracted the median value divided by the standard deviation. The final result gives a number between 0 and 2. Zero means that households have higher levels of social development including clean drinking water, sewage systems, education, and more etc.; 2 means that these conditions are completely absent.

References

Archivo Huatusco. 1915. Narración de los Hechos. Doc. AHD-1327.

Barickman, B. J. 1996. Persistence and decline: Slave labour and sugar production in the Bahian Reconcavo (Nicaragua), 1850–1888. *Journal of Latin American Studies* 28: 581–633.

Batterburry, Simon. 2001. Landscapes of diversity: A local political ecology of livelihood diversification in South-Western Niger. *Ecumene* 8, no. 4: 438–464

Bebbington, Anthony. 2001. Globalized Andes? Livelihoods, landscapes and development. *Ecumene* 8, no. 4: 414–436

Blaikie, P., T. Cannon, I. Davis, and B. Wisner. 1994. *At Risk: Natural Hazards, People's Vulnerability and Disasters*. Routledge.

Brandt, J., J. Primdahl, and A. Reenberg. 1999. Rural land-use landscape dynamics—Analysis of driving forces in space and time. In *Land Use Changes and Their Environmental Impact in Rural Areas in Europe*, ed. R. Kronert et al.. UNESCO and Parthenon Publishing Group.

Bryant, R. L., and S. Bailey.1997. *Third World Political Ecology*. Routledge.

Colombian Journal. 2004. Colombian History. www.colombianjournal.org.

Córdova, Susana. 2003. Historia de la Región de Huatusco. M.S. thesis, UNAM, México

Dicum, G., and N. Luttinger. 1999. *The Coffee Book: Anatomy of an Industry from Crop to the Last Drop*. New Press.

Duncan, James S. 2002. Embodying colonialism? Domination and resistance in nineteenth-century Ceylonese coffee plantations. *Journal of Historical Geography* 28, no. 3: 317–338.

Food and Agriculture Organization of the United Nations (FAO). 2003. Land Reform: Land Settlement and Cooperatives. www.fao.org.

Forman, R. T. T. 1995. *Land Mosaics: The Ecology of Landscapes and Regions*. Cambridge University Press.

Freidberg, Susanne E. 2003. Culture, conventions and colonial constructs of rurality in South-North horticultural trades. *Journal of Rural Studies* 19, no. 1: 97–109.

Global Environmental Foundation (GEF). 1995. Reporte Proyecto "Ordenamiento Ecologico en la Sierra de Santa Martha, Veracruz." Proyecto Sierra de Santa Marta A.C.

Gobierno de Veracruz. 2005. La Aventura del Café. Editora del Gobierno: Xalapa, Veracruz. www.veracruz.gob.mx.

Groppo, Paolo. 1996. Agrarian Reform and Land Settlement Policy in Brazil: Historical Background. FAO.

Groppo, Paolo. 1997. La FAO y la Reforma Agraria en América latina: hacia una nueva visión. Dirección de Desarrollo Rural de la FAO. www.fao.org.

Hopkins, T. K., and I. Wallerstein. 1994. Commodity chain in the capitalist world economy prior to 1800. In *Commodity Chain and Global Capitalism*, ed. G. Gereffi and M. Korzeniewicz. Praeger.

Instituto Nacional de Estadística e Información Geográfica (INEGI). 1995, 1996, 1997, 2005. Estadísticas Nacionales, México, D.F.

Jones, Clarence F. 1929. Agricultural regions of South America. Installment VI. *Economic Geography* 5: 390–421.

Kravis, Irving B. 1968. International commodity agreements to promote aid and efficiency: The case of coffee. *Canadian Journal of Economics* 1: 295–317.

Kristensen, L., C. Thenail, and S. P. Kristensen. 2004. Landscape changes in agrarian landscapes in the 1990s: The intersection between farmers and the farmed landscape. *Journal of Enviromental Management* 71: 231–144.

LeClair, Mark S. 2000. *International Commodity Markets and the Role of Cartels*. M. E. Sharpe.

Mansfield, Becky. 2003. From catfish to organic fish: Making distinctions about nature as cultural economic practice. *Geoforum* 34: 329–334.

Murdoch, Jonathan. 2000. Networks—a new paradigm of rural development? *Journal of Rural Studies* 16: 407–419.

Nightingale, Andrea. 2003. Nature, society and development: Social, cultural and ecological change in Nepal. *Geoforum* 34: 525–540.

Nolasco, Margarita. 1985. *Café y Sociedad en México*. Centro de Ecodesarrollo, México, D.F.

Perfecto, I., R. Rice, R. Greensberg, and M. Van der Voort. 1996. Shade coffee: A disappearing refuge for biodiversity. *BioScience* 46: 598–608.

Pindyck, Robert S. 1979. The cartelization of world commodity markets. *American Economic Review* 69: 154–158.

Renard, María C. 1999. Los intersticios de la globalización. Un label (Max Havelaar) para los pequeños productores de café. Embajada Real de los Países Bajos, México.

Rice, Robert A. 1993. New technology in coffee production: Examining landscape transformation and international aid in Northern Latin America. Smithsonian Migratory Bird Center.

Rochealeau, D., L. Ross, J. Morrobel, and T. Kominiak. 2001. Complex communities and emergent ecologies in the regional agroforest of Zambrana-Chacuey, Dominican Republic. *Ecumene* 8, no. 4: 465–492 .

Rodríguez Centeno, M. 2004. Paisaje agrario y sociedad rural. Tenencia de la tierra y caficultura en Córdoba, Veracruz (1870–1940). Ph.D. thesis, Colegio de México, México, D.F.

Roseberry, W., L. Gudmundson, and M. Kutschbach. 1995. *Coffee, Society, and Power in Latin America*. Johns Hopkins University Press.

Sklenicka, P., and T. Lhota. 2002. Landscape heterogeneity: A quantitative criterion for landscape reconstruction. *Landscape and Urban Planning* 58: 147–156.

Terkenli, Theano. 2001. Towards a theory of the landscape: The Aegean landscape as cultural image. *Landscape and Urban Planning* 57: 197–208.

Toledo, V., and P. Moguel. 1996. Searching for Sustainable Coffee in Mexico: The Importance of Biological and Cultural Diversity. Sustainable Coffee Congress, Smithsonian Institution. In *Proceedings of the First Sustainable Coffee Congress*, ed. R Rice, A. Harris, and J. McLean. Smithsonian Migratory Bird Center.

Trujillo, Laura E. 2000. Political Ecology of Coffee. Ph. D. thesis, University of California, Santa Cruz.

Whatmore, S., and L. Thorne. 1997. Nourishing networks: Alternative geographies of food. In *Globalising Food: Agrarian Questions and Global Restructuring*, ed. D. Goodman and M. Watts. Routledge.

Wickham, J. D., R. V. O'Neill, and B. Jones. 2000. A geography of ecosystem vulnerability. *Landscape Ecology* 15: 495–504.

Wilson, G.A. 2001. From productivism to post-productivism and back again? Exploring the (un)changed natural and mental landscapes of European agriculture. *Transactions of the Institute of British Geographers* 26: 77–102.

Zilli, José. 1981. *Italianos en México*. Editorial San José.

5

The Benefits and Sustainability of Organic Farming by Peasant Coffee Farmers in Chiapas, Mexico

María Elena Martínez-Torres

Although decades—if not centuries—of economic policies have entrenched poverty and environmental decline as the rule throughout rural areas of Mexico and Latin America, small-scale coffee producers of mostly Mayan origin in Chiapas are organizing themselves in search of new options to confront this stark reality. Their organizations may not only be paving the way toward alternative solutions to widespread rural poverty, but are also generating environmental recovery based on the frequent use of "organic" production practices. Despite the severe crisis that rural Mexico has undergone since 1982—which came to the world's attention with the Zapatista uprising in 1994—these peasant farmers, or *campesinos* as they are known in Spanish, have converted Mexico into the world leader in production of certified organic coffee.

The pioneer growers of organic coffee in the global economy have proven to be mostly indigenous peasants from the mountain ranges and ravines across the poor southern Mexican states of Chiapas and Oaxaca. During the 1980s and the 1990s these small-scale coffee growers became the most organized sector of Mexico's revitalized peasant movement. As part of the new movement, some groups (re)organized themselves into coffee cooperatives, combining traditional communal structures with elements of cooperative commercial enterprises. Some of them have been so successful that they provide a unique opportunity to analyze the elements that might be involved in more sustainable rural development policies and organization-building strategies for the future.

Coffee is the commercial crop best suited to the forested mountains with thin soils that are found in Chiapas. Coffee is also unique among Third World export crops in that the production and processing technologies involved were all developed in producing countries. This has meant that coffee has contributed greatly to state formation over more than a century of history and to the development of modern infrastructure, since the use of non-imported technologies allowed for far more capital accumulation inside producing countries than did other export crops. Today coffee continues to be an important source of foreign exchange for many countries, and both large- and small-scale farmers produce it. But recent years have seen the drastic reconfiguration of national and international coffee markets. While many of the changes hurt small farmers, they simultaneously generated new opportunities for those peasants that produce high-quality coffee, especially those producing organically, and thus many peasant coffee cooperatives have entered the growing market segments created by the differentiation of the coffee market into specialty and eco- or socially certified markets. As these market opportunities have appeared, small-scale farmers have proven able to take advantage of them, when they are well organized and produce with quality.

The degree to which they are well organized and networked is the key element that allows them to tap into market opportunities and to transform their production practices. In rural Mexico social capital grew extensively over the last few decades through a series of iterative cycles (Martínez-Torres 2005). Once farmers were well organized, they were able to make productive investments in their natural capital endowments—their land suitable for coffee, their soils, accompanying biodiversity, etc.—via organic farming practices, and thus tap into the market for organic coffee.

In this chapter, I briefly review the concepts of sustainable agriculture, agricultural intensification, and sustainable development, in order to situate the results of a case study of organic coffee production in Chiapas within the larger debates on these topics. I then present the results of the study of the economic, productive, and ecological benefits of this example of organic farming, which shed light on the viability of organic farming as an alternate path to intensification, and as an option for more sustainable development in rural areas.

Sustainable Agriculture, Agricultural Intensification, and Sustainable Development

The economic and ecological impacts of the ongoing small-farmer organic coffee boom in the southern state of Chiapas can be analyzed within the contexts of sustainable agriculture, agricultural intensification, and sustainable development. From data collected from Africa, Asia, and Latin America, Jules Pretty (1997, p. 249) concluded that farming communities using regenerative technologies have substantially improved agricultural yields, often using few or no external inputs. This is directly relevant to the ongoing debate in rural development circles concerning the intensification of agriculture. One perspective posits that increased agricultural output can only be achieved by using "science-based agriculture." In recognizing the mistakes of the first green revolution, this school proposes a "second green revolution" targeted at poorer, more marginal farmers (Borlaug 1992, 1994; Paarlberg 1994). The intention is to bring "modern technology" to bear on small-farmer agriculture:

improved seed varieties, monoculture, chemical fertilizer, pesticides, and perhaps biotechnology. The contrasting approach, where Pretty (1997) situates himself, is the so-called sustainable intensification of agriculture, which is based on investing in, and enhancement of, natural capital. Substantial productivity growth is possible, he argues, in currently unimproved or degraded areas, while at the same time protecting or even regenerating natural resources by using agroecological techniques. (See also Altieri 1995; Carruthers 1995; UNDP 1995.)

The intensification associated with the green revolution is based on the use of external inputs like chemical pesticides and fertilizers, machines and large-scale irrigation to boost food production. This technology generates economic concentration, social exclusion, the rise of expensive, patented "improved" seeds, and the depreciation of natural capital via compacted, eroded, and degraded soils, the loss of biodiversity, the pollution of groundwater, etc. (Rosset and Altieri 1997). In fact, the depreciation of productive resources like soils in many agricultural areas threatens the future sustainability of production.

Rosset and Altieri (1997) highlight the importance of breaking the monoculture structure of agricultural systems by introducing mixed cropping and by integrating crops and livestock. They argue that sustainable agriculture should avoid input substitution (i.e. using organic fertilizer instead of a chemical one), and focus rather on more complex systems that provide their own "services" like maintenance of soil fertility and reduction of pest damage. Pretty, Altieri, and others summarize empirical evidence of highly productive agriculture using low-input and regenerative techniques (Pretty 1995a,b; Altieri 1995, 1996, 1999; Hewitt and Smith 1995; Hazell 1995, Reijntjes et al. 1996).

Both routes to intensification are apparent among the coffee producers in Chiapas described in this chapter. One subset of small coffee farmers deployed agrochemicals to boost coffee yields; another subset adopted a series of organic, more agroecological techniques based on investment in and improvement of natural capital, what we might call "cultivated natural capital." While some (e.g. Avery 1995) argue that organic farming is the opposite of intensification, leading inevitably to lower yields, Pretty (1997) responds that there are many low-external input techniques which nevertheless represent net intensification or greater productivity per unit area, as was certainly the case for organic coffee farmers in Chiapas.

The term *sustainable agriculture,* of which organic farming may be thought of as an example (Pretty 1997, 1999), represents, in part, a reaction by civil society to 50 years of agricultural and rural development policies focused on the first kind of intensification of agriculture. Altieri (1995) and Pretty (1999) have provided the most comprehensive definition of sustainable agriculture—which the former calls agroecology and the latter equates with investment in natural capital, as summarized by Pretty (1999):

A more sustainable agricultural system systematically pursues the following goals:

a) A thorough integration of natural processes such as nutrient cycling, nitrogen fixation, and pest-predator relationships into agricultural production processes, so ensuring profitable and efficient production.

b) A minimization of the use of those external and non-renewable inputs with the potential to damage the environment or harm the health of farmers and consumers, and a targeted use of the remaining inputs used with a view to minimizing costs.

c) The full participation of farmers and other rural people in all processes of problem analysis, and technology development, adaptation and extension, leading to an increase in self-reliance amongst farmers and rural communities.

d) A greater productive use of local knowledge and practices, including innovative approaches not yet fully understood by scientists or widely adopted by farmers.

e) The enhancement of wildlife and other public goods of the countryside.

In comparison to conventional approaches to rural development, the concept of *sustainable development* is defined here as a development track that is more equitable, more ecologically sound, and economically viable for the poor. For example, development that is more sustainable should not be exclusionary, but rather should minimize social polarization and provide opportunities for the poor to make a decent living in ways that do not degrade productive resources—like soils—which are needed for the future. For the purposes of this chapter, the following synthetic definition drawing on Lele (1991) and Pretty (1998) is used:

Sustainable development consists of goals, strategies and processes which together provide more socially just, economically viable, and ecologically sound alternative tracks to conventional development pathways, offering improved livelihoods to the poor in ways which promote both their empowerment and the conservation or improvement of key natural resources so that the basis of productive activities can be maintained into the future.

Coffee-Production Technologies in Chiapas

In light of the discussion and debate on sustainable agriculture, intensification, and sustainable development, we can examine how coffee is grown in Chiapas. As noted above, practices are found in Chiapas which follow different paradigms of intensification, and which, as we shall see, have differing impacts on sustainability. It is simplest to classify the production technologies into the following four categories, though a lot of variation in practices is certainly encompassed within each one.

Traditional or Natural Coffee Cultivation

Until recently the only common method of coffee cultivation by small farmers was traditional, shade-grown coffee, which is also known as "passive organic" and which will be referred to here as *natural* (a common designation in Spanish). In this method, coffee trees are randomly planted with a diverse canopy of shade trees, many of which yield useful products themselves (Perfecto and Vandermeer 1994). The vegetative structure of a traditional, natural, or passive organic coffee farm is forest-like. Several layers of tree canopy provide shade to the coffee plants. The shade species may include legumes, fruit trees, banana plants, and/or hardwood species. The presence of all these species creates a stable production system, with protection from soil erosion, favorable temperature and humidity regimes, constant replenishment of soil organic matter via leaf litter production (Martinez and Peters 1991a), and an array of beneficial insects that act to keep potential pests under control (Perfecto and Vandermeer 1994). Traditional and/or passive organic coffee is considered to be the most ecologically sound agro-forestry system in Mesoamerica (Rice 1993, 2000; Perfecto et al. 1996).

Intensive Conventional or Technified Coffee: Chemical Cultivation

The green revolution came late to coffee production, but in the last few decades green revolution-style *technified production* has spread dramatically in Latin America's large and medium-size coffee plantations, as well as in other parts of the world (Rice 1993; Richter 1993a). This so-called modernization of the production process consists of the use of high-yielding varieties of seeds, agrochemical inputs, and significant reduction or outright elimination of shade (Rice 1993).

Typically the coffee plants and the shade trees are partially or completely removed, and the coffee is replaced with new sun-tolerant coffee hybrids planted in open-rows[1] along the slope, while the use of purchased chemical inputs like fertilizer, herbicide, and fungicide is initiated or intensified (ibid.). "Technification," "renovation," "rehabilitation," "revitalization," and simply "technology transfer" are among the terms that have been used to describe this process.

Technification has generally introduced low stature, compact varieties of *Coffea arabica*. The most common variety is *Caturra*, a mutant dwarf variety discovered in Brazil in the last century, which yields its first crop in the third year—almost two years earlier than the traditional varieties of *Tipica* or *Bourbón* (ibid.). A hybrid known as robusta, a cross between *C. arabica* and *C. canephora*, developed in Timor, was also introduced because of its resistance to rust diseases. Planting densities typically changed from 2 meters between plants in traditional coffee to only 35–40 centimeters on the densely planted technified plantations.

The new varieties have high response rates to chemical inputs such as fertilizer, and are relatively sun tolerant. Weeds proliferate without the shade, so the new systems are very intensive in their use of herbicides and/or manual labor for weeding. Part of this heavy use of agrochemicals has been a product of loan requirements, in which growers had to agree to purchase specific chemical inputs in order to get bank credit for technified production (Nolasco 1985). With technification, coffee went from being an exception among export crops, in that it had a low import coefficient because of its low reliance on imported chemical inputs, to a more typical export crop, which (like cotton and bananas before it) now depends heavily on agrochemicals.

Unlike traditional cultivation, technified production does not require that the farmer have any detailed knowledge of plant growth and ecology in particular microhabitats, because it is base on a "one size fits all" technological "package." Only large and medium coffee producers have fully adopted this intensive technology in Chiapas, and these are not the subject of this chapter. However, many small producers have adopted part of the technification "package," like agrochemicals for fertilization and weeding, and they have also reduced, though not eliminated, their shade, and may or may not have renovated their coffee varieties. This somewhat technified combination of natural technology with the use of chemical fertilizers and/or pesticides is termed *chemical technology* in this chapter.

Intensive Organic Coffee

Organic farming methods are similar to those of traditional or natural production in that substantial shade is used, though typically the density and diversity of shade is somewhat lower than in the "natural" technology. These methods also resemble those of more technified production in that there is more intensive use of inputs, although organic inputs and human labor are substituted for the agrochemicals used in chemical methods (Sánchez López 1990). One way to think of it, which will later be demonstrated with data in this chapter, is that there are two ways to boost—or "intensify"—the levels of production on traditional coffee farms. One is to technify; the other is to implement intensive organic farming practices. While chemical-intensive technification is environmentally destructive, as shall also be shown, organic, like natural production, is considered to be a relatively stable and ecologically sound agro-forestry system.

In organic production, plant care involves pruning (a first and second thinning) and the *recepa,* which is the practice of cutting back the whole tree, leaving just a small part from which the tree will regenerate itself. This pruning boosts berry production on the plant. Each plant is examined to determine the specific care it needs. Intensive organic farming is the only technology in which the laborious work of carefully selecting the branches that are going to be cut is done twice.

The construction of terraces is another unique feature of organic production, and is usually performed on a plant-by-plant basis, in which a flat area is built up around each plant to prevent erosion. Terraces demand a great deal of labor, and the time required to build them depends of the type of soil. The time required increases where there are steep slopes and rocky soils. This is perhaps the most difficult and labor-intensive activity of the organic coffee production technology and one which may not always be necessary, as shall also be shown in this chapter.

Transition to Organic

A large number of farmers are currently in the process of intensifying their traditional, natural production, using organic methods. Many of them have not yet qualified for organic certification, nor have they finished the technological transition. Therefore, in this chapter I refer to them as *transitional*.

Case Studies

In order to examine the productive, economic and ecological impacts of the techno logical choices that peasant farmers in Chiapas, I carried out a survey from 1997 to 2000. The participants in the study came from 36 communities in coffee-growing areas of Chiapas located on the mountain ranges running northwest to southeast, parallel to the coast across twelve municipalities and six regions of Chiapas.[2] (See figure 5.1.) With rich biodiversity and multiple microclimates, the Eastern Highlands, Western Highlands, Jungle, North, Sierra, and Soconusco regions of Chiapas are prime areas for coffee production.

The survey was designed to document productive practices and economic and environmental outcomes across a range of environments and organizational types. Questions were asked about land area and number of parcels, chemical use, agronomic techniques, other crops, family history, structure and ethnicity, yields and incomes. Ecological data was taken on soil erosion, ground cover with leaf litter and humus, and shade biodiversity. The goal was to be able to relate the outcomes to the productive practices, organizational membership and strategies, and geographic location of the families surveyed.

The Highlands is the most populated area of Chiapas, and one of the poorest. The greatest number of participants were from the ethnically diverse West and Eastern Highlands. The Tzotzil and Tzeltal indigenous families that participated in this study were from the Eastern Highlands region, including seven communities located in the municipalities of Chenalhó and Tenejapa. The Western Highlands participants were drawn from six Tzotzil communities distributed between the municipalities of El Bosque and San Andrés Larrainzar.

To the north of the Highlands is the North Region with low, mainly jungle lands. The mostly Tzeltal participant families in this region come from six communities located near the city of Ocosingo and the municipality of Chilón. This is an area of lower altitudes, 700–1,200 meters above sea level. Mixed forest and jungle are the main vegetation.

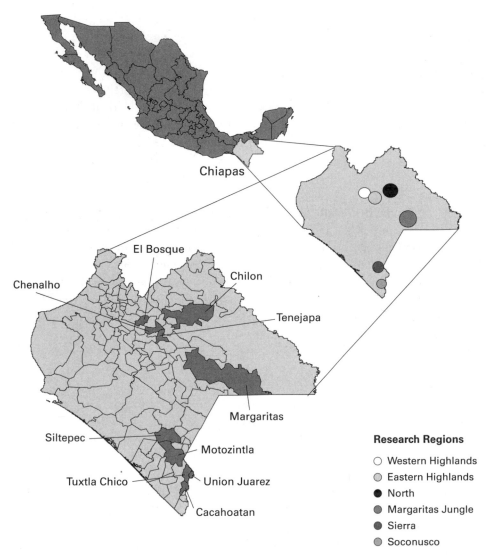

Figure 5.1
Coffee regions and municipalities in Chiapas. Source: GIS-IDESMAC, Mexico, 2001.
Designed by María Elena Martínez and Aurora Becerril.

A series of valleys and lower mountains oriented NW-SE, are located in the eastern section of Chiapas, and form the Jungle region. Ranging from sea level to 700 meters, with some mountains over 1,000 meters, the major part of the Jungle region is covered by the Lacandon Jungle. A smaller jungle in the southern part of the region is the Margaritas Jungle, from which the municipality of Margaritas takes its name. The soils of these areas can sustain jungle vegetation but once used for annual crop farming, the soils are good only for three or four harvests. The area visited in this survey was the Margaritas Jungle. This region encompasses the lowest-altitude coffee-producing lands in this study, ranging between 500 and 700 meters, and has a mostly migrant population originally from the Highlands and other parts of Mexico.

The Sierra region is located in the Sierra Madre mountains of Chiapas. Ranging from 1,000 to 3,000 meters above sea level, this fractured terrain has a dryer climate than other parts of Chiapas. The center of Sierra region, and the main commercial center, is the town of Motozintla.

In the Soconusco region, the participants come from eight communities in three municipalities. All of the participants in the Soconusco are Spanish-speaking *mestizo* families, mainly descendants from Guatemalan indigenous people. The Soconusco region encompasses a coastal plain and a mountain range with altitudes from sea level up to 4,000 meters. The research covered communities located on the slopes of the highest mountain in the region, the Tacana volcano (4,060 meters). It has the best type of soils for coffee production, and Soconusco is the number one coffee area, producing 65 percent of the coffee in Chiapas (Alvarez Siman 1996).

The 150 families surveyed in this study belong to 6 major organizations of small coffee farmers, and to several smaller ones, as shown in table 5.1. Details of the research methodology used in the survey can be found in Martínez-Torres 2003. The organizations were chosen to be representative of the larger variation among all such organizations in terms of their technology strategy (do they promote organic farming? Are their members mostly chemical-intensive producers?), their ethnic identification (does the organization represent itself to its members and to society as "indigenous"?), and their political affiliation (are they pro-Zapatista? are they aligned with the PRI? are they independent?). The following subsections are based on the responses to the survey questions that I administered, and on ecological data that I collected on the coffee farms.

Economic Benefits from Organic Farming

As described above, different technological approaches are used in coffee farming in Chiapas—natural, chemical, transitional, and organic. We want to know what the benefits—and costs—of these strategies are for the families that implement them.

Table 5.1
Profiles of organizations studied.

	Technological focus				Ethnic identification				Political stance			
	Organic	Natural	Chemical	Transition	Indigenous	Mixed[a]	Mestizo	Pro-EZLN	Pro-PRI	Independent/ Autonomous
ISMAM	X				X					X
Lazaro Cardenas		X					X		X	
La Selva	X					X				X
Majomut	X	X		X	X			X		X
MutVitz		X		X	X			X		
Tzotzilotic		X	X	X		X				X
Other	X	X	X	X		X			X	X

a. Includes indigenous and mestizo.

The focus here is on yield (the amount of coffee produced per unit of area) as a measure of the productivity of smallholder coffee production, and thus as one indicator of economic benefits for farm families. In general we expect that as farmers intensify production by investing more labor, capital and inputs in it, yield goes up. A caveat must be inserted however, which is that yield—which refers to the quantity of a single product (i.e. coffee) harvested per unit area—tends to underestimate the "true" productivity of smaller farmers with more diverse cropping systems (Rosset 1999). This is because small-scale farmers often produce many different crops, and even animals, yet yield refers to only one of these products, and fails to take into account the others produced in the same area.

Furthermore, intensification itself may occur by adding additional components to the mix (i.e. by intercropping, or, in the case of coffee, by adding shade-tree species that produce useful products), rather than by increasing the intensiveness of production of single crop. Nevertheless, the productivity of coffee *is* of prime importance to farmers for whom it is their major income source, and there exists a common tendency to intensify the production of coffee itself, in order to boost this income source. Thus in the following section of this chapter the focus is on coffee *yields* per se.

How Technologies and Geographies Affect Yields

The expectation one has in terms of technology and yield depends on our preconceived notions. The people who believe that organic farming is backward (e.g., Avery (1995))—a return to pre-modernity—expect that organic farmers will have much lower yields than chemical farmers. These organic farming skeptics tend to see "organic" as the antithesis of "intensification," and "chemical" as its realization. Proponents of organic farming, on the other hand, feel that both chemical and organic are different ways to intensify production (i.e. to boost productivity). In the former case, organic, and transition to organic yields would be expected to differ little from those in natural (non intensive) production, while chemical yields would be much higher than any of the others. This pattern might be represented as Chemical > Organic = Transitional = *Natural* (chemical greater than organic, transitional and natural, all of which are the roughly the same as each other).

In the case of the organic farming advocate, on the other hand, chemical and organic yields would be expected to be similar to each other, both being significantly higher than in natural production, with transitional to organic somewhere in between. This might be represented as Chemical = Organic > Transitional > Natural.

A simple analysis[3] of the 110 families from whom yield estimates were obtained, revealed that natural producers obtain an average of 7.54 quintales per hectare

Table 5.2
Comparison of yield (qq/hectare) by technology ($P = 0.1544$, $N = 110$).

Technology	Mean (± standard error)
Natural	7.54 (±1.68)
Transition	8.44 (±1.02)
Organic	10.26 (±0.78)
Chemical	11.40 (±1.31)

Table 5.3
Average yield by region ($P < 0.01$, $N = 110$).

	Mean	± Standard error
Jungle	5.68	1.43
North	9.86	1.30
Highlands, West	9.68	0.98
Highlands, East	8.86	1.07
Soconusco	13.07	1.62
Sierra	12.40	1.49

(qq/hectare)—the usual measure of coffee production (1 quintal = 100 pounds of green coffee beans)[4]—those in transition obtain 8.44 qq/hectare, those using organic technology 9.77, with 11.30 qq/hectare for those using chemicals (table 5.2). This pattern falls somewhere in between the two stereotyped expectations described above, with a pattern of Chemical > Organic > Transition > *Natural*; however, the data are not quite statistically significant, making it difficult to conclude that this is a real pattern and not just simply due to chance.

It is likely that the effect of using all four technologies in different regions, with altitudes, soils and weather patterns radically different in their suitability for coffee, statistically confounds the effect of technology, and that is why the results are not statistically significant. In fact, when analyzed alone, the geographic region in which a farm is located does very much affect yields. A simple analysis of how yield varies by region revealed significant differences (table 5.3). On average, the yield was highest in the Soconusco region (13 qq/hectare), followed by the Sierra (12 qq/hectare), the North (9.9 qq/hectare) the Western Highlands (9.7 qq/hectare), the Eastern Highlands (8.9 qq/hectare), and finally the Margaritas Jungle with the lowest yields (5.7 qq/hectare). The order of the yields was Soconusco > Sierra > Western Highlands ≥ North > Eastern Highlands > Jungle (table 5.3), which makes

Table 5.4
Yields by technology, corrected for effect of region ($P < 0.05$, $N = 110$).

	Mean yield (qq/hectare)	± Standard error
Natural	8.57	1.65
Transitional	8.87	1.92
Organic	9.85	1.80
Chemical	10.32	3.44

sense in that the Soconusco has the best coffee soils and the Jungle the worst. The Soconusco was the original area where coffee was introduced into Chiapas.

Given the significant effect of region on average yield, it is likely that the effect of technology in the earlier analysis was, indeed, confounded by the regions where the technologies were employed. In order to separate confounding effects (which could include a confounding effect of technology on region as well), it was necessary to use an analysis using multiple variables.[5]

The analysis for regions and technologies revealed significant effects on yields of both regions and technologies, and one significant interaction between them (a statistical interaction would be where the effect of a particular technology is different in different regions). Three regions had distinct effects in the statistical model (the Jungle, Soconusco and Sierra), as did all of the technologies, plus a very positive interaction between organic technology and the Sierra region. Perhaps this interaction is due to the effect of the ISMAM cooperative, which is dominant in the Sierra region, and has been producing organic coffee for many more years than any other organization. Its members may be more skilled at coffee-production practices. A simple statistical model relating yield (qq/hectare), Y, to regions and technologies was generated:

$$Y = 8.6 - 3.8(J) + 2.9(So) - 1.3(Si) + 0.3(T) + 1.3(Or) + 1.8(C) + 6.0(Or \times Si),$$

where J = Jungle, So = Soconusco, Si = Sierra, T = Transitional, Or = Organic, C = Chemical, and the variables for regions and technologies are what statisticians call "dummy variables" which can take on values of 0 or 1 (that is, they are either present or not, or "on" or "off").

This model allows us to estimate yields for each technology, corrected for the effect of regions. These corrected yields are presented in table 5.4 and shown graphically, in comparison to the raw, uncorrected yields, in figure 5.2.

It is worth noting the difference between the raw yields, which were not statistically different from one another, and the corrected yields, which do differ significantly from one another according to the statistical model. The neat "step-by-step"

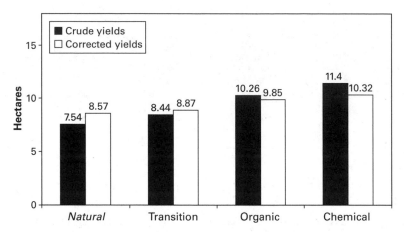

Figure 5.2
Comparison of raw yields by technology, with yield corrected for effect of region.

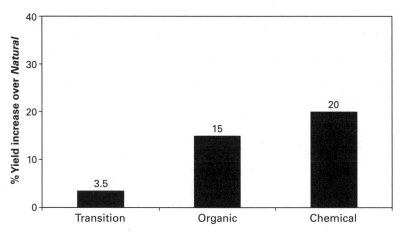

Figure 5.3
Average percent yield increase obtained with transitional, organic and chemical technologies, compared to *natural* technology (based on mean yields corrected for effect of region).

pattern of increases along the presumptive continuum of intensification from natural to transitional through organic and then chemical—which is observed with the raw yields, has been replaced by a pattern in which a slight increase between natural and transitional is followed by a jump to organic and chemical, which do differ from one another, though by less than in the non-significant raw data. In figure 5.3, the average percent yield increase—over natural technology—obtained with transitional, organic and chemical technologies is presented. While this finding is still intermediate between the expectations of our hypothetical organic farming advocates and organic farming skeptics, it is much closer to the expectation of the former, Chemical = Organic > Transitional > *Natural*, than to the latter, who expected Chemical > Organic = Transitional = *Natural*. In other words, it lends support to the hypothesis that investment in natural capital via organic farming practices is indeed a viable alternative route to the intensification of coffee production—i.e. an alternative to chemical intensification. While conventional wisdom holds that organic farming is the low-yield "opposite" of intensification, these data reveal that assumption to be false in this case.

Returning to the topic of yield correction for the effect of region, it indeed turns out to be the case with the raw data that region confounded the effect of technology

Table 5.5
Corrected mean yields for each combination of region and technology. Differences are significant ($P < 0.05$, $N = 100$).

Region	Technology	Corrected yield
Jungle	*Natural*	4.8
	Transition	5.1
	Organic	6.1
	Chemical	6.6
Soconusco	*Natural*	11.5
	Transition	11.8
	Organic	12.8
	Chemical	13.3
Sierra	*Natural*	7.3
	Transition	7.6
	Organic	14.6
	Chemical	9.1
Highlands (East, West, North)	*Natural*	8.6
	Transition	8.9
	Organic	9.9
	Chemical	10.4

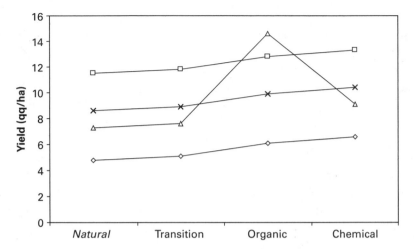

Figure 5.4
Corrected mean yields plotted by region and technology.

on yield, and vice versa, that technology confounded the effect of region on yield. The statistical model was used to generate corrected yields for each combination of region and technology, in order to view the full picture (table 5.5).

A plot of the corrected yields (figure 5.4) shows a uniform pattern for both technology and region, with the exception of organic coffee in the Sierra region. The Sierra region is where ISMAM dominates organic coffee production, and clearly they are doing something right—in that organic technology actually out-produces chemical technology in this region. While this may speak, on the one hand, to ISMAM and probably to agro-climatic conditions in the Sierra and their aptness for organic production as well, it also indicates the potential that organic production offers, as an intensification strategy, which is as yet not fully exploited in most regions.[6]

Income and Technology

When one is speaking of a cash crop, rather than subsistence production, yield advantages mean little if not translated into income gains. Income reflects a combination of yield, area harvested, quality and price, the latter of which is mitigated by a variety of factors such as sale to the organization versus sale to an intermediary, or *coyote*. This section analyzes average per hectare annual gross income from coffee.

Figure 5.5 presents gross annual income per hectare of coffee. Here there is a substantial step from natural (5,523 pesos/hectare) to transitional (8,867 pesos/hectare), and from transitional to organic (9,560 pesos/hectare), which is virtually the same as chemical (9,732 pesos/hectare), although the differences are not quite

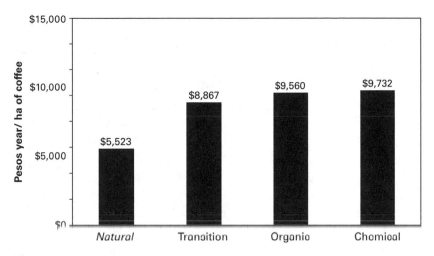

Figure 5.5
Average gross income from coffee per hectare of coffee.

statistically significant. This lends further support for the thesis being developed here that organic farming is a viable alternative to chemicals in terms of intensification—especially for cash poor families in an environment characterized by high underemployment and a low opportunity cost for extra family labor, typical of much of Chiapas—as the investment in organic technology is "cash cheap" and labor intensive, the opposite of chemical technology (Heinegg and Ferroggiaro 1996). A technology that is more labor intensive can be an advantage or a disadvantage depending on alternative employment opportunities, or lack thereof.

In many remote regions of Chiapas, such as the Jungle area, there are few opportunities to turn underutilized family labor into income. Families in such regions have little cash to invest in their coffee production, but they do have family labor available. Given the alternatives of a labor-saving, capital-intensive route to intensification—chemical production—versus a capital-saving, labor-intensive route—organic farming—the advantages of the latter are clear. Thus, organic farming may confer an advantage to families with the ability to "self-exploit" their own labor, in the sense of the phrase as used by Chayanov. (See Chayanov 1986 and Ellis 1988 for discussions of the peasant economy and self-exploitation.) The concept of self-exploitation refers to one of the competitive advantages that Chayanov attributed to family farming, namely that the use of family rather than hired labor means that the family farm often has a more committed, harder-working, and more responsible labor force than the capitalist farm that relies on alienated wage laborers. Migration is another alternative for excess family labor—thus it may be fairer to say, in the words of an informant from the Roberto Barrios community in the non-coffee-producing part of the Jungle area, that "organic farming is the alternative to migration if we want to hold our communities together" (indigenous agroecology promoter, interview, April 2002).

Of note is the fact that the income differences between organic and chemical production are much smaller than the yield differences, highlighting the price premium paid for organic coffee. It should be pointed out that during the years 1994–1998, for which these data were collected, coffee prices were much higher than they have been since. When coffee prices are high, the price differential between organic and conventional coffee is relatively slight compared to low-price years, when the relative premium for organic is much greater. Thus, it would be reasonable to assume that if similar data were available for subsequent years, it would show that organic production actually outperforms chemical in terms of gross income. This demonstrates the worthiness of conversion to organic production as an investment in natural capital, and the fact that it pays off in economic terms because certification allows the positive benefits for the environment to be "internalized," and thus realized in monetary terms, in the form of price premiums. Here we must point out that this internalization of benefits requires the prior formation of significant social capital, without which certification and price premiums are impossible to achieve.

Ecological Benefits of Organic Farming

If we are interested in organic farming as an approach that may contribute more to "sustainable development," then we must go beyond economic benefits and look at ecological ones as well. In general terms, farming of any kind has impacts on ecological variables like biodiversity and the quality of the soil (erosion and soil fertility). Not only are these important aspects of the environment from a purely ecological or conservationist point of view, but they may also have significant implications for the sustainability of production into the future. If all the topsoil is eroded away, for example, then future production will become impossible. Thus what are "ecological variables" today, often translate into "economic variables" for the future. In this sense, we postulate that when farmers make investments in their natural capital (i.e. their soil or their biodiversity), these investments bring a short-term ecological return, which over the medium term is very likely to also be an economic return, as we shall see below.

Ecological Indicators Measured in This Study

As ecological indicators of sustainability, soil erosion, leaf litter/humus, and shade biodiversity were selected. Soil erosion is probably the most direct short-to- medium-term threat to the sustainability of production, as this implies the direct loss of

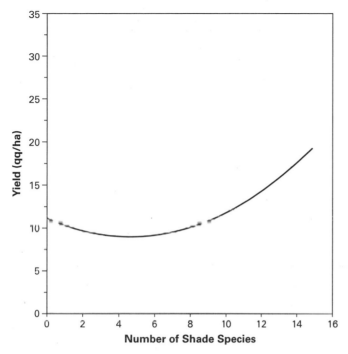

Figure 5.6
Yield by number of shade species. This is a curve statistically fitted to the data. The equation is Yield $= 11.1 - 0.9n + 0.1n^2$, where n is the number of shade species present in the coffee grove ($r^2 = 0.09$, $P = 0.03$).

natural capital. The area covered by leaf litter is a factor that may inhibit future erosion, as the soil is somewhat protected from the direct action of water and wind when it is completely covered by a layer of leaf litter. The depth of the leaf litter and humus layer reflects the organic matter that will contribute to future soil fertility via decomposition. Shade biodiversity is a strong determining factor for most other elements of biodiversity—which depend directly or indirectly on vegetational resources—ranging from birds, mammals, insects, and other organisms in the above ground environment to the soil biota which is critical to soil fertility and is most likely enhanced by a greater diversity of leaf litter being decomposed. For the details of how these indicators were measured in this study, see Martínez-Torres 2003.

Shade Diversity and Yield
An interesting finding from the ecological data collected on the farms of the participants in this study was the effect of the number of shade species on yield. There are some studies that find shade diversity to be an important element in determining

Table 5.6
Percentage of farmers for each type of technology, with up to the given number of shade species in their coffee grove.

	Up to 2	Up to 4	Up to 6	7+
Chemical	36	55	9	0
Transition	28	24	28	20
Organic	24	40	17	21
Natural	29	36	29	7

coffee yields (Soto 2001), and this study supports those findings. By statistically fitting a curve to yield versus the number of species of shade trees on each farm (figure 5.6), a significant positive relationship was found. Thus diversity has a positive effect on yield (although it may also be a case of mere association), perhaps by promoting decomposition and soil biology, or maybe by enhancing natural enemies of pests and diseases of the coffee trees. Whatever the underlying mechanisms, however, this suggests that planting a greater diversity of shade species in their coffee groves is a useful way for coffee farmers to invest in their natural capital.

Shade and Farming Technology
Table 5.6 presents the number of shade species that farmers using different technologies have in their coffee groves. It is interesting—and not surprising—to note the far lower shade diversity among chemical growers: 91 percent of them had fewer than five species of shade trees, while 36 percent of *natural*, 48 percent of transitional, and 38 percent of organic farmers had five or more species of shade trees in their groves.

Oddly, highly diverse systems were more common in transitional and organic settings than with natural technology, perhaps as a result of the strong promotion of diversity on the part of those organizations that are devoted to organic farming.

Erosion: Effects of Slope and Terraces
In the broken terrain of Chiapas, erosion is a critical indicator of the long-term sustainability of production. A highly significant positive linear relationship was found in this study between erosion and slope.[7] Because this relationship is well known, it has been common to construct terraces in coffee plantations on steep slopes. In fact, they are required for organic certification (Dardon 1995; IFOAM 1995). These are supposed to function to slow erosion and to promote better soil quality and water retention, though producers often complain that the labor required is excessive and they do not function well. There are two kinds of terraces: small ones for each individual plant and long ones, which follow contours. In this study I found

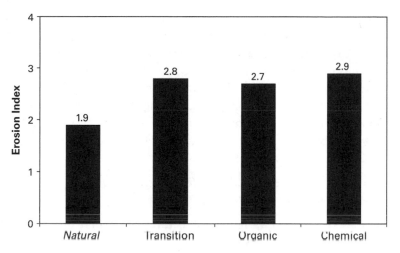

Figure 5.7
Erosion index by technology ($P = 0.13$, $n = 97$).

no statistical relationship whatsoever between either kind of terrace and erosion,[8] supporting the farmers' assertions and other studies. (See Pérez-Grovas 1996. Pérez-Grovas was the technical coordinator of an organic coffee coop.) This suggests that the laborious work of constructing these terraces is not a fruitful way to invest in improving natural capital.

For those farmers that used living barriers instead of terraces, the erosion was much lower, though they were not a sufficient number of cases for statistical analysis of firm conclusions.[9] Nevertheless, this is highly suggestive and worthy of further examination, especially as they are far less labor intensive to construct and maintain than terraces. This is the same conclusion reached by Pérez-Grovas (1996). Eliminating the requirement of terraces for certification might make conversion to organic both more attractive to farmers and more profitable (as it would lower labor costs), and living barriers should be explored as an alternative option. (For the same conclusion, see AICA 1997.)

Erosion and Farming Technology
A simple statistical analysis of erosion on farms with each type of technology revealed an almost significant relationship where natural technology had less erosion than the other systems (figure 5.7).[10] Here it should be noted that natural technology is the only system where the ground is relatively undisturbed, with no soil being moved in the process of weeding or terrace construction. Since moving soil increases the likelihood of erosion, this makes sense. As we shall see below,

there is also greater coverage of the soil with leaf litter in the natural system. This is an area that organic farming promoters and farmers need to work on, though it is also worth noting that the smaller-scale chemical farmers like those in this study have more shade and thus more leaf litter than do large-scale chemical farmers.

In order to discover as many of the causal factors behind erosion as possible, I carried out an exploratory statistical analysis[11] for all of the variables measured, to determine which would be related to erosion, and how. Those variables that were included in the highly significant model[12] were slope, technology, and the proportion of the ground covered by leaf litter and humus. The final model was

$$\text{Erosion Index } (1\text{--}4) = 1.6 + 0.4(\text{slope}) - 0.2(\text{leaf litter index})$$
$$- 0.9(\text{natural technology}),$$

where natural technology is a dummy variable with a value of either 1 or 0 (either the farmer uses natural technology or doesn't use it).

We can conclude that these are the important variables in determining the degree of erosion to be expected, with slope as the driving force, mitigated by keeping the ground covered with leaf litter and humus, and when natural technology is used.

Leaf Litter and Humus

It is precisely leaf litter/humus that was chosen as the final ecological indicator. In this study it was estimated both as an index of the proportion of the ground covered, and by the average depth of the leaf litter/humus layer. Leaf litter coverage varied significantly by technology, as shown in figure 5.8, where chemical technology had the least coverage, and natural the most, followed by organic. This reinforces the role of natural technology in stemming soil erosion, and probably is a positive indication for transitional and organic as well.

The truly remarkable results are those for the average depth of the leaf litter and the humus layer, a key indicator of future soil fertility. Here there was a very significant relationship with technology. The average depth on organic (7.2 cm) plots was nearly double that of *natural* and transitional (both 4.6 cm), which in turn had greater average depth than chemical plots (3.2 cm), as shown in figure 5.9. Here we could say that organic practices appear to be seriously improving the soil, lending support to the notion that the process of conversion to organic is a real investment in building natural capital for the future.

Organic Farming, Social Capital, and Sustainability

The data presented in this chapter clearly demonstrate how *investment* in natural capital, via conversion to organic, the introduction of shade biodiversity, and the

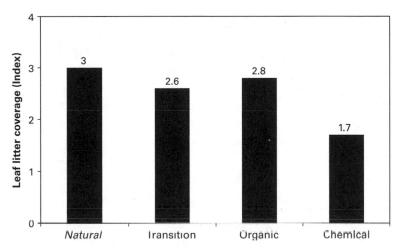

Figure 5.8
Leaf litter coverage by technology. Index (1–4) of proportion of ground covered by leaf litter and humus ($P < 0.01$, $n = 86$).

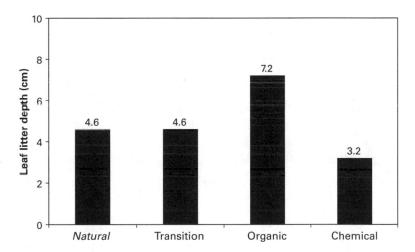

Figure 5.9
Average depth of leaf litter and humus layer, by technology ($P < 0.01$, $n = 86$).

build-up of leaf litter and humus, pay off in productive (yield) and economic (gross income) terms, and in terms of ecological indicators of the future sustainability of production (erosion prevention, future soil fertility). Clearly these factors make the strategy of investment in natural capital by organic farming a viable alternative to chemicals in terms of the intensification of coffee agriculture. Part of the economic payoff to these investments comes via the *internalization* of the environmental benefits of organic farming to society at large, via certification and price premiums. Under the conditions of Chiapas, organics provide a "Chayanovian" mechanism to turn under-utilized family labor into income—a factor, which might even provide an alternative to out-migration and the community breakdown it generates.

The payoffs from the internalization of benefits would be impossible without the earlier formation of significant social capital, because only effective organizations are able to obtain and maintain certification in the organic market. Given the amount of technical assistance provided by these organizations to the farmers who are undergoing the transition to organic, by either professional staff or by campesino-promoters, and the organizational learning involved, it is unlikely that even the natural capital investment of converting to organic could have been undertaken without having had sufficient social capital already in place.

Sustainable development is supposed to be economically viable, socially just and ecologically sound, in the sense both of not damaging the external environment and of conserving or enhancing the resource base for future production. While "socially just" in the larger sense may mean a broad distribution of the benefits of development, in the case of coffee-production technologies, we might evaluate "socially just" as encompassing those options which are appropriate for smaller and poorer farmers. In that context, the low investment embodied in natural technology, and the route to intensification based on applying more labor and enhancing natural assets with organic technology, would be favored over the more capital-intensive chemical approach (at least under conditions like those of much of rural Chiapas, where there is a low opportunity cost for family labor). The data on the outcomes of the different technological options evaluated here allow us to examine the remaining two dimensions of sustainability. We can use yield (productivity) and gross income as proxies for economic viability, and erosion, ground cover, depth of leaf litter and humus, and shade biodiversity as indicators of ecological sustainability.

In figure 5.10 the values of these parameters are plotted on radar or kite graphs as a visual display of the relative sustainability of the four technological approaches. In this graph, the larger the area covered by the diagram, the more "sustainable" the system is (Monzote et al. 2002). One can see the advantages of the organic approach quite clearly in this way. It is equal or superior to the other three in all

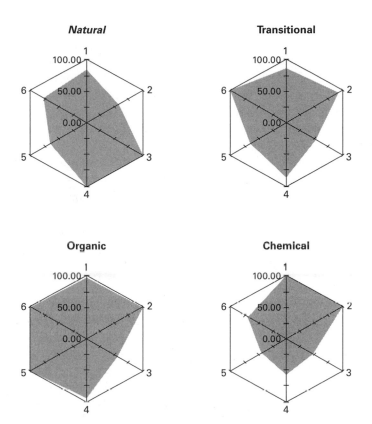

Figure 5.10
Measures of indicators of sustainability by technology. 1 = Corrected yield; 2 = Gross income/hectare; 3 = Erosion index (inverse); 4 = Ground cover index; 5 = Depth of leaf litter/humus; 6 = Average number of shade species. All indicators have been placed on a relative scale of 0 to 100.

dimensions except soil erosion—where natural technology is superior—and even here one might imagine that its superior ground coverage with leaf litter might, over time, give it an advantage. Seen this way, organic farming of coffee bests natural technology in economic terms and is superior to chemical technology in ecological terms, providing the best overall combination of productivity today, plus the likely sustainability of that productivity into the future. If we add to that the advantages mentioned above in terms of socially just criteria, then it would be fair to conclude that overall, this approach—which is based on investing in natural capital and internalizing benefits to society—is the most sustainable option.

Notes

1. This approach uses greater air flow and penetration of sunlight to combat coffee rust—a fungus called *roya* in Spanish—that usually affects the arabica varieties.

2. For detailed information on participant families, see Martínez-Torres 2003.

3. In this case a simple analysis of variance for the 110 cases where yield estimates were obtained in the interviews (averaged over years to reduce variation). For details on the statistical analyses utilized in each case, see Martínez-Torres 2003.

4. qq/hectare is the standard measure of coffee yield. qq stands for quintales (hundredweight). One quintal equals 100 pounds (57.5 kilograms) of green coffee.

5. Stepwise multiple regression. For the statistical methodology used in this and all other analyses, see Martínez-Torres 2003.

6. For data on yield and the other variables reported in this chapter analyzed by organization, see Martínez-Torres 2003.

7. Erosion Index (1–4) = 0.27 + 0.41(index of slope), $P < 0.001$, $r^2 = 0.33$.

8. $P > 0.05$.

9. Living barriers are perennials planted along contour lines to slow erosion.

10. In a later analysis in this chapter, where more variables are controlled for, this relationship is found to be significant.

11. Stepwise multiple regression. For details of this and the other statistical analyses presented here, see Martínez-Torres 2003.

12. $P < 0.001$; $r^2 = 0.44$.

References

AICA. 1997. Evaluación del programa de producción de café orgánico de Las Margaritas, Chiapas, Mexico. Informe final: Mexico.

Altieri, Miguel A. 1995. *Agroecology: The Science of Sustainable Agriculture*. Westview.

Altieri, Miguel A.1996. Hacia un concepto de salud agroecológica. In *Ecología aplicada a la agricultura: Temas selectos de Mexico*, ed. J. Trujillo Arriaga et al. Universidad Autónoma Metropolitana-Xochimilco, Mexico, D.F.

Altieri, Miguel A.1999. Enhancing the Productivity of Latin American Traditional Peasant Farming Systems through an Agroecological Approach. Paper presented at International Conference on Sustainable Agriculture: An Evaluation of New Paradigms and Old Practices, Bellagio Center.

Alvarez Simán, F. 1996. *Capitalismo, el estado y el campesino en México un estudio sobre la región del Soconusco en Chiapas*. Universidad Autónoma de Chiapas.

Avery, Dannis. 1995. *Saving the Planet with Pesticides and Plastic*. Hudson Institute.

Borlaug, N. 1992. Small scale agriculture in Africa: The myths and realities. *Feeding the Future* 4, no. 2.

Boserup, Ester. 1965. *The Conditions of Agricultural Growth: The Economics of Agrarian Change under Population Pressure*. Aldine.

Carruthers, David V. 1995. Agroecology in Mexico: Linking environmental and indigenous struggles. Paper presented at Conference on the Politics of Sustainable Agriculture, University of Oregon.

Chayanov, A. V. 1986. *The Theory of the Peasant Economy*. American Economic Association.

Dardon, José A. 1995. *Conferencia internacional sobre café orgánico: memorias*. Mexico: IFOAM-AMAE-UACH.

Ellis, Frank. 1988. *Peasant Economics: Farm Households and Agrarian Development*. Cambridge University Press.

Hazell, P. 1995. Managing Agricultural Intensification. IFPRI.

Hewitt, T. I., and K. R. Smith. 1995. Intensive Agriculture and Environmental Quality: Examining the Newest Agricultural Myth. Henry Wallance Institute for Alternative Agriculture.

IFOAM. 1995. Basic Standards for Organic Agriculture and Food Processing.

Lele, Sharachandra M. 1991. Sustainable development: A critical review. *World Development* 19, no. 6: 607–621.

Martinez, Eduardo, and Walter Peters. 1991. El sistema agroforestal de la Finca Irlanda. Descripción y actividades. Technical report, Tapachula, Mexico.

Martinez, Eduardo, and Walter Peters. 1994. Cafeticultura ecológica en el estado de Chiapas: Un estudio de caso. Tapachula, Mexico.

Martínez-Torres, María Elena. 2003. Sustainable Development, Campesino Organizations and Technological Change Among Coffee Producers in Chiapas, Mexico. Doctoral dissertation, University of California, Berkeley.

Martínez-Torres, María Elena. 2006. *Organic Coffee: Sustainable Development by Mayan Farmers*. Ohio University Press.

Monzote, Marta, Eulogio Muñoz, and Fernando Funes-Monzote. 2002. The Integration of Crops and Livestock. In *Sustainable Agriculture and Resistance: Transforming Food Production in Cuba*, ed. F. Funes et al. Food First Books.

Nolasco, Margarita. 1985. *Café y sociedad en Mexico*. Centro de Ecodesarrollo, Mexico.

Paarlberg, R. L. 1994. Sustainable farming: A political geography. IFPRI.

Perezgrovas Garza, Victor. 1996. Evaluación de la sustentabilidad del sistema de producción de café orgánico en la Unión Majomut, en la región de los Altos de Chiapas. Master's thesis, University of Chapingo, Mexico.

Perfecto, Ivette, and John Vandermeer. 1994. Understanding biodiversity loss in agroecosystems: Reduction of ant diversity resulting from transformation of the coffee ecosystem in Costa Rica. *Entomology-Trends in Agriculture* 2: 7–13.

Perfecto, Ivette, Robert Rice, Russell Greenberg, and Martha E. Van der Voort. 1996. Shade Coffee: A Disappearing Refuge for Biodiversity. Draft Report, School of Natural Resources, University of Michigan, Ann Arbor.

Pretty, Jules. 1995a. Participatory learning for sustainable agriculture. *World Development* 23, no. 8: 1247–1263.

Pretty, Jules. 1995b. *Regenerating Agriculture: Policies and Practices for Sustainability and Self-Reliance.* Earthscan.

Pretty, Jules. 1997. The sustainable intensification of agriculture. *Natural Resources Forum* 21, no. 4: 247–256.

Pretty, Jules. 1998. Sustainable agricultural intensification: Farmer participation, social capital, and technology design. Paper prepared for World Bank's Rural Week, Washington.

Pretty, Jules. 1999. Current challenges for agricultural development. Paper presented at Kentucky Cooperative Extension Service Conference, Lexington.

Reijntjes, Coen, Bertus Haverkort, and Ann Waters-Bayer. 1996. *Farming for the Future: An Introduction to Low-External-Input and Sustainable Agriculture.* Macmillan.

Richter, Michael. 1993. Ecological effects of inappropriate cultivation methods at different altitudes in the Soconusco region of Southern Mexico. *Plant Research and Development* 37: 19–44.

Rice, Robert A. 1993. New Technology in Coffee Production: Examining Landscape Transformation and International Aid in Northern Latin America. Smithsonian Migratory Bird Center.

Rice, Robert A. 2000. Managed biodiversity in shade coffee systems: The non-coffee harvest. Paper presented at XXII International Congress of Latin American Studies Association, Miami.

Rosset, P. 1999. The multiple functions and benefits of small farm agriculture in the context of global trade. Food First Policy Brief No. 4, Institute for Food and Development Policy.

Rosset, P., and M. Altieri. 1997. Agroecology vs. input substitution: A fundamental contradiction of sustainable agriculture. *Society and Natural Resources* 10, no. 3: 283–295.

Sánchez López, Roberto. 1990. *Manual práctico del cultivo biológico del café orgánico.* ISMAM.

Soto-Pinto, Lorena. 2001. Características ecológicas que influyen en la producción de café. Ph.D. dissertation. Colegio de la Frontera Sur, Chiapas.

UNDP (United Nations Development Programme). 1995. Agroecology: Creating the Synergism for a Sustainable Agriculture.

6

A Grower Typology Approach to Assessing the Environmental Impact of Coffee Farming in Veracruz, Mexico

Carlos Guadarrama-Zugasti

Scholars have long debated the path, the degree, and the pace of agricultural change, rural social differentiation, and development. In many accounts, techno logical change is identified as the main catalyst of the whole process, and it has been assumed that heterogeneity among farmers can explain their conflicts of interest, their varying responses to economic policies and different agricultural landscapes. Recognizing the heterogeneity within rural groups is the main issue when addressing the nature of structural change and social differentiation in agricultural settings (Shulman and Garret 1990). Recently, the inclusion of the sustainability paradigm has added to the complexity of rural development models (Altieri 1993; Thrupp 1993) and raised concerns as to the appropriate theoretical perspectives needed to study changing agricultural practices. As sustainability comes to the fore, researchers are developing new methods and concepts to assess the environmental impacts of farm practices in rural settings where the heterogeneity and social differentiation of growers are significant.

Farmer typologies have both contributed theoretical frameworks to help explain rural heterogeneity and provided empirical evidence that has influenced agricultural and rural development policies. Most studies have focused on identifying the methodological problems inherent in the techniques used to separate farmer types and the criteria that classify them as a particular group (Darnhofer et al. 2005). A review of these studies can be found in Shulman and Garret 1990.

Sustainability is a relatively new dimension that scholars have attempted to incorporate into farmer typologies. However, little work has been done to clarify the connections between farmer types, sustainable agriculture, and technological change. Nevertheless, several recent studies have begun to delineate this new terrain. Lighthall (1995, p. 506) sought "to develop insights on the interrelationships between environmental degradation, technological change, and social welfare in agriculture and other sectors of capitalist production." He also developed useful conceptual tools such as "operational structure" and "production window" to understand the barriers to a full adoption of sustainable practices among organic and conventional corn growers. Comer et al. (1999) addressed the issue of what accounts for the slow and differential rates of adoption of sustainable practices by investigating cultural, sociological, and ideological components of attitudes and beliefs among sustainable and conventional farmers in Tennessee. They emphasized the need to go beyond economic determinism when selecting criteria to explain grower behavior. Focusing on another dimension of sustainability, scale, Köbrich and Rehman (1998) discussed the problems involved in constructing farmer typologies when the level of spatial resolution is analytically inadequate. For example, highly aggregated indicators at the national scale can mask farmer heterogeneity and, on the other hand, farm-level analysis can fail when dealing with policies of national scope. Therefore, they propose an intermediate-level analysis, which they refer to as micro-regional. In this way, by linking farm types to micro-regional-level analysis, a more meaningful assessment of sustainability can be achieved.

The present chapter constructs farmer typologies to evaluate sustainability and analyze the relationships between coffee grower types, farming choices and environmental impacts that are occurring within the context of ongoing technological transformation of coffee farming in Veracruz, Mexico. Lighthall's (1995) conceptual insights are used to develop a more disaggregated approach, which helps to compare sustainability in the separate components of coffee agroecosystems. The distinction between the whole farm and its component parts is critically important to a sustainability analysis because a practice that increases sustainability in one part of the system may reduce it in another part. By using the notion of "operational structure," the nature of hybrid technologies used in coffee farming and their fragmented sustainability can be addressed. Cluster analysis was used to construct heuristic typologies based on empirical field data. As Köbrich and Rehman (1998) correctly indicate, the decision criterion when choosing farmer groups derived from multivariate analysis is a qualitative one and depends upon theoretical assumptions.

Types of Coffee Growers

Table 6.1 presents three typologies, the criteria used to define them, and the main variables of those criteria. The different farmer types illustrate the evolution of this concept through time. Nolasco's (1985) typology, which is considered a turning point in the coffee literature, is mainly based on the economic traits of growers in a political economy framework. It focuses on the nation as a whole and its main innovation was the assumption that farmers follow different logics of reproduction depending on their combination of land access, credit availability, market orientation, and type of labor used. A positive correlation between market-oriented pro-

Table 6.1
Typologies of coffee growers.

	Nolasco (1985)	Méndez and Benoit (1994)	Díaz (1996)
Defining criteria	Farm economic orientation	Age of grower	Available land
	Labor type	Age of coffee trees	Paid labor
	Land access	Size of farm	Crop specialization
	Capital sources		Main income source
	Agroecosystem techniques		Belongs to farmer organization
Selected variables	Cash/staple crop	Age of grower	Available land per grower stratification
	Family or waged labor	Education	Hired labor ratio
	Land tenure	Experience as grower	Crop percentage
	Credit and financial sources	Coffee area	Percentage main income
	Shade tree type	Other crops area	Regional percentage of organized farmers
		Agricultural practices Yield	
Coffee grower types found	Simple commodity	Capitalized	Specialized
	Commoditized	Young	Peasant
	Entrepreneurial	Diversified	Traditional
	Agro-industrial entrepreneur	Senior	Transitional
		Traditional	Farm worker
		Semi-proletarian	Entrepreneurial
			Organized

duction and technological intensification is implied. The drawbacks of this typology are an over-reliance on economic variables, leading to the exclusion of social and cultural farmer traits and too little attention given to regional particularities. The absolute positive correlation between small farmer and low-input use, and big farmer and high-input use, may also be uncertain as the field data obtained for my study will demonstrate. Recently, some other coffee grower typologies have been constructed in Centroamerica, but they do not fully address the relationships considered here (Bonilla and Somarriba 2000).

Table 6.2
Coffee cultivators' constraints and strategies, after Méndez and Benoit-Cattin (1994).

Type code	Type name	Major constraints	Strategy/objective
A	Capitalized	Land availability	Income maximization
B	Young	Labor	Labor efficiency
C	Diversified	Education	Output optimization
D	Senior	Education	Family cycle
E	Traditional	Aged plantation	Not identified
F	Semi-proletarian	Land availability	Subsistence

The typology developed by Méndez and Benoit-Cattin (1994) introduced two interesting elements regarding farmer type criteria: the age of grower and the age of the plantation. The former is important for understanding the family cycle, a crucial factor in peasant household evolution; the latter centers on the fact that the coffee tree is a perennial plant with a long life span, an attribute that strongly affects farmer decisionmaking when innovative practices are proposed. This typology also reveals how the plantation size is linked with intensification, but, at the same time, mediated by cultural and social factors that preclude the possibility of making an automatic and direct correlation between scale and the use of high-tech inputs.

The third typology, created by Diaz (1996), attempted to illustrate the difference between a regional farmer type structure and a national one, such as Nolasco's. It includes degree and type of farmer organization. This is important because different modes of organization affect the capacity of growers to influence change instead of being merely passive actors in the transformation of regional agriculture.

Table 6.1 summarizes the key theoretical and methodological findings concerning attempts to better understand coffee farmer heterogeneity. The comprehension of farmer type differences has evolved from economic agency to complex interrelationships among social, cultural, economic, political and technological components; from large-scale typologies to the particularities of micro-regional levels; from exogenous determination of farmer social mobility to the local and regional collective action through organization.

Table 6.2 summarizes the major findings of Méndez and Benoit-Cattin (1994) and highlights the linkages between farmer type, factors limiting production, and the strategies followed to overcome such limitations. This study was carried out among smallholders from coffee-producing communities in Guatemala and it sought an explanation for why these farmers were not adopting a plantation renovation program launched by the government. The code in the first column is a letter used to identify a group of farmers that corresponds with specific types. The second column shows the economic, social, cultural or technological feature that best describes the group; the next column synthesizes the major limitations the group has that prevent the renovation of their coffee plantations; the fourth column displays the strategy or objective function underlying the logic of reproduction of each group. For example: the coffee grower type A is termed "capitalized" because these farmers experience higher productivity compared to the other groups, use an intensified management system and have more land. Their objective is to maximize income by increasing productivity since there is minimal additional land available. It makes sense that this group was much more sensitive to labor saving and other productivity enhancing technologies than all of the other groups. Group F, the semi-proletarian, is in a much different situation than group A, but paradoxically it exhibits the same restraint as group A—land availability—but for subsistence purposes rather than to maximize income.

Although the productive possibilities in both groups are constrained, they respond differently to exogenous pressures for technological innovation: the capitalized group will look to adopt intensive technologies whereas the semi proletarian type will tend to reject or resist them because these technologies are not affordable. The family cycle in the type B group, designated as "young," suffers from a shortage of family labor, because it is composed mostly of small nuclear families, which are still raising their children. This set of farmers might be responsive to innovation as long as they can afford the new technologies. The other farmer type behaviors can be contrasted in a similar way in order to illustrate how an apparently homogeneous group of smallholders actually consists of multiple groups with structural differences that affect decisionmaking and the farming objectives. The specific decisions that arise from different farming choices will have particular impacts on resources use. These impacts can then be assessed in terms of sustainability.

Coffee Grower Typology of Central Veracruz

In this section, I develop a coffee farmer typology that elucidates (1) how new technologies are being adopted, and / or contested, (2) how agroecosystem management is affected by these technological changes that make up "hybrid technologies," and (3) how these farming choices are generating different impacts on the microregional landscape, impacts that can be assessed using sustainability indicators. In order to address agroecosystem sensitivity to innovation, I rely on Lighthall's (1995) concept of operational structure, which maintains that technological adoption

Table 6.3
Variables used in developing a coffee grower typology in Veracruz, México.

Economic	Socio-cultural	Technological
Farm size	Grower age	Plantation age
Acreage with coffee	Experience	Shade-tree type
Other farming activities	Education	Number of coffee varieties used
Non-farming activities	Family size	Weeding practices
Coffee berry processing	Land tenure	Number of fertilizer applications
Agro-industrial facility ownership		Compost use
Yield/hectare		Soil-conservation practices
		Pest-control practices

will take place as long as it is compatible with the farm's current overall operating practices. I assume that techniques will be incorporated into the agricultural production process if they do not drastically affect the farmer's routine and their expected profits. It is hoped that this assumption will help to uncover the barriers and incentives to technological innovation. It will be argued that the most prevalent outcome of such a process is the assembly of hybrid technologies whereby sustainable practices are pursued in some parts of the coffee agroecosystem, but not in others. I develop farmer typologies based on the studies reviewed earlier placing emphasis on economic, technological and social variables. The connections between farming choices and resource use are appraised as a further step in the analysis.

Fieldwork was carried out during 1998 and 1999 in twelve communities of four counties in Central Veracruz, Mexico. A research team collected data from 87 coffee farmers using a combination of structured, open and conversational interviews. Field transects and site observations were performed to gain complementary information on coffee plantation management. Table 6.3 summarizes the 21 variables that were used in the first clustering; two of them were discarded because of their high correlation with the others. Interviews consisted of 20 questions divided into economic, technological, and socio-cultural categories. The data included a complete description of each practice employed in the coffee plantation, including the time of year and the farmers' rationale for using this practice.

Using nineteen of the variables in table 6.3 (acreage with coffee was discarded because of its high correlation with farm size), an ascendant hierarchical cluster analysis was run using the Statsoft statistical software package. Deep knowledge of each grower case and visual inspection of the cluster tree were used to establish the coffee grower types.

Table 6.4
Types of coffee growers in Central Veracruz, México.

Grower type	Main farming choices	Socio-cultural traits	Economic features
A Newcomer	High agro-chemical use, two coffee varieties, no compost or soil conservation	Urban, middle-aged, nuclear family, private property, no experience	Small farms (4–5 ha), medium yield, non-dependence on coffee farming
B Coffee and corn cultivator	Low external input use, three coffee varieties, varied shade trees, composting and soil conservation	Elementary school, 35–65 years old, large family, long time in farming, *ejido* land tenure	2 ha total acreage, occasional farm worker, low yields, corn for subsistence
C Farm worker– coffee cultivator	Young and old farms, many coffee varieties, weeding and fertilization practices	Elementary school average, large nuclear family, ages 30s–60s, long time in farming	Livelihood depends on wages, very small plantations (<1 ha), low yields
D Medium coffee grower	Farms 15 years old, mostly one shade-tree species, two coffee varieties, hand-tool and chemical weeding, high fertilizer use	Middle-aged, high school, small nuclear family, more than 20 years farming experience, private property	Middle-size farms (15 ha), main income from coffee, medium-high yields, coffee berry processing
E Large diversified grower exporter	All types of shade trees, many coffee varieties, very high fertilizer use, chemical pest control, terracing and contour line	30–58 years old, junior high school, extended family composing an enterprise, private property	Big coffee farms, sugar cane and cattle (180 ha); large-scale agro industrial activities and coffee exporting; high yields
F Agro-industrial family enterprise	Old farms, mostly one shade-tree species, 3–4 coffee varieties, chemical plus hoe weeding, fertilizer use, some use of compost, varied pest control	Junior high school average, extended family composing an enterprise, long time farming, average age 59, private property	Medium–large coffee farms (20–45 ha), low–medium yields, depending on coffee farming; small coffee berry processing facilities
G Coffee–sugar cane producers	Old farms, 1–2 shade-tree species, 1–2 coffee varieties, chemical plus hoe weeding, high fertilizer use, biological pest control	32–52 years old, long time farming, small nuclear family, junior high school, private property	10–25 ha total acreage, low yields, small coffee berry processing facilities; only farming income

Table 6.4
(continued)

Grower type	Main farming choices	Socio-cultural traits	Economic features
H Small growers	Young and old farms, all range of shade-tree species, many coffee varieties, weeding and fertilization practices; no pest control	Young, mature and old growers, short or long time farming, elementary school, private and *ejido* land tenure	1.5–5 ha farm size, occasional farm workers, no other crop, low yields

Definition of the types was solved in the third level of the ascendant hierarchical clustering resulting in eight types of coffee farmers, which together make up the typology of coffee producers in four counties of Central Veracruz.

Table 6.4 presents the main features of these eight grower types. They were given names using the combinations of variables that were most responsible for each cluster grouping and best described their main traits. The first column pictures the farming choices for each type describing their technological characteristics. In the second and third columns, the associated attributes of each type such as family size, age, education level, farming experience, income sources and farm scale are shown. A farm size regional standard was used to name farmer types according to their acreage: it is considered a small grower when he/she cultivates no more than ten hectares, a medium grower owns between 10 and 30 hectares, and a big grower term includes all those that possess more than 30 hectares.

As table 6.4 shows, the variables that defined each grower type were different in every case. Type A, Newcomers, are defined primarily by age, education level, experience and non-farming activities. This group has a very homogeneous management system, which relies on high agro-chemical use, no resource conservation practices, and consists of new plantations with two coffee varieties. Farmers in this group can be considered an emergent type of grower (only 7.5 percent of the total) that is taking advantage of the economic crisis in coffee since they are buying land from indebted small and medium-size growers. This consolidation of farms provides a vivid illustration of social differentiation processes operating in central Veracruz.

Type B, Coffee and Corn Cultivators, make a living by combining coffee and corn farming on small plots of land. The key variables that defined this type were education level, other farming activities, land tenure, and farm size. These farmers own small coffee farms and occasionally need to look for temporary work as farm laborers. The fact that they are still able to grow corn is related to the ejido land tenure system and the remote access to their communities. This type of grower is actually disappearing since fewer farmers are growing corn in the region due to land scarcity, the ejido transformation from collective to private property ownership and population growth.

The biggest type is C, Farm Worker Coffee Cultivators, whose main trait is the lack of dependence on coffee farming in order to make a living. The interviews revealed that these farmers are focused on maintaining a full-time job as farm laborers and coffee farming is an activity performed in their spare time on the weekends. Therefore, their holdings are less intensively managed. This is also the most technologically heterogeneous type of the eight groups, with a wide range of shade trees, coffee varieties, and weeding and fertilization techniques. Since farm households in this group are not completely dependent on income from coffee farming, it seems that they are willing to test more techniques than the other groups. Because these farmers work full-time off the farm, the techniques most likely to be adopted will be affordable and labor saving. Even though this is the biggest cluster, the farm worker-coffee cultivator type is the group least understood by politicians, non-government organizations, and extension consultants, and thus their impacts on natural resources are often underestimated.

Type D, Medium Coffee Growers, possess most of the features developed in the INMECAFE model promoted during the 1970s. These farmers are defined by their farm size, main source of income, and family size and type. Coffee farming is performed in old plantations with specialized (few planted) shade trees, two coffee varieties, a high use of agrochemicals, and soil conservation and composting are rarely practiced. Since these growers rely entirely on coffee farming and processing to make a living, new practices will not be fully adopted until they are first completely tested in small trial plots within the plantation. Two farmers in this group tried composting and rejected the practice, because it proved unprofitable. Contrary to the idea that small-scale producers are the most conservative in their farming styles, this so-called entrepreneurial group also proved to be conservative in their management practices.

Also defined by farm size is type E, Big Diversified Grower-Exporters. Within this group, other activities, including cattle ranching and the management of sugar cane and forest plantations, accompany large-scale coffee farming. Surprisingly, these farmers are experimenting with many techniques or crops that might prove to be more profitable than the ones already in practice. They are testing new coffee varieties, all types of shade trees, composting, terracing and contour lining, macadamia and blackberry cultivation, and even eco-tourism. The environment is a concern as long as it can make them money. There are two cases within this group where some of their coffee groves are grown under native oak trees, but with high agrochemical use, showing again the management of hybrid agroecosystems. They process and export un-roasted coffee beans and one of the two farmers surveyed participates in the coffee stock exchange market.

Type F, the Agro-Industrial Family Enterprise, has as its main feature extended family enterprises which own small coffee berry processing facilities. Their coffee plantations are medium to large by regional standards and their management mainly follows the INMECAFE model but with more techniques and coffee varieties incorporated. It is noticeable that on average they do not have high yields and this might be the reason why they are trying new techniques like composting or biological control that might increase productivity without increasing their costs.

The cultivation of two cash crops at the same time distinguishes type G, Coffee–Sugar Cane Producers. The technology involved in the management of these two agroecosystems is primarily conventional, although biological control is used against the coffee berry borer. This is a small group because coffee farms are located in the geographical transition zone at mid-level altitudes between the highlands where coffee is the primary crop and the lower lands where sugar cane is the main cash crop. Sugar cane is generally a conventionally managed crop and it is reasonable to assume that this "agrochemical thinking" is translated to coffee farming as shown in table 6.4.

Type H, Small Growers, is the type with the second-largest (21 percent) share of the total. As the technological column in table 6.4 indicates, they have a wide array of management systems, including different plantation ages, shade trees, coffee varieties, weeding and fertilization practices. Composting and soil conservation practices are uncommon, and pest control is almost completely absent. This is the most varied group in terms of education level, age, land tenure, experience, and family size and maybe this is the reason why all sorts of technological arrangements are being implemented. What they have in common is farm size, low yields, occasional part-time jobs off the farm, and no other farming activity. Because of these economic conditions, this group is probably the most vulnerable to low coffee prices.

Overall, this typology of coffee growers that I have developed and applied at the micro-regional scale is an empirical tool intended to detect the main traits that help classify growers by their farm-management decisions and socio economic conditions. Table 6.4 summarizes the farming choices growers are imposing on the agricultural landscape. Paradoxically, the groups of farmers traditionally expected to be the most conservative in a technological sense because of their peasant cultural roots, such as farm workers and small growers, are the most apt to incorporate some new techniques into their systems, while farmers with a greater ability to buffer economic risk when adopting new technology, such as medium-size growers, proved to be more conventional in their management practices.

This section illustrates the variety of technological responses that can be related to different farmer typologies. If these responses can be codified into environmental impacts, then we will be able to identify where those impacts are most likely to occur, what their magnitude will be and what the reasons are behind the management decisions associated with these impacts. In the following section, I develop quantitative and qualitative indices for this purpose, and one index for each part of the operational framework of coffee agroecosystem management for each type of grower.

Farming Choices and Environmental Impact

In Central Veracruz most coffee is produced under shade trees. Coffee farm classification systems that use shade trees as their primary indicator, including the work by Toledo and Moguel, might assume that the presence of diverse shade trees is related to low external inputs and reflects reliance on traditional management techniques (Toledo and Moguel 1996; Escamilla et al. 1995; Fuentes-Flores 1979). However, during interviews farmers revealed different degrees of agrochemical use. The intensity of agrochemical applications correlated closely with the farmer typology of growers and their agricultural management practices presented in the previous section. The nursery, where coffee seedlings are produced, is the stage of production which accounts for the greatest quantity and variety of agrochemical use. Bio regulators, containing hormones combined with macro and micronutrients, are used to boost growth and rooting in the nursery and in the field when planting coffee seedlings in the process of plantation renovation. Foliar sprays that contain macro and micronutrients and fulvic acids (a soil organic matter compound) in their formulae, promote vegetative growth in the early stages of young coffee plants. The nature of foliar spray is clearly representative of hybrid technologies, since they include components from industrial agriculture, such as nitrogen, phosphorus, and potassium, and components derived from composting processes, such as fulvic acids. Agrochemical use is only one of many impacts that result from coffee farming. Topsoil and biodiversity are largely affected by the agricultural routines practiced by growers. In order to detect and quantify such effects, the operational framework of agroecosystem management was split into eight basic agricultural practices: shade-tree management, weeding practices, fertilization, plantation renovation, coffee variety use, nursery, sapling planting, and pest control.

The effect of each of these practices was measured for each type of grower. Figures 6.1–6.7 rank these effects on scales from 0 to 10, in which the higher the number the greater the impact (Trujillo 1997). When appropriate, quantitative data were converted to a 1–10 scale. Ordinal qualitative data were converted to this same

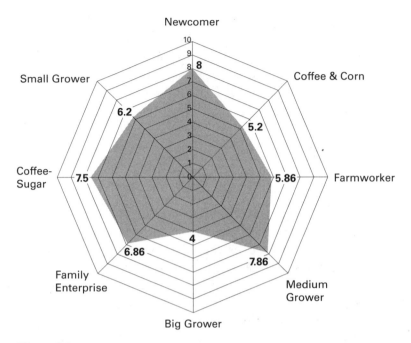

Figure 6.1
Shade-tree management impact by grower type. Scale values: 0 = cloud forest, 1 = nine shade-tree species, 2 = eight shade-tree species, . . . , 10 = full sun.

numeric scale for the purposes of a visual comparison (Ericksen and McSweeney 1999). The process of scoring certain practices like weeding or pest control was based on farmer and consultant opinions, literature review, and field observations.

Shade Management
Figure 6.1 shows the shade-tree species richness associated with different grower types. Farmer interviews served as the source for this data. I estimated impact by comparing the number of different coffee shade-tree species per hectare between the farms, assuming that the lowest impact on biodiversity was associated with the highest number of shade-tree species (a value of one was assigned to the farms with the highest average figure; a value of zero would correspond to the local natural ecosystem, the cloud forest). Conversely, the lowest number of shade-tree species was ranked as having the strongest negative effect on biodiversity (a value of nine for one shade species and a value of ten for full-sun plantations). Two points can be clearly drawn from figure 6.1: (1) None of the grower types have a low impact on diversity. (2) Big growers have the lowest impact of all grower types. The relationship between

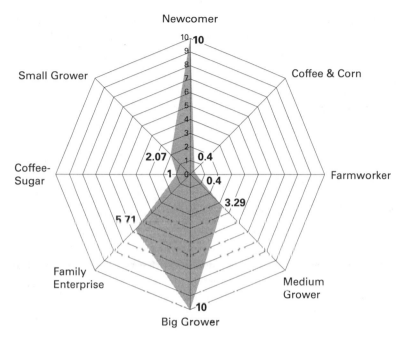

Figure 6.2
Intensity of pest control by grower type, ranked from 0 (no control) to 10 (extremely hazardous control measures).

farmer typologies and shade-tree management is an understudied area of the operational framework characterized by rapid changes (Albertin and Nair 2004).

All growers were in the process of reducing shade diversity. This is because they were increasing the density of coffee bushes within their plantations by clearing shade trees that do not fit into the new required row alignments. There has been a switch in the recommended plantation design from the traditional 3-meter space intervals between coffee plants to 2 m or even 1.5 m. Land scarcity has also pushed growers to put more plants in a given plot; at the same time, government and research institutions like INIFAP were encouraging the renovation of old plantations and the removal of big shade trees that occupy precious space, which is especially valued by smallholders. On the other hand, big growers have more land and thus seem more likely to experiment with shade-tree species. In one large plantation there were nine plots with many different shade trees including oaks and second-growth cloud forest.

Shade-tree diversity assessment is difficult because growers tend to have their plantations distributed in three or more coffee groves that are sometimes separated by several kilometers; shade-tree composition often varies between these groves. Moreover, the extreme patchiness of coffee farms in a short transect (50–100 m) due to land fragmentation, indicates that further research on assessing shade-tree biodiversity is necessary. Ideally, studies would also measure within plot diversity before comparing between farms. The former is called "alpha" diversity and the latter "beta" diversity (Gliessman 1998).

Pest Control

Figure 6.2 illustrates a ranking system that assesses the environmental impact of the different pest-control practices associated with different coffee grower types. Each practice was assigned a ranking on a scale of 0–10, "no control" is rated 0 impact, biological control as 2.5, the combination of chemical and biological control as 5, and "chemical control" with Endosulfan (extremely hazardous agrochemical) as an impact equivalent to 10. I then calculated each grower's environmental impact by averaging the value assigned to the management practice of all growers within each grower type. The differences among grower types are clear. The Newcomer and Large Grower apply the most pesticides. Small Growers show a greater intensity than Coffee–Sugar Cane Cultivator, a group expected to be much more conventional. Comparing these groups also highlights the use of hybrid technologies. Coffee–Sugar Cultivators rely on biological controls to cope with coffee berry borer attacks, which are more intense in the lower-elevation lands where sugar cane is the main cash crop, but in the other parts of the operational framework agrochemicals are more heavily used.

The Family Enterprise and Medium Growers also exert a strong impact on the environment through the intensity of chemical pest control. These farmers often spray extremely hazardous insecticides and nematicides such as methyl parathion, endosulfan, and promecarb. The government does not enforce restrictions or monitor their use on these farms. Small Growers, Coffee and Corn Cultivators, and Farmworkers stated that they do not use chemical pest control because they consider their farms free of pests and disease. They did not explain their reduced agrochemical usage based on concern for the environment. These statements refer to mature coffee trees, whereas seedlings and saplings are the object of careful attention and chemical input as long as it is affordable. This distinction reveals a behavior pattern in management among these smallholders: once the coffee sapling is established and safe, no further care is needed. This distinction highlights the differential environmental impacts derived from the agricultural practices as a result of the agroecosystem management, which for the purposes of this analysis has been broken down into pieces of the farm's operational structure. Thus, while one part of the system (mature coffee trees) is managed by some grower types in a more

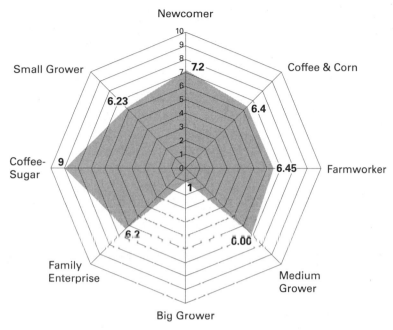

Figure 6.3
Impact of coffee variety use by type of grower, ranked from 0 (ten coffee varieties) to 10 (one coffee variety).

sustainable manner, other parts (production of saplings and seedlings) are managed conventionally.

Number of Coffee Varieties Used

Structural diversity in coffee plantations is also affected by farmer decisions on the kinds of coffee varieties to be planted. In figure 6.3, as in figure 6.1, I estimated the environmental impacts using a scale based on the number of coffee varieties planted in the grove. Since more varieties provide greater structural diversity due to size, color, phenology, and canopy variation, the scale was built according to the average of varieties planted by all of the members of each grower type, with more varieties signifying a lower impact. Figure 6.3 closely resembles figure 6.1 with some slight variations. As in shade-tree management, decisions about what coffee varieties to plant greatly affect structural diversity, which in turn affects other properties of the agroecosystem such as temperature stability (L. E. Trujillo, personal communication, 2004). Coffee trees are perennials, which like the shade trees have long life spans, but with one essential difference: coffee trees produce the source of grower's livelihoods. Therefore, figure 6.3 reflects the decision of growers to keep production levels constant by relying on two or three tested varieties rather than doing trials with a greater number of them. Again, big growers have the capability of experimenting with the best fit for their plantations (up to nine varieties were found in a mosaic array in one of their coffee groves).

In order to relate the number of varieties to quality, it is necessary to point out that the historical development of coffee varieties can be divided into three phases: a first generation of tall varieties was used before the introduction of the INMECAFE model; a second generation, promoted by INMECAFE, that included high-yielding and short-stature coffee trees; and a third generation composed of the newest releases of short-stature, high-yielding, and disease-resistant coffee plants. Complementary data for figure 6.3 show that the Small Growers, Coffee and Corn Cultivator and Farm Worker types are using mostly the *Tipica*, *Bourbón*, and *Mundo Novo* varieties (first generation); for the Medium, Coffee–Sugar Cane and Family Enterprise types *Caturra*, *Catuai*, and *Garnica* varieties prevail (second generation); and, the Big Growers rely much more on *Costa Rica 95* and *Colombia* types (third generation). That some types of growers are bound to certain types of varieties means that each group has differential adoption strategies, which are driven by a set of cultural-economic variables. The third-generation varieties, for instance, are expensive and scarce, which limits their adoption. In addition, a network of relations is needed that can provide connections with high-tech nurseries and information about the characteristics of these new coffee plants.

Besides the environmental impact associated with reducing structural diversity, there is the problem of diminished coffee quality attributed to the newest coffee varieties. In 1998, ICAFE issued a warning about the possible inferior aromatic and taste properties of coffee beans produced by the *Costa Rica 95* variety. The ICAFE advised against its production until this concern is adequately addressed. This shows that the replacement and reduction of varieties not only influences ecological sustainability, but may also affect economic sustainability. In the aftermath of the International Coffee Agreement's collapse, declining quality threatens price premiums in the world coffee market (Stewart 1992; Reardon and Farina 2002).

Weeding

Weeding practices are an important part of the operational structure of the agroecosystem and have been addressed as a key issue related to environmental and economic costs (Semidey et al. 2002). Practices were classified according to their potential erosion effects and the toxicity level of the herbicides. In figure 6.4, the area shaded in the center, with a value of one, signifies minimum negative environmental impact, in this case the use of "machete." The values, which range from 2 to 9, are a ranking

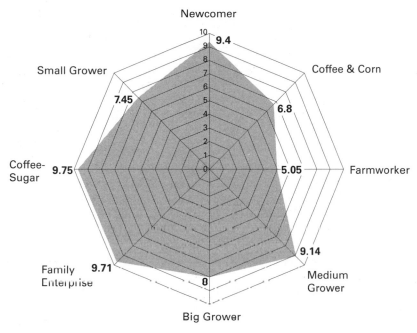

Figure 6.4
Impact of weeding by grower type.

system from a lesser to a greater environmental impact. These impacts are associated with different weeding methods, including a combination of machete, hoe, herbicide and portable lawn mower. The most damaging method, which received a score of 10, is a combination of a hoe and herbicide use. Weeding is the agricultural practice that has the highest indices of environmental impacts for all types of growers. Despite concerns about soil erosion stated by the majority of growers interviewed, the hoe is still widely used. Big and medium growers prefer to combine hoe use with herbicides, because coffee groves under renovation are dominated by weedy grasses, which are more difficult to control with a machete. Small growers have cultural reasons to keep using the hoe: a "clean" hoed farm gives prestige to the owner as it represents a "good farmer." Moreover, "machete" weeding increases labor costs because with this method it is necessary to weed two or three mores times than with a hoe or an herbicide. Thus, due to cost pressure and community status, coffee farmers of all types are utilizing a variety of weeding methods still based on or combined with the hoe. This can cause significant damage to soil quality. As long as plantation renovation continues to intensify, weeding will be a critical issue to address when trying to reverse unsustainable and high-impact practices.

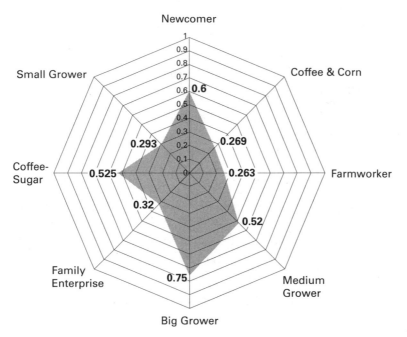

Figure 6.5
Impact of fertilization use by grower type (kilograms per plant per season).

Fertilization

Figure 6.5 presents the rates of chemical fertilizer use in kilograms per coffee plant per season for each grower type. Rates vary from 0.263 kg in Farm Workers to 0.75 kg in Big Growers. The index was constructed by multiplying the amount of fertilizer applied to one plant by the number of times applied during the season. Because of its rapid effects in promoting higher production, fertilization is one of the main concerns for all type of growers.

Low levels of fertilizer use among Small Growers, Coffee and Corn Cultivators, and Farm Workers are related to their lack of money. Many of these farmer types and some of Medium Growers, who did not report fertilizer application at the time of the survey, stated that they would apply it as soon as they could obtain some money. One of the consequences of this monetary scarcity was that several of them are buying by-products of fertilizer manufacturing (actually "spillovers" in the mixture process at the fertilizer factory), which are cheap but of low quality. Other farmers buy fertilizer formula that are cheaper than the popular brand names, such as "ammonium sulfate," which has a soil acidification tendency, rapidly lowering soil pH (Brady and Weil 2000), and has been related to nutritional disorders in plants

(Phongpan et al. 1988). The Agro-industrial Family Enterprise type, whose average is low (0.32), are applying fertilizer only once a year because of the increasing costs. Although compost is a cheaper alternative than fertilizer per unit land, small and medium-size land owners have not adopted this practice primarily because of high labor and transportation costs. Furthermore, no social information network is available to advise farmers about how to best apply it. Farmer prestige is another reason many do not use compost; two Small Growers that experimented with compost were considered "to be nuts" by their peers for using an uncertain technique instead of commercial fertilizer for such a critical part of the coffee farming process. Among the type C growers, time was the reason given for not engaging in composting since using fertilizer is quicker and allows them to keep their off farm jobs.

All the factors that limit the adoption of composting suggest that fertilization has a very narrow "time production-window" (Lighthall 1995), where availability of the fertilizer at the farm gate, timely application (after a rainfall moistens the soil enough), and an affordable price combine to establish one of the most critical moments within the coffee agricultural calendar. There is little chance that a practice that barely fits in only one of these requirements will be widely adopted, not to mention the prestige of the farmer, which may be at risk in a highly experienced coffee grower environment.

Plantation Renovation

Measurement of soil sediments, runoff, and nutrient exportation under shade-grown coffee are components of an important field of research (Jaramillo-Robledo 2003), but, paradoxically, there is very little research on plantation renovation, which has a major impact on soils and soil quality. Digging holes to plant coffee saplings that will renew the plantation, which can be for the purposes of replacing dead plants or constitute a whole new coffee grove, is a process that removes and loosens great amounts of soil. Establishing a new coffee plantation, even under shade management, requires a large clearing of coffee and shade trees so large that it rivals the potential erosion of full-sun coffee plantations. A study conducted in Venezuela by Ataroff and Monasterio compared three sun coffee systems and pointed out the difficulty of renovation:

During the first year of the establishment phase of a sun coffee plantation, the disturbance to the system resulted in an important increase in erosion. Firstly, original vegetation was cut down (which in our case was a shade coffee plantation), secondly, holes were made and small coffee plants were transplanted. . . . The disturbance caused by these activities (especially opening the holes) lasted for several months. More than four times the average loss of the mineral fraction < 4 mm in a 7–10 years old sun coffee plantation was recorded for the first year of plantation. (1997, pp. 103–104)

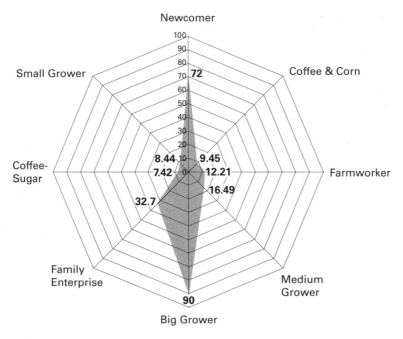

Figure 6.6
Effects of farm renovation on soil by grower type (cubic meters of soil removal per hectare per year).

The ongoing renovation in coffee shade systems showed no difference in its first year with the sun coffee plantations mentioned in this quotation. Figure 6.6 illustrates the impacts measured in cubic meters of loosened soil per hectare for each grower type. The larger the amount of soil removed the greater the impact and the higher the number. This figure was calculated by multiplying the average number of saplings planted by grower type by the amount of soil loosened and removed when digging a 30 × 30 × 30-cm hole. The helix-like shape of figure 6.6 is due to the fact that two farmer groups are renovating coffee trees at high rates: Newcomer and Big Grower impacts of 72 and 90 cubic meters respectively. As a means of comparison, a truck loaded with sand for construction carries approximately 7 cubic meters. In summary, these growers annually remove and loosen 10–13 truckloads of soil per hectare. Both types, particularly the latter, generally replace coffee trees in an entire grove at one time; this is called "whole renovation." The other grower types renovate by either replacing ill, dead or non-productive plants or planting coffee saplings between mature coffee plants in existing rows; this is called "gradual renovation." All grower types are renovating their farms, albeit at different paces, which is related to the presence of more nurseries and increased agrochemical applications associated with tree nursery management strategies.

Potential erosion increases during the rainy season when planting is performed. The "whole renovation" management strategy includes clearing all coffee bushes and shade trees; this greatly increases the potential for erosion. Some growers also practice terracing and contour planting, which also removes and loosens more soil (Blaikie 1985). The soil remains bare and exposed for three years (while the shade trees and coffee bushes grow). Although this practice can cause a one-time impact, many farmers are implementing it simultaneously and this can exacerbate landscape-level erosion processes. Moreover, new coffee varieties have a shorter life span and demand shorter renovation periods than the older varieties. In summary, plantation renovation relies on and promotes very unsustainable practices.

Biocide Index

As mentioned earlier, the sapling and nursery phases entail the use of large amounts of agrochemical inputs in order to control damping-off and die-back seedling diseases, nematode pests, and to achieve healthy and strong plants that are able to be productive through their lives. To estimate the environmental impacts resulting from these phases, a Biocide Index (BI) was calculated for each grower type based on the one proposed by Jansen et al. (1995). This index was calculated as follows: the amount of the agrochemical product sprayed was multiplied by the number of applications in the season and weighted by its hazard level. For example, one spray in coffee saplings of a farmer of the Big Grower type needed 25 ml of Nemacur; the saplings were sprayed eleven times in the season ($25 \times 11 = 275$); this number was multiplied by the hazardous level of the active ingredient of Nemacur (fenamiphos) scored with a value of 4 (extremely hazardous). The result, 1,100, was converted to its logarithm value, 3.04 in the case of the sapling phase. These calculations were performed for each one of the farmers interviewed and grouped in a grower type and then averaged for that group. The resulting figure is the BI for each type of grower plotted in figure 6.7.

Figure 6.7 presents the resulting BI for the sapling phase by grower type; the scale is logarithmic, to facilitate visual differentiation among farmer types. This practice has the heaviest use of agrochemicals in the operational framework of coffee agroecosystem management. In the sapling planting phase, all types of grower are involved. Figure 6.7 shows that this stage is critical since most of the grower types try to fulfill the plant requirements by using external inputs during this stage regardless of their socio-economic conditions.

The nursery is a part of the farms' operational management structure, and it is the place where many growers start and access new varieties and plant the quantities

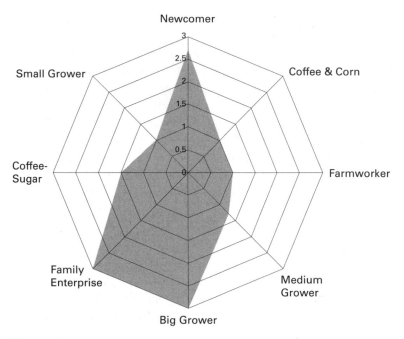

Figure 6.7
Biocide index by grower type at sapling phase.

of coffee seedlings required to complete the renovation process. It is also a step that regulates the rate of adoption of new varieties; many grower types do not have access to the seed or cannot afford the agrochemicals commonly associated with introducing new seedling varieties.

While most scholars studying coffee farming changes have focused on the replacement of shade systems by sun systems, or on the use of external inputs in the regular maintenance of established plantations, the process of renovation in traditional coffee shade plantations goes largely unnoticed. The assessment of the environmental impacts and sustainability of shade plantations should be reframed to consider the current transformation of the components of a farm's operational structure, including the seedling nurseries. This disaggregated analysis helps to inform the larger political-economic context including changing demand in international coffee markets. Thus, an exploration of these relationships questions current policies, such as Mexico's Consejo de Café's claim that coffee farm renovation will lead to the improvement of Mexico's competitive advantage within the world coffee market (Consejo Mexicano del Café 1998). Plantation renovation, overall, is the driving force behind external input intensification and soil degradation. Both trends undermine the traditional view that shade-grown coffee is inherently sustainable.

Environmental Impacts and Operational Structure by Grower Type

This analysis has distinguished the separate practices that comprise the annual operational structure on a coffee farm. I have reviewed each practice and its subsequent environmental impact by grower type, elucidating how different components of the farm's annual operational structure are related to technological change and sustainability (Guadarrama-Zugasti 2000). It is now possible to integrate all the effects of the agricultural practices depicted in figures 6.1–6.7 into a more complete perspective for each type of grower.

In figure 6.8, each multi-axial graph includes the eight practices clockwise starting at the top center (shade-tree, pest control, coffee varieties, weeding, fertilization, renovation, nursery biocide index and, planting biocide index) and their environmental impact ranking for each one of the eight grower types (the letters A–H refer to the grower types listed in table 6.4). In order to understand the point of this visual approach it is necessary to imagine first how the graphs would look if there were no impacts, and second, how they would look if there were maximum impacts. In the first case, the graph would not have any shaded area, and in the second the entire graph would be shaded. These two extremes represent the idealized divide between "conventional" and "sustainable" growers. As shown in figure 6.8, such an extreme situation does not exist for any of the grower types. Instead, figure 6.8 illustrates very clearly how hybrid technologies are manifested in each grower type's operational framework. The variety of shapes reflects different degrees of hybridization and the difficulty in identifying a particular type of farmer as "sustainable" or "conventional."

Some growers affect the natural resource base more than others. The larger the shaded area the greater the impact exerted by that type of grower. As shown, there are important differences in the magnitude and quality of these impacts. The Newcomer (A), Big Diversified Grower Exporter (E), and Agro-industrial Family Enterprise (F) types account for the largest shaded areas. However, there are three practices in which all of the types display a strong environmental impact: weeding, shade-tree management, and number of coffee varieties. These are environmentally sensitive areas that need to be addressed in the context of sustainable management of the coffee agroecosystem. Figure 6.8 synthesizes the linkages between farmer types, farming choices, and environmental impacts. Throughout this section, economic as well as socio-cultural incentives and barriers to carrying out agricultural practices were discussed as a way of explaining why certain growers farm in a particular style.

Shade-grown coffee is in a stage of transformation that is affecting its sustainability. Contrary to the view that shaded coffee plantations managed by smallholders are

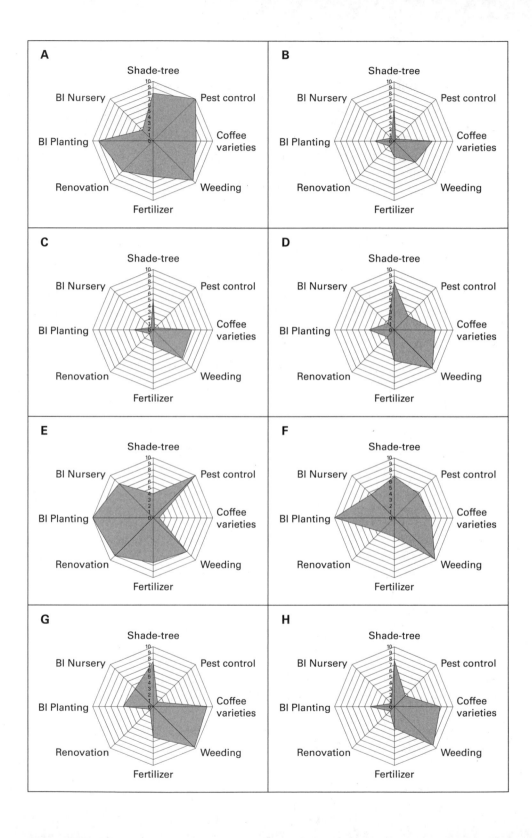

necessarily sustainable, the results presented here show that smallholders, like large holders, greatly impact the environment. Hybrid technologies embody the choices that coffee farmers have in a regional context of continuous depression of coffee prices, increasing environmental regulation and farmer's contested cultural heritage.

Conclusions

A close analysis of the shaded areas in figure 6.8 highlights the uneven environmental impacts associated with different grower types. The differences in the size of the shaded areas among grower types can be seen as a subsidy that certain grower types grant to the regional environment. Figure 6.9 illustrates such differences through a comparison of the impacts associated with coffee agroecosystem management practices pursued by Big Diversified Exporter Growers and the Coffee and Corn Cultivators. Their overall impacts are plotted in the same multi-axial graph at the same scale, this shows the Coffee and Corn Grower's limited environmental impact as compared to the impacts associated with the Big Diversified Exporter's management strategies.

The fact that most small growers, such as the Coffee and Corn type, do not intensify their farming practices diminishes their negative environmental impacts. Their low-impact farming practices can be considered an environmental subsidy. Previous studies have also highlighted small growers' invisible and unpaid environmental subsidy. For example, Low (1994) advanced this argument concerning farm-households in Africa. In this case, a low-external-input agriculture posed a dilemma for growers, because this sustainable agriculture translated into unsustainable livelihoods. In Veracruz, small-scale coffee growers' low income impinged on their economic sustainability, prevented their social reproduction, and caused many farm families to abandon agricultural activities and often the land (Guadarrama-Zugasti 2000).

Figure 6.9 shows that the sections of agroecosystem management largely considered to be of very low impact among traditional smallholders actually have relatively large impacts when compared to other practices such as shade-tree management, coffee variety use and weeding practices. These unexpected outcomes suggest that the smallholders' environmental subsidy, which arises from their reluctance to intensify their operations, is about at its end. Unless farmers receive compensation for these practices, perhaps through coffee labels like shade-grown, Bird-Friendly,

Figure 6.8
Environmental impacts and operational framework by grower type. (A–H refer to grower types listed in table 6.4.)

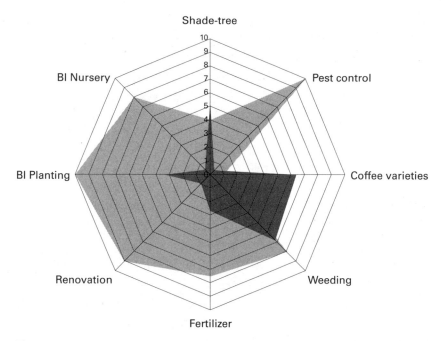

Figure 6.9
Environmental subsidy by coffee grower. (Light area = Big Diversified Grower; dark area = Coffee and Corn Cultivator).

and organic, the environmental subsidy provided by these growers will fade out completely in the near future.

References

Albertin, A., and P. K. Nair. 2004. Farmers' perspectives on the role of shade trees in coffee production systems: An assessment from the Nicoya Peninsula, Costa Rica. *Human Ecology* 32, no. 4: 443–463.

Altieri, M. 1993. Sustainability and the rural poor: A Latin America perspective. In *Food for the Future*, ed. P. Allen. Wiley.

Ataroff, M., and M. Monasterio. 1997. Soil erosion under different management of coffee plantations in the Venezuelan Andes. *Soil Technology* 11: 95–78.

Blaikie, P. 1985. *The Political Economy of Soil Erosion in Developing Countries*. Longman.

Bonillla, G., and E. Somarriba. 2000. Tipologías cafetaleras del Pacífico de Nicaragua. *Agroforestería en las Américas* 7, no. 26: 27–29.

Brady, N., and R. Weil. 2000. *Elements of the Nature and Property of Soils*. Prentice-Hall.

CEPAL. 1982. *Economía Campesina y Agricultura Empresarial: Tipología de Productores del Agro Mexicano*. Siglo XXI, México.

Comer, S., E. Ekanem, S. Muhammad, S. Singh, and F. Tegegne. 1999. Sustainable and conventional farmers: A comparison of socio-economic characteristics, attitude, and beliefs. *Journal of Sustainable Agriculture* 15, no. 1: 29–45.

Consejo Mexicano del Café. 1998. Plan de Renovación y Rehabilitación de la Caficultura Mexicana. Manuscript.

Darnhofer, I., W. Schneeberger, and B. Freyer. 2005. Converting or not to organic farming in Austria: Farmer types and their rationales. *Agriculture and Human Values* 22: 39–52.

Diaz, C. S. 1996. Estratégias de Participación de los Caficultores de la Región de Huatusco. Ver. M.S. thesis, Universidad Autónoma Chapingo, México.

Ericksen, P. J., and K. McSweeney. 1999. Fine-scale Analysis of soil quality for various land uses and landforms in Central Honduras. *American Journal of Alternative Agriculture* 14, no. 4: 146–156.

Escamilla, P., V. Licona, C. Díaz, C. Santoyo, R. Sosa, and R. Rodríguez 1994. Los sistemas de producción de café en el Centro de Veracruz, Mexico: Un análisis tecnológico. *Revista de Historia* 30. 41–67.

Fuentes-Flores, R. 1979. Coffee production systems in Mexico. In *Workshop on Agroforestry Systems in Latin America*, ed. F. De las Salas. CATIE, Costa Rica.

Gliessman, S. R. 1998. *Agroecology: Ecological Processes in Sustainable Agriculture.* Ann Arbor Press.

Guadarrama-Zugasti, C. 2000. The Transformation of Coffee Farming in Central Vera Cruz, Mexico: Sustainable Strategies? Ph. D. thesis, University of California, Santa Cruz.

INMECAFE. 1979. Tecnología Cafetalera Mexicana. 30 Años de Investigación y Experimentación. Dirección Adjunta de Producción y Mejoramiento de la Caficultura: Xalapa, Veracruz.

Instituto del Café de Costa Rica. 1998. *Manual de Recomendaciones para el Cultivo del Café*: San José, Costa Rica.

Jansen, D., Stoorvogel, J., and R. Shipper. 1995. Using sustainability indicators in agricultural land use analysis: An example from Costa Rica. *Netherlands Journal of Agricultural Science* 43: 61–82.

Jaramillo-Robledo, A. 2003. La lluvia y el transporte de nutrimentos dentro de ecosistemas de bosque y cafetales. *Cenicafé* 54, no. 2: 134–144.

Köbrich, C., and T. Rehman. 1998. Sustainability, farming systems and the mcdm paradigm: Typification of farming systems for modeling. In *Technical and Social Systems Approaches for Sustainable Rural Development*. Dirección General de Investigación y Formación Agraria: Andalucía, Spain.

Lighthall, D. R. 1995. Farm structure and chemical use in the corn belt. *Rural Sociology* 60, no. 3: 505–520.

Low, A. R. C. 1994. Environmental and economic dilemmas for farm-households in africa: When "low-input sustainable agriculture" translates to "high-cost unsustainable livelihoods." *Environmental Conservation* 21: 220–224.

Méndez, J. C., and M. Benoit-Cattin. 1994. Intensificación de la caficultura de los pequeños productores de Guatemala: Una topología. *Café, Cacao, The* 38, no. 2: 125–133.

Nolasco, M. 1985. Café y Sociedad en México. Centro de Ecodesarrollo, México, D.F.

Phongpan, S., S. Vacharotayan, and K. Kumazawa. 1988. Efficiency of urea and ammonium sulfate for wetland rice grown on an acid sulfate soil as affected by rate and time of application. *Fertilizer Research* 15: 237–246.

Reardon, T., and E. Farina. 2002. The rise of private food quality and safety standards: Illustrations from Brazil. *International Food and Agribusiness Management Review* 4: 413–421.

Semidey, N., E. Orengo-Santiago, and E. Más. 2002. Weed suppression and soil erosion control by living mulches on upland coffee plantations. *Journal of the Agricultural University of Puerto Rico* 86, no. 3–4: 155–157.

Stewart, R. 1992. *Coffee: The Political Economy of an Export Industry in Papua-New Guinea.* Westview.

Shulman, M., and P. Garret. 1990. Cluster analysis and typology construction: The case of small-scale tobacco farmers. *Sociological Spectrum* 10: 413–428.

Thrupp, L. A. 1993. Political ecology of sustainable rural development: Dynamics of social and natural resource degradation. In *Food for the Future: Conditions and Contradictions of Sustainability*, ed. P. Allen. Wiley.

Toledo, V., and P. Moguel. 1996. Searching for sustainable coffee in Mexico. In *Proceedings of the First Sustainable Coffee Congress*, ed. R. Rice, A. Harris, and J. McLean. Smithsonian Migratory Bird Center.

Trujillo, L. E. 1997. Análisis de la Sostenibilidad de la Sierra de Santa Marta, Ver. Proyecto Desarrollo Regional Sustentable, Tomo 5. SEMARNAP, México, D.F.

7

Confronting the Coffee Crisis: Can Fair Trade, Organic, and Specialty Coffees Reduce the Vulnerability of Small-Scale Farmers in Northern Nicaragua?

Christopher M. Bacon

Activist pressure and the expanding specialty coffee market have provoked a small, but growing, percentage of those that daily drink 1.5 billion cups of coffee to remember the 20–25 million families that produce and process this valuable bean (Conroy 2001; Dicum and Luttinger 2006). Small-scale family farms produce over 70 percent of the world's coffee in 85 Latin American, Asian, and African countries (Oxfam 2001). Most coffee producers live in poverty and manage agroecosystems in some of the world's most culturally and biologically diverse regions.

Changing patterns in global coffee commodity chains including the disintegration of the international coffee agreement in 1989, market liberalization, corporate consolidation, increasing production, and a worldwide coffee glut have plunged commodity prices to their lowest levels in a century (Ponte 2002a,b). However, increasing consumer awareness regarding issues of quality, taste, health, and environment have created a growing demand for specialty and eco-labeled (i.e., organic, Bird-Friendly, and Fair Trade) coffees (Goodman 1999; Rice 2001). Specialty and eco-labeled coffees offer price premiums. The volumes of coffee moved through specialty, organic, and Fair Trade commodity chains remain relatively small and must be set within the context of changing global coffee markets.

In the period 1999–2003. the price of a pound of green coffee fell from US$1.20 to between US$0.45 and US$0.75. Low prices continue to devastate rural economies and threaten the biodiversity associated with traditional coffee production (CEPAL 2002; IADB 2002). Permanent employment in Central America's coffee sector has fallen by more than 50 percent and seasonal employment by 21 percent (IADB 2002). In Matagalpa, Nicaragua, falling coffee prices have accelerated migration to urban poverty belts. A walk through a coffee farming community in Coto Brus, Costa Rica, reveals eroded hillsides where farmers recently replaced coffee agroforestry systems with treeless cattle pastures. Since the 1999–2000 harvest, the value of Central American coffee exports has fallen from US$1.678 billion to US$938 million in 2000–01 and an estimated US$700 million for the 2001–02 harvest

(IADB 2002). Declining export revenues have created debt that exceeds US$100 million. As debt in the coffee sector increases, banks have foreclosed on farms and export companies (Díaz 2001).

This chapter examines how changes in the global coffee market and falling coffee commodity prices affect small-scale farmers' livelihood vulnerability in northern Nicaragua. The first section is a synopsis of the changing tendencies in the global coffee trade. The second section briefly reviews theories linking price shocks to livelihood vulnerability and then applies this framework to a farmer typology revealing the consequences of the coffee crisis. The third section presents the results of research that investigated the hypothesis that farmers selling Fair Trade and organic coffees are less vulnerable than those linked only to conventional coffee markets. In the final section, I discuss strategies to reduce vulnerability without reproducing the same structures that created the coffee crisis.

Changing Commodity Chains

Booms and busts punctuate the history of international commodity prices. The driving forces behind the 1999–2003 decline in green coffee commodity prices suggest this cycle will continue, and prices may remain low for the coming years (CEPAL 2002). The disintegration of the International Coffee Agreement (ICA) and market liberalization contributed to increasing global coffee production. The increasing coffee supply led to rising inventories in consumer countries and coincided with sluggish demand and market concentration in the roasting and trading industries (Ponte 2002a).[1] Among the consequences are shifts in power to the roasting and retailing end of the commodity chain and falling prices paid to producers (Talbot 1997).

The ICA was a set of international agreements that set production and consumption quotas and governed quality standards for most of the coffee industry from 1962 to 1989. A combination of processes, including increasing fragmentation in the geographies of production and consumption, shifting geopolitical conditions as the United States perceived less of a threat from the Latin American left, and the changing development models as Indonesia and Brazil moved away from import substitution toward export led growth contributed to the disintegration of the agreement (Ponte 2002b). Free from international quotas, green coffee prices initially fell, briefly rebounded during the period 1994–1998, then plummeted before rebounding slightly in early 2004. The two primary coffee varieties are arabicas and robustas. Farmers in Latin America, Ethiopia, and Kenya have historically cultivated mostly arabica beans, which are generally considered of higher quality and sold to specialty markets at slightly higher prices than robustas. Brazil, Vietnam, and Uganda

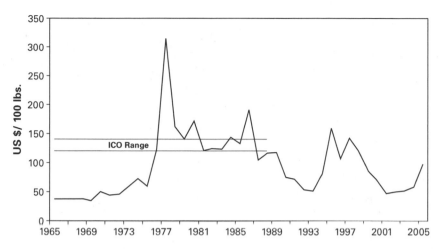

Figure 7.1
Monthly averages of New York Coffee Futures prices for 100-pound sacks, 1965–2005. Data from ICO 2005; figure from Petchers and Harris, this volume.

produce most of the world's robusta coffees. Two tendencies are eliminating the previous competitive advantages held by countries producing arabica coffee varieties. In the 1990s, Brazil more than doubled its production of arabica coffees; it now produces close to half of the world's arabica coffee. Furthermore, many roasting companies can substitute between robusta and arabica beans in their blends; thus, the price differential between robusta and arabica coffees is rarely more than 10 cents per pound. The price reported below is for other mild arabica beans grown outside Colombia.

The disintegration of the ICA coincided with geopolitical shifts, including the fall of the Soviet Union and the state's declining role in commerce. As many national agricultural ministries dramatically decreased their role in coordinating coffee production, commercialization, and quality control, governments lost international negotiating power. Producers and exporters gained flexibility and more direct market access to roasters and retailers. Large-scale transnational trade and roasting companies were quick to enter the spaces opened by the retreating state. The combination of market liberalization and increased coffee production coincides with high rates of transnational corporate concentration. By 1998, Philip Morris, Nestlé, Sara Lee, Procter & Gamble, and Tchibo controlled 69 percent of the market for roasted and instant coffees (van Dijk, van Doesburg, Heijbroek, Wazir, and Wolff, 1998, cited in Ponte 2002a). Eight transnational export-import companies control 56 percent of the coffee trade (van Dijk et al., 1998, cited in Ponte 2002a).

The changing structure of the global coffee commodity chain has led to declining prices paid to producers. Since the fall of the ICA producers' share of the final retail price has fallen from 20 percent to 13 percent (Talbot 1997). Historically, coffee-producing countries in Latin America, Asia, and Africa captured close to 55 percent of the coffee dollar, significantly more than many other tropical export crops, such as bananas and cacao. However, power shifts and production trends in the coffee commodity chains have decreased producing countries' share to an estimated 22 percent (Talbot 1997).

These are the trends in the conventional green coffee market which in 1999–2000 moved an estimated 102.5 million 60-kilogram sacks of coffee with a wholesale value of US$14 billion (SCAA 1999). These dominant trends mask the growth and emergence of specialty and certified coffees.

The Rise of Specialty Coffee
The North American specialty coffee market grows 5–10 percent per year, and it reached an estimated retail value of $7.8 billion by 2001. This rapid growth contrasts to slow demand growth for bulk commercial-grade coffees. Unheard of 30 years ago, the specialty or gourmet market segment represents 17 percent of US coffee imports by volume and 40 percent of the retail market by value (Giovannucci 2001). The United States purchases 25 percent of the internationally traded coffee in the world (Giovannucci 2001). In 1982, a small number of small-scale coffee roasting companies joined together to form the Specialty Coffee Association of America (SCAA). The mission of the SCAA is to promote high-quality gourmet coffee and sustainability (SCAA 2002). The SCAA's 2,600+ members are primarily small-scale roasting companies, traders, and sellers of coffee-related accessories, but the membership also includes larger companies (Starbucks and Folgers), farmer organizations, and producing country representatives. Commercial-grade coffees do not have equally strict quality requirements, are commonly sold in tin cans, and often cost the consumer half the price. In addition to claims to superior taste, specialty coffee companies celebrate the craftsmanship of coffee roasting and preparation; they employ more specialized roasting processes, focus on product freshness and use large marketing expenditures to differentiate their product from the bulk commercial-grade coffees. The specialty roasters depend on a higher-quality coffee bean and are generally willing to pay producers price premiums for better beans.

Eco-Labels and Alternative Coffee Markets
Small specialty roasting companies pioneered the introduction of organic and Fair Trade coffees into the United States and helped the specialty coffee market become

the most active space for eco-labeling in the food sector. Nearly all eco-labeled coffees are also considered specialty coffee. The North American retail market for certified organic, Fair Trade, and shade-grown coffee is approximately US$188 million. The estimated worldwide retail value of these coffees is roughly US$530 million (Giovannucci 2001). Despite their relatively small market share, coffee roasters and retailers anticipate rapid and sustained growth for certified coffees.

Certified organic coffee currently accounts for 3–5 percent of the US specialty coffee retail market and remains the most widely recognized eco-label (Giovannucci 2001). Most consumers in the United States and Europe recognize the organic label from its widespread usage on fresh fruits and vegetables. The International Trade Center estimated the worldwide retail market value of all organic food and beverage products at US$21 billion in 2001 (International Trade Center 2002). Price premiums and 10–20 percent growth in retail markets have contributed to an increasing number of acres entering certified organic production. In workshops, Nicaraguan farmers almost always list price premiums as their primary reason for converting to certified organic production, however, the full list of motivations also includes the following: It is safer for their families and children without agro-chemicals on the farm, it lowers expenditures for synthetic inputs, it is better for the environment, and it helps protect the water. In Latin America, thousands of coffee, cocoa, vegetables, and fruits farmers have solicited and received organic certification.[2] Mexico exported the first organic coffee and remains a pioneer in the organic industry (Nigh 1997). While health remains consumers' primary motivation for purchasing organic products, development agencies, environmental activists, and many farmers' associations also support the certification for the ecological benefits gained from eliminating synthetic pesticides and fertilizers. Coffee covers an estimated 2.8 million hectares in Mexico, Colombia, Central America, and the Caribbean. While some of this coffee is produced without shade trees, farmers grow more than 60 percent under the shade of native and exotic trees. These shade coffee landscapes conserve biodiversity, soil, and water (Méndez 2004; Perfecto, Rice, Greenberg, and Van der Voort 1996).

In contrast to organic certification, which is a set of standards that regulates inputs and practices in the production process, Fair Trade-certified coffees certify the trade process.[3] Advocates of the Fair Trade movement believe that trade has the potential to either exploit or empower producers in the global South. These advocates refute the basic neo-liberal assumption that expanded trade will increase social and environmental benefits for everybody, and assert that North–South trade relations are plagued by power inequalities and exploitation. Four international Fair Trade associations define Fair Trade as follows:

Fair Trade is a trading partnership based on dialogue, transparency and respect that seeks greater equity in international trade. It contributes to sustainable development by offering better trading conditions to, and securing the rights of, marginalized producers and workers—especially in the South. Fair Trade organizations (backed by consumers) are engaged actively in supporting producers, awareness raising, and in campaigning for changes in the rules and practice of conventional international trade. (IFAT 2004)

Fair Trade markets find their roots in more than 50 years of alternative trade relationships. Long before certification existed, churches, disaster relief organizations, and solidarity groups had formed more direct trade relationships with refugees and marginalized groups. They paid producers better prices, offered market access, and provided technical assistance. These Northern organizations distributed Fair Trade crafts and foods through religious and solidarity networks. However, the volumes of Fairly Traded goods remained small and the development impact limited. In 1988, a church-based organization in the Netherlands teamed up with a Mexican smallholder coffee cooperative to launch the Max Havaalar Fair Trade product certification (IFAT 2004). The certification started a Fair Trade mainstreaming process that permitted wider participation by industry actors. This initiative grew quickly; Northern countries formed national Fair Trade labeling organizations, and more Southern producer groups accessed these networks. In 1997, these organizations joined to form Fairtrade Labelling Organizations International (FLO), which promotes Fair Trade products and practices by establishing standards and coordinating an international system of product monitoring and certification. This system now includes hundreds of companies, and more than 800,000 producers in over 40 countries are involved in Fair Trade networks (FLO 2003).

The Fair Trade standards stipulate that traders pay a price that covers the costs of sustainable production and livelihoods, provide a premium for social development, sign contracts that encourage long-term planning and stability, and help provide pre-harvest credit (FLO 2003). In the case of both coffee and cocoa, only small-scale producer organizations are eligible for certification. However, the FLO's certification organization also certifies large agricultural businesses producing bananas, tea, and fruit. Certification standards vary between crop and social organization (large farm or cooperative), but they all share minimum standards for social and economic development supplanted by anti-exploitation clauses. The expanding list of Fair Trade certified products includes coffee, cocoa, tea, fruits, wine, sugar, honey, bananas, rice, crafts, and some textiles (EFTA 2003). Coffee was the first certified Fair Trade product and remains "the backbone" of the system, accounting for the majority of the Fair Trade retail sales (Raynolds 2002b). Analysts estimate that roughly 1–2 percent of the global coffee trade is certified Fair Trade (Oxfam 2001). A livelihood vulnerability framework will help understand how participation in Fair Trade and organic coffee networks impacts vulnerability to the coffee crisis.

Livelihood Vulnerability, Farmer Typology, and the Coffee Crisis in Nicaragua

Livelihood Vulnerability

The regional impact of the coffee crisis can be considered an example of the frequent economic crises that affect the global South. These economy-wide shocks have many possible trigger events, including hurricanes, earthquakes, rapid devaluations, recessions, market shifts, declining terms of trade, and commodity price crashes (Skoufias 2003). From 1980 to 1999 the Latin American and Caribbean regions experienced at least 38 major natural disasters and over 40 episodes when the GDP per capita fell by 4 percent or more (IADB 2000). Scholars and development professionals have considered these phenomena in a special issue of *World Development* that examined the interplay between household vulnerability, coping strategies, economic crises, natural disasters, and household well-being (Skoufias 2003).

The livelihood vulnerability framework offers a common approach for both economic crisis and natural disasters (Combes and Guillaumont 2002; Moser 1998). This approach examines causes, impacts on household well being and mechanisms to cope with and buffer damage (Blaikie, Cannon, Davis, and Wisner 1994; Skoufias 2003). Vulnerability contains an external source of stress or shock and an internal component describing the exposure and response to this shock as it is interpreted through the socio-ecological relationships that shape individuals or group's livelihood assets. These descriptions of livelihood vulnerability respond to critiques of a narrow focus on income-based definitions of poverty and draw from Sen's pioneering work on assets, entitlements and famines (Moser 1998; Reardon and Vosti 1995; Scoones 1998; Sen 1981, 1997; Shankland 2000).

'Livelihood' refers to the means of gaining a living, including the tangible and intangible assets that support an existence (Chambers and Conway 1992). Bebbington added a cultural component to the material and economic focus behind livelihood assets, simply defining livelihoods as the way people make a living and how they make it meaningful (Bebbington 2000). The addition of meaning into the definition of livelihoods provides a theoretical space for including farmer perceptions and narratives, and an entry point for beginning to understand the subjective feelings of well-being and empowerment. In this way, livelihood vulnerability = livelihoods (material and intangible assets) + (exposure to) a stress or shock.

When vulnerable livelihood assets are exposed to a stress, the stress can diminish the asset's productivity or quality and/or limit access; the consequences are declining resource flows to the households. Intangible assets, such as kin and friendship networks, are often the most important relationships that households mobilize to reduce vulnerability. Household livelihood projects that are exposed to a stress will likely reallocate their assets to cope with the declining quality of life (Skoufias

2003). Previous studies have documented a wide variety of coping mechanisms to reduce damages and survive crises; many of these mechanisms such as pulling children out of school to avoid expenses can diminish long-term development potential and maintain households in a "poverty trap" (Skoufias 2003; Varangis, Siegel, Giovannucci, and Lewin 2003). Other common coping mechanisms include migration, increased borrowing, crop substitution, and decreasing inputs. Households will decide to reallocate their assets according to their perceptions and capabilities.

A Farmer Typology for the Central American and Nicaraguan Coffee Sector

Different farmers produce coffee in different ways, under different agroecological conditions, and in a variety of positions vis-à-vis the commercialization chains leading to the market. Farm size provides a good general indicator to describe the different forms of coffee production and commercialization (CEPAL 2002). An estimated 85 percent or 250,000 of Central America's coffee farmers are micro and small-scale producers. The family is the primary source of labor on these farms. In contrast to the micro-producers, who own less than 3.5 hectares of coffee land, most small-scale farmers employ day laborers during the coffee harvest. The small-scale farmers I surveyed in Nicaragua grow more than half of the food they eat. These farmers intercrop bananas, oranges, mangos, and trees for firewood and construction within their coffee parcels. Households measure annual yields in coffee and associated crops. Medium- and large-scale coffee plantations maintain a permanent labor force. Most large-scale farms are also agro-industrial plantations that have integrated processing facilities on the farm, occasionally exporting their own coffee. (See table 7.1.) These farms usually provide living quarters and food to farm worker families. Rural landless workers continue to live in extreme poverty. During the coffee harvest, the large plantations employ and house hundreds, sometimes thousands of coffee pickers. Like most countries in Central America, Nicaragua's coffee farm ownership is highly concentrated. In Central America, the largest plantations and agro export businesses account for 3.5 percent of the farms, 48.6 percent of the total land in coffee production, and an estimated 57.8 percent of the region's coffee production (CEPAL 2002). As of 2001–02, 5.4 percent of the country's largest farms accounted for 42 percent of the land in coffee production and roughly 75 percent of production (UNICAFE 2003).

The Coffee Crisis in Nicaragua

It is difficult to isolate the impacts of the coffee crisis from the series of negative shocks (Hurricane Mitch, drought, declining commodity prices) that continue to affect Central America (Varangis et al. 2003; Wisner 2001). In Nicaragua, the 1999–2001 droughts added further stress to low coffee prices. In the tropical dry

Table 7.1
Typology of coffee producers in Nicaragua (adapted from UNICAFE 2003). Estimated total harvest levels for 2001–2002. Average productivity statistics were generated from previous studies.

	Producers		Area		Production		
Farm size	Number	Percent	Hectares	Percent	Quintals[a]	Percent	Yield (quintals/ ha)
Small (< 12 ha)	28,745	94.56	52,719.4	57.41	356,766	24.28	4
Medium (13–30 ha)	1,492	4.91	29,183.5	31.78	576,483	39.23	11.6
Large (> 30 ha)	163	0.54	9,926.47	10.81	536,325	36.5	11.8
Total or average	30,400	100	9,1829.4	100	1,469,574	100	9.4

a. 1 quintal = 101.2 pounds.

regions, including the northern districts of Estelí, Madriz, and Nueva Segovia, the farmers did not harvest their subsistence crops. In focus groups, small-scale farmers told us how they lived on mangos, yucca, bananas, and the other subsistence crops that they intercrop with their coffee.

People's vulnerability to the falling prices depends upon their location in the coffee commodity chain and their access to assets such as land, credit, employment, and social networks. The coffee crisis is felt by most of the country's estimated 45,334 micro and small-scale farmers. These smallholder households sell coffee as their primary source of cash income. Farmers talk about pulling their children out of school, migration, and increased heath problems. The microproducers often work as day laborers on large plantations because their small parcels and current management practices are not sufficient to support the family. In the late 1990s coffee annually contributed US$140 million to the national economy and provided an equivalent of 280,000 permanent agricultural jobs (Bandaña and Allgood 2001; CEPAL 2002). Researchers estimate that Nicaraguan laborers have lost more than 4.5 million days of work during the first two years of the coffee crisis (CEPAL 2002). The rural landless coffee workers are more vulnerable than smallholders. The large plantations that employ these workers have high monetary costs of production (US$0.74–1.08 per pound) due to dense cropping patterns, dependence on paid labor, and intensive chemical inputs.[4] In 2001, the banks stopped offering credit for coffee and foreclosed on debt-ridden farms.

In the mountains north of Matagalpa, banks and plantation owners stopped paying and later stopped feeding their workers. Hungry and without work, thousands of families marched down from their individual parcels and large plantations. People grouped together along roadsides and in public parks where they lived in miserable conditions surviving on food donations. They demanded food, work, health care, and land (Calero 2001; Gonzalez 2001). I interviewed one woman who had camped by the road for the last three days with her children, and she stretched out the palm of her calloused hands and said this:

You see these hands. These hands are for working not for receiving donations." The aid agencies have responded with food for work programs, providing packages of donated rice, beans, sugar, and oil to plantation owners who can supplement their lower wages with food and entice the rural laborers back for this season's crop. As a recent World Bank study notes the Central American governments have largely failed to address the structural problems underpinning the crisis. (Varangis et al. 2003)

However, after 5 years of protests and three years since signing an agreement with the government, the rural landless workers union recently won titles for more than 2,000 hectares of land for some 3,000 rural workers.

The Effects of Participation in Organic, Fair Trade, and Specialty Coffee Trade Networks

Study Design and Methods

The research I conducted in Nicaragua started after I had spent more than 15 months evaluating a coffee quality improvement project with coffee cooperatives. I developed a set of indicators combining my research interests with criteria suggested by the cooperatives' administrative directors and elected leadership. After designing and field testing a survey, we scheduled a training workshop attended by the cooperatives' agricultural extension agents. Following the training, extension agents decided to either perform a complete census or I randomly selected 12–15 farmers from their cooperatives' membership lists. The larger unions of cooperatives designated a representative first-level cooperative from which farmers were randomly sampled.

The survey primarily contained structured closed ended interview questions and a walking assessment of the farmer's principal coffee parcel. While the extension agents conducted the survey, I followed up with multiple visits to each research site. During these visits, I evaluated data quality and ensured comparative methods. I also worked with a gender specialist to conduct 10 focus groups separated by sex. I drew focus group participants from the same list of farmers that participated in the sample and used these results to help triangulate responses given in the surveys.

Table 7.2
ANOVA results comparing altitude and certification with price as dependent variable. The average prices received at the farm gate were calculated by multiplying the (volume sold to each market)(by price for that market) + (volume market 2)(price sold to market 2) = (totals revenue)/total volume sold, these average totals were calculated for all farms. Source: Participatory survey, 2001.

	DF	Sum of squares	Mean square	F	P	Lambda	Power
Certification	1	1640169.310	1640169.310	78.945	<0.0001	78.945	1.000
Altitude	1	21131.332	21131.332	1.017	0.3144	1.017	0.162
Certification and altitude	1	34775.200	34775.200	1.674	0.1972	1.674	0.237
Residual	209	4342184.525	20776.003				

Finally, I interviewed the cooperatives' elected leadership and professional staffs and reviewed the cooperatives' internal documents regarding coffee sales.

The 228 farmers that participated in this survey are from a diverse social and ecological terrain. The social landscape includes first-level cooperatives (20–50 members) and regional cooperative unions (1,500+ members). Although the distribution of farm sizes in this sample resembles the percentages described in table 7.1, this sample differs from the national census because 180 farmers sold coffee to organic, Fair Trade, or Bird-Friendly markets.

Findings

The question of how coffee first came into farmer livelihood projects in northern Nicaragua precedes a discussion of the impact of participation in alternative and conventional markets.[5] Farming is part of a dynamic and mutually constructed relationship between households and agroecosystems (Gliessman 1998). Households engage their farms for multiple purposes few of which are captured in a single survey. Despite dependence on an export commodity for currency, a strong subsistence ethic survives among many small-scale Nicaraguan coffee farmers. Of the 225 farmers that responded to this survey question, 137 or 61 percent stated that they grow half of more of the food they eat, 58 farmers claimed that they grew about ¼ of the food they annual consume in the household, and only 30 farmers grew less then ¼ of the food they ate in 2000. Many coffee farmers also produce corn, beans, bananas, fruits, chayote, and yucca. The list of foods purchased on the farm, generally included salt, sugar, oil, and meat. In the focus groups, we asked how men and women allocate the total harvest of different crops. Both men and women allocate the first 80–90 percent of their corn and beans for household consumption, and they sell the surplus. Milk and cheese were sometimes divided evenly between resources for the household and those for sale. In contrast, farmers sold 80–90 percent of the coffee harvest, generally keeping only the lowest quality beans for their own consumption.

Although coffee is exotic to Nicaragua, for many farmers, this seed now contains a different story. Coffee agroecosystems and farm households coevolved as coffee slowly wove its way into the culture and landscape.[6] In this research population, 143 heads of household are second-generation coffee farmers, of which 50 or 23 percent of the total are third-generation producers, and only 36 percent (79 of 222 farmers) reported that they are the first generation to cultivate coffee. Coffee producers are also increasingly an older group. Ages of the 221 survey respondents ranged from 78 to 20, and the average farmer was 44 years old. In the focus groups, we asked, "what does coffee mean in your daily life?" A male coffee farmer in Estelí said "Coffee is the hope of a better future," a female coffee farmer in Matagalpa said "It provides sustenance to our family," another woman from Jinotega said "Coffee gives value to our land," and two men from the district of Madriz called coffee "the best crop to improve our lives and find an equilibrium with the environment." These quotes suggest some of the cultural values associated with coffee cultivation. Coffee dollars build houses, send children to school, and provide hope for the future.

What are the determinants of prices paid at the farm gate? The general assumption underlying coffee quality improvement projects is that higher-quality coffee receives a better price. Higher-quality coffee is sold quicker and earns higher yields when the coffee is processed from the parchment to exportable green beans. The results of a professional coffee tasting and the number of physical defects are the best quality measures; however, altitude is an easily accessible and commonly used proxy indicator. A systematic comparison of price and altitude reveals a statistically insignificant correlation between altitude and price. I used the average prices in local currency to run a two-way ANOVA comparing the impact of altitude and certification on coffee prices. The results support the conclusion that access to certified markets leads to significantly higher prices paid to farmers. Certification has a greater influence on price than altitude (quality). These relationships between price, quality, certification, and cooperative membership merit further research.

The cooperative is the primary intervening variable affecting prices received at the farm gate. Small-scale farmers, not organized into a cooperative or a marketing association, do not produce the volumes of coffee necessary to fill a container (275 sacks) and access the certified markets or sign contracts with importers.

The export cooperative manages external relationships that move coffee to certified markets and organizes an internal price structure that determines prices received

Table 7.3
Average prices reported at the farm gate for the 2000–01 harvest. Source: Participatory farmer survey conducted from July to August 2001.

Where did you sell the coffee?	Price paid per pound green coffee[a]	How long until you were fully[a] paid?	How many farmers sold to each market?
Cooperative—direct to roaster	US$ 1.09 (0.04)	33 (6.1) days	11
Cooperative—Fair Trade[b]	US$ 0.84 (0.07)	41 (86.6) days	36
Cooperative—Organic[b]	US$ 0.63 (0.11)	73 (78.4) days	61
Cooperative—conventional	US$ 0.41 (0.04)	46 (62.9) days	84
Agro export company	US$ 0.39 (0.04)	24 (50.3) days	51
Local middleman	US$ 0.37 (0.02)	9 (27.3) days	72

a. Numbers in parentheses are standard deviations
b. Although some coffee was certified as both Fair Trade and organic, most farmers understood this and reported that they were commercializing either Fair Trade or organic. They did not give a single price for both certifications.

at the farm gate. All cooperatives that commercialize coffee penalize farmers for defects, but none that I observed provides clear incentives for high-quality coffee. As coffee roasters increase their push for higher quality and cooperatives increase their knowledge and infrastructure for measuring quality, an incentive system will likely emerge.

The cooperatives allocate a portion of the higher prices offered by sales into certified Fair Trade and organic markets to invest in productive infrastructure, pay debts, provide credit, provide technical assistance, cover administrative and certification costs, and fund housing and education projects in farming communities. In two cooperatives, half of the Fair Trade and/or organic price differential was used to pay outstanding debt. These practices lower the final price of coffee to producers. Table 7.3 summarizes the average prices received by producers at the farm gate for sales through different commodity chains.

Most farmers sell their coffee to multiple markets. Nicaraguan cooperatives linked to organic and Fair Trade markets sell up to 60 percent of their coffee through conventional markets. Thus, the average price for all the coffee sold by the farmer may be significantly less than prices paid in the different alternative markets. For example, although the eleven cooperative members received US$1.09 per pound for the portion of their coffee sold directly to the roaster, the average price for all their coffee was US$0.58 per pound.[7] Thirteen members of a cooperative linked to organic and Fair Trade markets averaged US$0.56 per pound. In comparison, farmers selling to conventional markets averaged US$0.40 per pound. Many of these average farm gate prices are below smallholders' estimated monetary production costs, which are between US$0.49 and US$0.79 per pound.[8] The consequences of low farm gate prices are further exacerbated by long delays between depositing the dehulled coffee beans at the processing mill and receiving the final payment.

Most cooperatives pay farmers in stages: first as credit for the harvest and wet-milling, next a payment when they bring the wet coffee parchment to the dry processing facility and a final adjustment when all has been exported and actual prices are calculated. A few export cooperatives treat farmers and cooperatives as clients who own the inventory, and thus bear the risk, until the importer buys the coffee. If their coffee does not sell, the farmers receive no payment. Farmers waited an average of 73 days before receiving the full payment for their organic coffee. Most farmers sell their coffee to two or more market channels. Farmers generally sell some of their coffee to low-paying middlemen to satisfy the immediate need for cash as they wait for higher prices in the specialty markets.

These smaller producer cooperatives have joined together to form unions of cooperatives that can manage the economies of scale, pool their resources, and export coffee. Export cooperatives need access to larger credit lines to pay the farmers before their physical product is actually exported. Banks, roasting companies, and importers are increasingly reluctant to provide this credit to these cooperatives. Even well-established export cooperatives with over US$300,000 in working capital must rely on fewer than ten foundations and one roasting company for pre-harvest financing.

Vulnerability and Changes to the Quality of Life

Farmers selling to a cooperative connected only to conventional markets are four times more likely to perceive a risk of losing the title to their land due to low coffee prices than members of cooperatives connected to alternative coffee markets. In the survey, 224 farmers answered the following questions: "Is there a risk you will loose your farm this year?" "If there is a risk, why?" Of the 180 farmers who commercialized a portion of their coffee to organic, Fair Trade, or roaster-direct market channels, eight farmers perceived a risk that they could lose their farm this year due to low coffee prices. Eight of the 44 farmers who belong to cooperatives selling only to conventional markets also indicated a risk of losing their farm due to bank foreclosures and low coffee prices.[9]

When I asked leaders from each cooperative to design project evaluation indicators, they suggested I consider health, environment, education, and community development in addition to coffee price and quality. Measuring quality of life is a difficult task. A small-scale farmer from a cooperative in Jinotega said: "Well being is to have health, food, education and tranquility in the family." Farmers articulated the relationships between low coffee prices and their quality of life in focus groups. Their own words tell the story. A female coffee farmer from Jinotega said: "We can't buy our clothing, shoes. . . . We are surviving on bananas." Two other farmers, also from Jinotega, said that they were not investing enough fertilizer or labor in their coffee farms. The same farmers continued to say that the difficulties related to the coffee crisis included deteriorating social relationships within their families.[10] A farmer from the district of Madriz said "We have a little help, a little room to breath, with the 50 percent the coop buys as Fair Trade."

The evidence from this survey suggests that participation in alternative coffee trade networks reduces exposure and thus vulnerability to low coffee prices. The farmers linked to cooperatives selling to alternative markets received higher average prices and felt more secure in their land tenure. However, 74 percent of the farmers surveyed reported a decline in their quality of life during the last few years. The responses to this question about quality of life showed no significant difference between farmers participating in conventional and alternative trade networks. This finding and the results of the focus groups suggest that income from coffee sales to alternative markets is not enough to offset the many other conditions that have provoked a perceived decline in the quality of one's life.

Learning from Alternatives to Reduce Vulnerability

What can the livelihood vulnerability framework reveal about the coffee crisis? In contrast to the narrowly focused income-based approaches to poverty, the livelihood approach provides a more detailed description that explores how people make a living and how they make it meaningful. It provides a theoretical space for incorporating the multiple household and collective coping strategies, including subsistence production, kinship networks, barter, migrations, increased labor time, political mobilization, and protest. Linking vulnerability to livelihood projects and trade networks begins to suggest why some households are more vulnerable than others. The approach will lead to an integrated response to the coffee crisis well beyond the current program of debt relief, quality improvement programs, and food donations.

Diversification to Reduce Vulnerability

International development projects designed around a livelihood approach will work with small-scale producers and laborers to increase access to land, build stronger producer organizations, promote access to alternative markets, increase government investments in rural health and education, and diversify income sources. Development actors can learn from and support local coping mechanisms and diversified farming practices. I asked farmers in the focus group to identify their activities and strategies to address the coffee crisis. Their responses reveal a few coping mechanisms: "Planting more bananas and citrus." "Redoubling the labor that we put in as a family in order to survive." "Work organically to obtain better prices and lower the costs of production, because chemical fertilizers are very expensive." Diversifying the crops on a farm, such as planting additional fruit and/or the continued subsistence cultivation of corn and beans has long been a key strategy to maintain food sovereignty and manage risk within the household (Ellis 1998; Reardon 1997).

Coffee farmers continue to produce much more than coffee, both within the coffee parcels on their farm and on the rest of their total farm area. In this survey, 111 of the 221 farmers that responded to this question reported that 50 percent or more of their total farm area was used for other agricultural purposes. The tendency of small farms to survive price crashes by exploiting their own labor has been documented since Chayanov first investigated agrarian transitions in the former Soviet Union (Chayanov 1966 [1925]). The third quote represents two reasons—lower costs and price premiums—for moving toward organic agriculture. These are only a few coping mechanisms that farmers mobilize to negotiate with the coffee crisis, other observed activities include sharing resources through kinship networks, local migration, and increased barter. All these activities may reduce vulnerability without reproducing the same structures that created the coffee crisis. Diversification beyond coffee is important, and much can be learned from the failures of previous export-oriented diversification projects (Sick 1997). However, the following discussion investigates the role of diversifying into alternative coffee production and trade networks. What interventions can help expand coffee production and trade models that reduce vulnerability and move toward long-term sustainability?

Making Markets: The Promise and Peril of Coffee's Alternative Trade and Production Networks

Nicaragua has the potential to emerge as a world leader in the production and trade of specialty, organic, and Fair Trade coffee. In the last ten years, cooperatives, technical assistance organizations, and the donor community have worked with Nicaraguan farmers and their organizations to increase participation in specialty coffee markets. Although 80 percent of Nicaraguan coffee is potentially specialty coffee, only about 10 percent of the 2000–01 harvest was exported as specialty coffee (Bandaña and Allgood 2001; USAID 2002). To increase participation in the specialty coffee markets, including sales into the Fair Trade and organic segments, producers and their organizations must invest in coffee quality improvement infrastructure and training (table 7.4).[11]

Although the global demand for Fair Trade certified products grew by 42 percent during the period 2002–04, Fair Trade remains a small segment of the global market (FLO 2003). Due to low demand and high quality requirements, many Fair Trade certified cooperatives must sell close to 70 percent of their coffee into the lower paying conventional markets. Assuming that one accepts Fair Trade as one model that can help reduce vulnerability, the next question is how to scale up.

Markets are institutions that reflect the collective results of socially agreed upon rules and practices. The North American public is increasingly aware of sustainable coffee marketing messages, and mainstream news has covered the coffee crisis. Specialty roasting companies are forming campaign alliances with civil-society organizations and producer cooperatives. A few roasting companies, such as Equal Exchange, have teamed up with the faith community (Lutherans, Quakers, Catholics, etc.) and civil-society organizations (Oxfam) to build campaigns promoting Fair Trade. Student activists have recently formed the United Students for Fair Trade to coordinate a national student Fair Trade movement in more than 100 universities across the United States. People and their organizations are making markets. Fair Trade and organic certifications are two examples of attempts to build alternative production and consumption networks. To the extent that Fair Trade networks create a working model of their principals in practice, they help coffee drinkers align their tastes to specialty coffee with their social justice values.

Table 7.4
Nicaraguan production of specialty, organic, and Fair Trade coffee, 1987–2007. Source: Cooperative League of the United States of America (CLUSA) 2002.

	Area in production		Farmers	
	Hectares	Percentage[a]	Number	Percentage[a]
Pre 1994	420	0.5	156	5.1
1994–2002	6,089	6.7	3,927	12.9
2002–2007[b]	10,959	12.0	7,070	23.3

a. Both area and farmers percentage calculations use the data from 1997–98 harvest. However, as the coffee crisis continues the national area in coffee production will likely decrease, so these are conservative estimates. It is still unclear what the future holds for the total number of farmers involved in coffee production.
b. CLUSA technicians derived these projections by multiplying the current numbers by 15% per year for the first four years and 10% for the two final years.

Seen from this perspective, Fair Trade offers a technology that can help re-embed economic relationships into a set of social values (Polanyi 1944; Raynolds 2000c).

The promise of re-embedding trade into a social value system is matched by the challenges and contradictions involved in attempts to infuse 21st century capitalism with social and ecological justice. In the United States, most Fair Trade and organic products are considered specialty items and sold at prices significantly above their conventional competitors; this links affluent consumers in the North to livelihood struggles in the South. Further research, action, and exploration would investigate and address these class differences and escape the confines of this market niche. Market-based approaches accept consumers as stakeholders in the international development process and downplay their role of citizens (Goodman and Goodman 2001). Certification as a tool for producer empowerment is further challenged by the proliferation of certifications, such as Rainforest Alliance and Utz Kapeh, which offer lower social standards than Fair Trade and lower environmental criteria than organic certification. However, in the short term, many of the questions concern how to grow these markets.

Changing markets and power shifts to the roaster and retailer end of the commodity chain suggest a set of demand side interventions that compliment more innovative supply-side projects. The US Agency for International Development, the European Union, the World Bank, and the Ford Foundation are funding projects to address the coffee crisis. While some foundations have funded innovative approaches partnering business and civil-society organizations to expand alternative markets, most of the multilateral funding remains narrowly focused on production practices for niche markets. If multilateral funding does not also promote consumer education and expand alternative markets, these actors risk pushing too many people toward a small exit (Oxfam 2002).

Conclusions

A few farmers also offered their strategies for the long-term resolution of the crisis. Byron Corrales said "We need to apply agroecological coffee-production practices and sell to a just market." Jose Saturnino Castro Peralta of the La Providencia Cooperative said "We need to maintain and strengthen the cooperative and improve the quality to get better prices." These quotes represent individual responses to the changing structures of the global coffee markets and international development agendas. Eight Nicaraguan cooperatives collectively representing more than 7,000 small-scale farmers created a collective response. They formed CAFENICA, the Nicaraguan association of small-scale coffee farmer cooperatives. CAFENICA provides technical assistance helping member cooperatives coordinate and execute their own development projects. It also provides political representation for small-scale producers and coordinates collective marketing strategies.

Alternative models can help reduce livelihood vulnerability to the crisis in conventional coffee markets. As the crisis deepens and alternative models become mainstream, they will encounter increasingly large obstacles and contradictions. Addressing these issues requires a more diverse, committed, and critical dialogue that engages historical ideals and existing trading practices. This dialogue could stimulate Fair Trade praxis and the continued evolution of a process intended to increase social justice in our food systems.

Acknowledgments

Pedro Haslam, Byron Corrales, Nick Hoskyns, and Paul Katzeff are coffee and development professionals who taught me through experience the meaning of quality coffee, Fair Trade, and small-scale farmer cooperatives in northern Nicaragua. Henry Mendoza and I worked with a group of 16 agricultural technicians to help design, field test, and conduct the survey. Nicholás Arróliga at GeOdigital did an excellent job managing the mountains of data created by these surveys. This article benefited from ideas and commentary from Daniel Press, Jonathan Fox, Roberto Sanchez-Rodriguez, V. Ernesto Méndez, Stephen Gliessman, and David Goodman. Funding to cover my expenses came from the Switzer Foundation, a Center for Global Conflict and Cooperation International Field Dissertation Scholarship, PASA-DANIDA, and the Department of Environmental Studies at the University of California, Santa Cruz.

Although I have included more survey results and analysis, most of this article is reprinted with permission from Elsevier from an earlier piece that appearing in *World Development*: C. Bacon, Confronting the coffee crisis: Can Fair Trade, organic and specialty coffee reduce small-scale farmer vulnerability in northern Nicaragua? (33, 2005, no. 3: 497–511).

Notes

1. See Ponte 2002a for a detailed discussion of the causes, mechanisms, and consequences of the coffee crisis. Ponte also provides a good summary of global coffee commodity chain theory. He uses this approach to carefully demonstrate how shifts in consumption patterns and commodity chain governance structures have led to declining revenues to producing countries and increased profits to the international roasting companies.

2. Millions of peasant farmers around the world produce food without using synthetic inputs. Thousands of coffee producers continue to manage their coffee trees applying the minimum amount of work and no inputs from outside the farm. They may simply manually

remove the weeds once or twice per year, and harvest the cherries when they ripen. While these farmers may meet the basic requirements for organic certification, the fact that they do not actively manage their farms, and have not filled out necessary documentation, nor solicited third-party inspection legally prohibits them from selling certified organic products. Many classify this as passive organic production.

3. For additional research on Fair Trade movements and markets, see Whatmore and Thorne 1997, VanderHoff 2002, Raynolds and Murray 1998, Raynolds 2000, and Raynolds 2002a. For more recent work, see Goodman 2004 and Jaffee 2007. Raynolds is among the earlier sociologists to publish papers about Fair Trade bananas and coffee. Early works on Fair Trade coffee include Brown 1993, Renard 1999a, Renard 1999b, and Rice 2001. For a more comprehensive summary of the alternative trade organizations and the fair trade of crafts as well as food products, see Leclair 2002..

4. Estimates of the costs of production vary widely. Many of the costs incurred on family-labor farms do not show up in monetary values. It is clear that large technified farms have higher dollar expenditures. An internal report at National Union of Farmers and Ranchers (UNAG), estimates that monetary production costs on large farms are double those on passively managed small-scale farms (Corrales and Solorzano 2000).

5. Méndez (2004) describes the multiple roles that shade trees play in the livelihood strategies of small-scale producers in El Salvador. For detailed descriptions of the social dimensions of organic coffee production in Mexico, see Nigh 1997, Hernández Castillo and Nigh 1998, and Bray et al. 2002.

6. In Matagalpa, cooperatives, municipal authorities, exporters, and businesses recently sponsored the first fair to celebrate the beginning of the coffee harvest. People have long celebrated the end of the harvest. All festivities were canceled during the first three years of the coffee crisis. But the recent fair reflects the regions determination to keep planting this once golden bean. Folkloric dances often depict *campesino* families cultivating corn and picking the red coffee cherries.

7. The prices in table 7.3 are average prices for each market reported by the farmer. These prices are received on the farm after deducting costs for dry processing, organic certification, debt service and export, other costs including transportation to market, land, labor, and capital have not been deducted.

8. See note 5.

9. Land ownership in Nicaragua has been highly contested for more than 25 years. These perceptions are not ill founded. CEPAL estimates that 500–3,000 Nicaraguan coffee farms have been lost due to the coffee crisis (CEPAL 2002). Follow-up research found that members of the cooperative selling all of their certified organic coffee to Fair Trade markets were able to purchase additional land and the cooperative membership continued to expand, while two of 18 members of a cooperative selling to conventional markets sold their land (Bacon 2005).

10. This direct translation refers to increased stress, more arguments, and likely more abuse as the poor farm households try to make do with less. One leader of rural peasants has clearly linked falling coffee crisis to increased abuse and discrimination against women.

11. Paul Katzeff (personal communication) and Daniele Giovannucci (2001) concur that flavor is the most important factor in the specialty coffee roaster's buying decision. Flavor

is identified in tasting laboratories. Eight cooperatives in Nicaragua recently teamed up with Thanksgiving Coffee Company and used funds from USAID and other donors to build cupping labs as part of the coffee infrastructure controlled by the cooperatives. The project has led to an improved reputation for Nicaraguan coffee, better prices to the cooperatives, and an estimated 25 containers (valued at more than one million dollars) in additional coffee sales (Bacon 2001, 2005).

References

Bacon, C. 2001. Cupping what you grow: The story of Nicaragua's coffee quality improvement project. Unpublished.

Bacon, C. 2005. Confronting the Coffee Crisis: Nicaraguan Smallholders' Use of Cooperative, Fair Trade and Agroecological Networks to Negotiate Livelihoods And Sustainability. Doctoral dissertation, University of California, Santa Cruz.

Bandaña, R., and B. Allgood. 2001. Nicaraguan coffee: The sustainable crop. Unpublished.

Bebbington, A. 2000. Reencountering development: Livelihood transitions and place transformations in the Andes. *Annals of the Association of American Geographers* 90, no. 3: 495–520.

Blaikie, P., T. Cannon, I. Davis, and B. Wisner. 1994. *At Risk: Natural Hazards, People's Vulnerability, and Disasters*. Routledge.

Bray, B., J. Plaza Sanchez, and E. Murphy. 2002. Social dimensions of organic coffee production in Mexico: Lessons for eco-labeling initiatives. *Society and Natural Resources* 15, no. 6: 429–446.

Brown, M. B. 1993. *Fair Trade*. Zed Books.

Calero, E. C. 2001. Intentan frenar éxodo campesino. *La Prensa* (Managua), July 23.

CEPAL (Comisión Económica para América Latina y el Caribe). 2002. Centroamérica: El impacto de la caída de los precios del café. LC/MEX/L.517. Mexico, DF.

Chambers, R., and G. Conway. 1992. Sustainable Rural Livelihoods: Practical concepts for the 21st Century. Discussion paper 296, Institute of Development Studies, University of Sussex.

Chayanov, A. V. 1966 [1925]. *The Theory of the Peasant Economy*. University of Wisconsin Press.

CLUSA (Cooperative League of the United States of America). 2002. Nicaragua Specialty coffee production 1987–2007. Internal database, Managua.

Combes, J.-L., and P. Guillaumont. 2002. Commodity price volatility, vulnerability and development. *Development Policy Review* 20, no. 1: 25–39.

Conroy, M. 2001. Can Advocacy-Led Certification Systems Transform Global Corporate Practices? Evidence and Some Theory. Working Paper DPE-01-07, Political Economy Research Institute, Amherst, Massachusetts.

Corrales, B., and J. Solórzano. 2000 Costos de Producción: Aporte al informe del consejo nacional. Memorandum, Unión Nacional de Agricultura y Ganadero, Managua.

Díaz, R. P. 2001. Situación y perspectivas de la caficultora en Centro América: La crisis internacional de precios. United Nations Development Program.

Dicum, G., and Luttinger, N. 1999. *The Coffee Book: Anatomy of an Industry from Crop to the Last Drop*. New Press.

Ellis, F. 1998. Household strategies and rural livelihood diversification. *Journal of Development Studies* 35, no. 1: 1–38.

EFTA (European Fair Trade Association). 2003. *The Fair Trade Yearbook: Challenges of Fair Trade 2001–2003*.

FLO (Fairtrade Labelling Organizations). 2003. Standards in general. http://www.fairtrade .net.

Giovannucci, D. 2001. Sustainable Coffee Survey of the North American Specialty Coffee Industry. Summit Foundation.

Gliessman, S. R. 1998. *Agroecology: Ecological Processes in Sustainable Agriculture*. Ann Arbor Press.

Gonzalez, D. 2001. A coffee crisis' devastating domino effect in Nicaragua. *New York Times*, August 29.

Goodman, D. 1999. Agro-food studies in the "age of ecology": Nature, corporeality, biopolitics. *Sociologia Ruralis* 39, no. 1: 17–38.

Goodman, D., and M. Goodman. 2001. Sustaining foods: Organic consumption and the socio-ecological imaginary. In *Exploring Sustainable Consumption: Environmental Policy and the Social Sciences*, ed. J. Murphy and M. Cohen. Pergamon.

Goodman, M. 2004. Reading fair trade: Political ecology imaginary and the moral economy of fair trade foods. *Political Geography* 230: 891–915.

Hernández Castillo, R., and R. Nigh. 1998. Global processes and local identity among Mayan coffee growers in Chiapas, Mexico. *American Anthropologist* 100, no. 1: 136–147.

IADB (Inter-American Development Bank). 2000. *Social Protection for Equity and Growth*.

IADB, US Agency for International Development, and World Bank. 2002. Managing the competitive transition of the coffee sector in Central America. Background report for conference, April 3, Antigua, Guatemala.

ICO (International Coffee Organization). 2005. Trade Statistics—Monthly Imports.

IFAT (International Federation for Alternative Trade). 2004. What is Fair Trade? http://www .ifat.org.

International Trade Center, UNTAD/WTO. 2002. Overview of World Markets for Organic Food and Beverage, Estimates.

Jacoby, H., and E. Skoufias. 1997. Risk, financial markets, and human capital in a developing country. *Review of Economic Studies* 64, no. 3, 311–335.

Jaffee, D. 2007. *Brewing Justice: Fair Trade Coffee, Sustainability and Survival*. University of California Press.

Leclair, M. S. 2002. Fighting the tide: Alternative trade organizations in the era of global free trade. *World Development* 30, no. 6: 949–958.

Méndez, V. E. 2004. Traditional Shade, Rural Livelihoods, and Conservation in Small Coffee Farms and Cooperatives of Western El Salvador. Doctoral dissertation, University of California, Santa Cruz.

Moser, C. O. N. 1998. The asset vulnerability framework: Reassessing urban poverty reduction strategies. *World Development* 26, no. 1: 1–19.

Nigh, R. 1997. Organic agriculture and globalization: A Maya associative corporation in Chiapas, Mexico. *Human Organization* 56, no. 4: 427–436.

Oxfam (O. Brown, C. Charavat, and D. Eagleton). 2001. The Coffee Market: A Background Study.

Perfecto, I., R. Rice, R. Greenberg, and M. Van der Voort. 1996. Shade coffee: A disappearing refuge for biodiversity. *BioScience* 46, no. 8: 598–609.

Polanyi, K. 1944. *The Great Transformation: The Political and Economic Origins of Our Time.* Beacon.

Ponte, S. 2002a. The "Latte Revolution"? Regulation, markets and consumption in the global coffee chain. *World Development* 30, no. 7: 1099–1122.

Ponte, S. 2002b. Brewing a bitter cup? Deregulation, quality and the re-organization of coffee marketing in East Africa. *Journal of Agrarian Change* 2, no. 2: 248–272.

Raynolds, L. 1997. Restructuring national agriculture, agro-food trade and agrarian livelihoods in the Caribbean. In *Globalizing Food: Agrarian Questions And Global Restructuring*, ed. D. Goodman and M. Watts. Routledge.

Raynolds, L. 2000. Re-Embedding global agriculture: The international organic and fair trade movements. *Agriculture and Human Values*, 17, no. 3, 297–309.

Raynolds, L. 2002a. Poverty Alleviation through Participation in Fair Trade Coffee Networks: Existing Research and Critical Issues. Ford Foundation.

Raynolds, L. 2002b. Consumer/producer links in fair trade coffee networks. *Sociologia Ruralis* 42, no. 4: 404–424.

Raynolds, L., and D. Murray. 1998. Yes, we have no bananas: Re-regulating global and regional trade. *International Journal of Sociology of Agriculture and Food* 7: 7–43.

Reardon, T., and S. Vosti. 1995. Links between rural poverty and the environment in developing countries: Asset categories and investment poverty. *World Development*, 23, no. 9: 1495–1506.

Reardon, T. 1997. Using evidence of household income diversification to inform study of the rural nonfarm labor market in Africa. *World Development* 16, no. 9: 735–747.

Renard, M.-C. 1999a. The interstices of globalization: The example of fair trade coffee. *Sociologia Ruralis* 39, no. 4: 484–500.

Renard, M.-C. 1999b. Los Intersticios de la globalización: un label (Max Havelaar) para los pequeños productores de café. D.F., Mexico: Universidad Autónoma de Chipingo, Embajada Real de los Paises Bajos, ISMAM, CEPCO.

Rice, A. R. 2001. Noble goals and challenging terrain: Organic and fair trade coffee movements in the global marketplace. *Journal of Agricultural and Environmental Ethics* 14, no. 14: 39–66.

SCAA (Specialty Coffee Association of America). 1999. 1999 Coffee Market Summary.

SCAA. 2002. Strategic Plan 2002–2004.

Scoones, I. 1998. Sustainable Rural Livelihoods: A Framework for Analysis. Working Paper 72, Institute for Development Studies, Brighton, UK.

Shankland, A. 2000. Analysing Policy for Sustainable Livelihoods. Research Report 49, Institute for Development Studies, Brighton, UK.

Sick, D. 1997. Coping with crisis: Costa Rican households and the international coffee market. *Ethnology* 36, no. 3: 225–275.

Talbot, J. M. 1997. Where does your coffee dollar go? The division of income and surplus along the coffee commodity chain. *Studies in Comparative International Development* 32, no. 1: 56–91.

UNICAFE (Unión Nicaragüense de Cafetaleros). 2001. Cosecha cafetera ciclo 2000/2001, Managua.

USAID (US Agency for International Development). 2002. Description of the Coffee Sector in Nicaragua. Unpublished internal report.

van Dijk, J., D. van Doesburg, A. Heijbroek, M. Wazir, and G. Wolff. 1998. *The World Coffee Market*. Rabobank International.

VanderHoff Boersma, F. 2002. Poverty alleviation through participation in Fair Trade coffee networks: The case of UCIRI, Oaxaca, Mexico. http://www.colostate.edu.

Varangis, P., P. Siegel, D. Giovannucci, and B. Lewin. 2003. Dealing with the Coffee Crisis in Central America: Impacts and Strategies. Policy Research Working Paper 2993, World Bank.

Whatmore, S., and L. Thorne. 1997. Nourishing networks: Alternative geographies of food. In *Globalising Food*, ed. D. Goodman and M. Watts. Routledge.

Wisner, B. 2001. Risk and the neoliberal state: Why post-Mitch lessons didn't reduce El Salvador's earthquake losses. *Disasters*, 25, no. 3. 251–268.

8

Coffee Agroforestry in the Aftermath of Modernization: Diversified Production and Livelihood Strategies in Post-Reform Nicaragua

Silke Mason Westphal

In contrast to other agricultural areas in western Nicaragua, the coffee-growing area of the densely populated Meseta region is dominated by green and lush vegetation, trees of many different kinds and coffee plants grown in the shady environment. Contemplating the landscape at the time of the study, it was hard to believe that a radical elimination of virtually all trees in the coffee fields had taken place in the 1980s under an ambitious coffee modernization program.

Entering the village of Fátima, one encounters the coffee farm of Don Alejandro,[1] who received the small plot of farmland during the Sandinista agrarian reform. Alejandro lives on the farm with his wife Magdalena, who works in a sewing factory in a nearby town, and two of their children.

Before the agrarian reform, Alejandro had worked as a farm laborer for a large coffee planter in the area, who belonged to the group supporting the then dictator Somoza. Alejandro had started working as a farm laborer when he was still a boy, accompanying his father. Farm wage labor was necessary to make ends meet at that time as the small plot of land that the family lived on was not sufficient to maintain the household. When his father died, this small property was left to Alejandro's stepmother and children while he himself did not inherit.

At the beginning of the 1980s, when he was working as a farm laborer, the coffee plantation that Alejandro worked on was abandoned by its owners. Because they had been affiliated with the Somoza regime they fled the country after the Sandinista revolution. The plantation was turned into a cooperative that Alejandro became a member of. The cooperative held the title to the land, and the members were organized as a collective for acquiring credits, purchasing chemical inputs and technical assistance. Coffee production was modernized with capital-intensive technology and managed according to national production plans.

Toward the end of the decade, Alejandro was sent out to fight against the *contras*[2] in the civil war for a few years and then returned to his cooperative in Fátima. Soon after, in the elections of 1990, the Sandinista party was voted out of office and a liberal coalition headed by Violeta Chamorro took over government. Along with a severe structural adjustment program the new government started a rapid dismantling of the agrarian reform. In the course of the decollectivization process, Alejandro's cooperative was split up and he received an individual plot of 3 mz.[3] He thus became a *parcelero*, as the individual beneficiaries of agrarian reform land are called in Nicaragua.

Having received his own farmland for the first time in his life, Alejandro began remodeling the coffee-production system he had taken over from the cooperative. Whereas previously the extensionist had decided what and how to plant and had prescribed the types and quantities of chemicals to be applied, Alejandro now adapted production practices to his own ideas. One of the first changes was to plant and let grow a variety of trees on the coffee plantation: orange, lemon, avocado, mango, timber trees, and bananas. He also reduced input levels considerably. Due to the new government's liberalization policies, access to credits and chemical inputs had become difficult for small producers like Alejandro. Within a few years, Alejandro's coffee management practices came to resemble the "traditional" coffee-production systems such as that of Gustavo, whose family had owned a small coffee farm nearby for generations. On the whole, the similarity in production methods, livelihood strategies and the households' socio-economic characteristics between the two historically different groups of former cooperative members and private small-scale coffee producers was striking, only 10 years after decollectivization.

The Nicaraguan Context

With a national economy based on agro-export, a long history of economic polarization reaching back to the colonial period, an agrarian reform, and in recent years a strong political move toward market liberalization and deregulation, the Nicaraguan case epitomizes several important historical trends characterizing the Mesoamerican region, albeit in a somewhat condensed form. The problems and challenges arising from the social context of rural households are also well known throughout the region: poverty, population growth, increasing pressure on agricultural land, limited employment opportunities in other sectors, and withdrawal of the state from agricultural and social service systems.

Coffee has historically had a central position in the Nicaraguan economy. Amounting to one-fifth of national export, coffee was still an important product at the time of the study, not only from a national economy perspective, but also as an income source for small-scale producers. Small-scale farmers[4] represented about 89 percent of the coffee producers and cultivated almost 40 percent of the area under coffee (UNICAFE 1997, 1998).

As the case of Alejandro indicates, several shifts in the wider political and economic context influenced the production and livelihood strategies of small-scale coffee producers importantly between the 1980s and the 1990s. Among these were the agrarian reform and the coffee modernization program CONARCA,[5] shifts in credit policies and international coffee price trends.

Modernization of coffee production was increasingly promoted in Nicaragua during the second half of the twentieth century, culminating in the carrying out of the CONARCA program in the Meseta region in the 1980s. The CONARCA program targeted coffee modernization based on the technological model of Brazilian large-scale coffee plantations, with monocultural design, extensive use of chemical inputs, and full sun exposure (Rice 1990). Among other measures, modernization entailed wide-ranging elimination of the existing coffee plantations, or in other words, clear felling coffee bushes and the abundant shade trees that used to be part of the coffee growing systems in the area, and substituting them with new varieties in unshaded plantations. The goal of the CONARCA program was to replant 12,000 mz of coffee, and practically all the coffee cooperatives in the study area had been replanted. Modernization efforts focused mainly on the state farms and cooperatives of the collective sector, while strategies for small family managed coffee farms were rather to convert them to fruit and vegetable production for the domestic market (ibid., pp. 63–65). Modernization policies moreover included facilitation of good credit possibilities, low input prices and promotion of "technified" production methods through the agricultural extension system.

In the 1980s, agricultural credit was mainly given by the State bank. Although not to the same degree as in the collective sector, credit was widely available to small-scale producers and the terms were favorable, which meant that many producers started to depend on this source of finance. In the 1990s, it became increasingly difficult for small-scale producers to obtain credit in Nicaragua, especially formal credit at reasonable interest rates, and banks virtually ceased to be a source of capital for small-scale producers. When the liberal government came into power, the banking system was privatized, interest rates liberalized and the volume of credit was reduced drastically. Between 1991 and 1992, credits given to coffee production fell by 72 percent, and small and medium-scale producers were left with a share of less than one-fifth of the total credit volume (Romero and Hansen 1992, pp. 98–99). Apart from difficult access, conditions such as short terms and high interest rates were also a constraint, and, as land was now required as collateral, credit entailed a serious risk to producers.

For the producers of export crops such as coffee, severe moments of risk consisting of fluctuations in world market commodity prices and climatic effects marked production conditions during the 1990s. Since 1989, international coffee prices have fluctuated markedly around a general downward trend, due to a breakdown of the international price agreements under the ICO[6] (Ponte 2001, pp. 5–6). For the small coffee producers of Nicaragua the resulting loss in this period was estimated to be one-third of previous income levels (Romero and Hansen 1992, pp. 17, 117). The drastic social, economic and political changes and disruptions during the past three decades influenced the outlook of the Nicaraguan farm households, emphasizing uncertainty as a major consideration in their lives and their planning for the future. This is the context of the present chapter, which examines the ways in which coffee producer households have changed and adapted their production and livelihood strategies to face these new conditions.

The Shade Debate: Modern vs. Traditional Coffee Production

The use of shade trees is a leading issue in the debate on technological development in coffee cultivation in general and in the study area in particular. In Nicaragua, the tension in the empirical debate has been played out between modernized (*tecnificado*) and traditional (*tradicional*) coffee production, with shade trees as the most important signifier of the latter. When efforts to modernize coffee cultivation in the Meseta region started, shade cover was considered detrimental to raising productivity. The CONARCA program with its focus on crop specialization and high-input technology can be understood as an expression of the modernization approach to agricultural development.[7] The introduction of a technological model from a geographical and social context quite different from that of Carazo and Masaya reflects the universalist technology perception characteristic of the modernization approach. The priorization of production collectives during the agrarian reform to some extent also can be attributed to a belief in the economic advantages of large-scale production units and capital-intensive technologies.[8]

More recently, the so-called traditional agroforestry systems such as shaded coffee have attracted renewed interest among researchers and practitioners in view of the continuing social and environmental challenges faced by the rural areas of Central America and other developing regions (see e.g. Beer et al. 1998; Llanderal 1998; Nielsen 1998; Bonilla 1999; Escalante 1999; Fernández and Muschler 1999; Vaast and Snoeck 1999; Guharay et al. 2000). Instead of seeing them as backward, a different and more optimistic interpretation of the potential of small family farms and traditional farming practices is offered in the recent current of literature on small farms, sustainable agriculture and indigenous knowledge (Bunch 1985; Altieri, Trujillo et al. 1987; Chambers 1989; Nair 1993; Netting 1993; Pretty 1995; Maldidier and Marchetti 1996; Rosset 1999; Röling and Brouwers 1999). Within this strand, the use of household labor, agro-ecological farming practices and adaptation to local natural conditions are seen as crucial elements for successful small-holder production, at the same time as promoting environmental sustainability.

Regarding its focus on family labor and intra-household demographic dynamics the small-farm literature has been inspired by the historical populist debate on peasants.[9] Some commonly accepted ideas are that of special smallholder or peasant rationalities diverging from the common understanding of the logic of the market economic system, e.g. the principle of risk minimization as opposed to surplus maximization, and considerations linked to the special structure of the peasant household as a unit of both consumption and production based on family labor. (See e.g. Sahlins and Service 1960; Scott 1976; Lipton 1989.) Netting (1993) is a more recent, prominent, and theoretically well-grounded proponent of neo-populist small-farm and sustainable-agriculture approaches. With his analytical focus on the individual farm household and its internal dynamics, Netting's approach can be characterized as one of theoretical-methodological individualism (Kearny 1996, p. 105). Opposed to the modernization approach, proponents of smallholders' sustainable agricultural practices do not see modern high-input farming methods as the key to increase productivity of small farms. Rather than externally induced technology, knowledge of the local environment and farming methods adapted to these are seen as the basis for intensification. Instead of universal technology models, explanations for changes in production systems are generally sought within the local setting, including the natural conditions and internal demographic dynamics of the household (Netting 1993, p. 57).

The present chapter discusses how the methodological focus of the small-farm literature on local agro-ecological conditions and household characteristics can contribute to explain the evolution of diverse coffee agroforestry systems in the study area—especially the change of the *parceleros'* modernized coffee-production systems to diverse agroforestry systems—and what were some of its limitations to understand the change processes at stake.

Data and Sources

The analysis refers to different types of information and sources, comprising literature of theoretical, thematic and geographical relevance. The primary data for this chapter were collected during field studies in Nicaragua (July 1998–May 2000). The empirical analysis is based on a combination of qualitative and quantitative data and methods of interpretation. The core of the fieldwork was made up of farm and household-level studies carried out in the villages of Fátima (Department of Carazo), San José de Monteredondo and San Juan de la Concepción (Department of Masaya). A survey with a sample of 62 respondents and a follow-up survey with

39 of these were undertaken, including interviews and measurements in the coffee agroforestry systems. The respondents were selected at random from the most comprehensive list of coffee farmers available, aiming at 20 in each village. In addition, six in-depth case studies, a series of individual key person and group interviews and a focus-group workshop were carried out. The primary information also comprised local and national-level interviews with individuals and representatives of public and private organizations, administrators, researchers, extensionists, development workers, as well as informal conversations and personal observations.

Villages Studied

The coffee-growing region within which the study area was located is known as the Meseta de los Pueblos. The natural conditions of the Meseta with its fertile volcanic soils and elevation of 450–700 meters are generally considered appropriate for coffee cultivation, although not optimal due to the relatively hot dry climate and short rainy season.

With 120 inhabitants per square kilometer, population density in the region is among the highest in the rural areas of Nicaragua. By comparison, the national average was 15 inhabitants/km^2 in 1993–94 (OIM 1999, p. 18). An extensive grid of dirt roads and paths connects villages and settlements, while paved roads mostly link municipal centers. Public transport is widely available between towns and cities, and also to some extent to the villages, though more sporadically and in more rustic forms.

The area has a heterogeneous structure with some large and many very small landholdings. The principal economic activity is agriculture with cultivation of beans, corn, coffee, citrus, other fruits, bananas, plantains, and some vegetables. A wide range of other economic activities complemented farm income including small-scale home-based craft production and home industry (wood, leather, textile, basket weaving and food), petty commerce oriented toward the markets of the nearby towns of Jinotepe, Diriamba, Masaya, and Managua and farm and urban wage labor, often involving seasonal and semi-permanent migration.

Coffee Farms and Producer Households

The average number of family members living in the sample households was six, comprising two children under 16 years, two adult women, and two adult men. According to the survey data, 48 (79 percent) of the respondents considered the farm the most important income source of the household, plus 5 (8 percent) who considered the farm and another income source equally important. Other income sources, however, played a considerable role for the producer households, and two-thirds of the households engaged in economic activities apart from farm production.

Of the 62 households included in the survey, 27 (43 percent) had received land through the Sandinista agrarian reform. The former cooperative members who had received individual plots of land are called *parceleros*. The remaining 35 (57 percent) producers had obtained their farms by inheritance or purchase and are here labeled "historically private producers'

The average size of the 62 farms of the sample was 6.9 mz. However, the range was quite wide (0.35–115 mz), which means that the median value of 3 mz gives a better impression of the typical coffee farm in the study area. The average area planted with coffee was 3 mz, corresponding to 64 percent of the farm. Ten households had rented additional land to cultivate annual crops, mainly beans and maize.

When asked about the importance of the different crops they cultivated, the majority of the respondents considered coffee their principal product, but fruit also played an important role. The importance assigned to the different crops reflected their function as products for sale, the most frequent cash crops being coffee, mentioned by all respondents, and citrus fruits, mentioned by 34. Other marketed crops were avocados (mentioned by 24), other fruits (mentioned by 15), plantains and bananas (mentioned by 21), food grains (mentioned by 13), and vegetables (mentioned by 3).

Homogenization of Production Strategies of *Parceleros* and Historically Private Coffee Producers

Coffee Agroforestry Systems

In order to undertake an agro-ecological characterization of the coffee and tree components in the coffee agroforestry systems, measurements and inventories of shade and coffee plants were carried out. Data were taken from a 1,000-m² sample plot within the coffee agroforestry system of each of the 62 farms. These data included an assessment of coffee plant density and state of productivity, measurement of shade percentages, and inventories of trees by species, numbers and size categories. Moreover, on three of the case-study farms a complete mapping of shade trees was undertaken in order to get a picture of the horizontal structure of the coffee agroforestry systems.

The most common coffee varieties grown in the study area were of the arabica species (*Coffea arabica*). The most frequent variety was *Caturra*, which had been promoted under the CONARCA program, followed by the older *Bourbón*, and a few producers cultivated *Catuai rojo* and *Catuai amarillo*. In most cases, the coffee plantations comprised a mix of these varieties. Average plant density in the coffee fields was 2,900 plants/mz. Average plant age was 12 years.

Table 8.1
Shade density and composition (*N* = 62).

	Mean	SD
Shade percentage	47	12
Number of species/1,000 m²	9	4
Population (ind./mz)	576	49
plantain/banana	368	39
fruit trees	65	8
timber trees	38	7
other trees	104	12

The trees in the coffee plots could be divided into different categories by their main function. Many of the trees were multi-purpose trees, providing at least one additional product apart from their function as shade trees. In the following discussion, the trees are divided into the categories of fruit trees, timber trees, other trees and bananas and plantains.[10] In practice, all trees in the coffee agroforestry system contributed to shade cover, but the tree species had quite different qualities in terms of creating a favorable agro-ecological environment for coffee production. Some trees, for instance the *madero negro*,[11] were primarily grown for their beneficial effect on the coffee, while others, such as citrus or timber trees, were mainly grown for their products. Table 8.1 shows the average shade percentages, number of species and populations of shade trees, classified by functional categories, found in the 1,000-m² sample plots.

The shade canopy in the coffee plantations was generally dense. Based on the measurement of shade percentages during the dry season in each of the 62 sample plots, an average of 47 percent was calculated.[12] The tree inventory demonstrated a wide range of species with an average of 9 species per sample plot (range 1–18) and a total of 80 tree species in the 62 sample plots. The most frequently occurring shade species were different banana and plantain varieties and *madero negro* trees, although it should be emphasized that the most prominent feature of the agroforestry systems was the mixture of different types of trees.

Comparison of shade strata of *parcelero* and historically private farms. As described above, the study deals with two types of coffee farms: One type had historically been privately owned. The second type had first formed part of the large estates belonging to Somoza and the group supporting his dictatorship until 1979, which then were turned into cooperatives during the Sandinista agrarian reform of the 1980s, before finally being divided up among the cooperative members in the process of decollectivization in the 1990s.

Table 8.2
Shade characteristics of historically private and parcelero farms (N = 61).

	Private farms: mean (n = 35)	Parcelero farms: mean (n = 27)
Trees (ind./1,000 m^2)	52.31	92.33
Tree species (ind.)	7.97	9.00
Shade density (%)	47.96	45.58
Coffee area	4	2

The *parceleros'* coffee plots had been replanted under the CONARCA program in the 1980s and the shade trees had been removed. One of the working hypotheses here is that the almost complete elimination of shade trees would be reflected in the shade structure of the coffee plantations at the time of the study. Thus, it was expected that lower shade percentages and fewer trees and tree species would be found on the farms of *parceleros* than on the historically family owned coffee farms that had not been modernized by CONARCA. To test this hypothesis, the two groups of coffee farms were compared regarding shade characteristics, shade cover, number of trees and species. The results are shown in table 8.2.

Contrary to the hypothesis that more trees and more diverse shade would be found among the farms of historically private producers, the table shows that the group of farms owned by *parceleros* had higher average values for abundance of trees, as well as for species. A *t* test indicated that the difference was statistically significant in the case of number of trees,[13] but not significant with regard to numbers of species. The difference between the shade percentages, which were slightly higher in the group of historically private farms, did not prove statistically significant and so did not support the hypothesis either. The fact that the higher number of trees in the group of *parcelero* farms was not reflected in a correspondingly higher shade percentage could be explained by the size of the trees. The canopy of an old *guanacaste*[14] tree, for instance, can be equivalent to the shade of several *madero negro* trees planted some 10–15 years earlier. Hence, the number of trees with canopy diameters measuring less than 5 meters was 76 among the group of *parceleros*, compared to 45 in the other group.

The data from the study support the conclusion that a comprehensive homogenization has taken place in the use of shade trees in the coffee-production strategies of the *parceleros* and the historically private producers. These results show that, despite the introduction of unshaded coffee technology under the CONARCA program, the producer households who received their land through the Sandinista agrarian reform had converted their coffee-production systems to so-called traditional shade-grown coffee by the end of the 1990s. The reestablishment of shade canopies in the coffee plots of the *parceleros* is a radical change in these production systems, considering that it takes a lot more time to grow trees than to fell them as was done under the CONARCA program. This radical change in the coffee-production strategies of the former cooperative farmers suggests a further question: how can this return to shade-grown management practices be explained? In the conversations with *parceleros* it was obvious that the growing of trees in coffee plantations had been a common point of disagreement between the extensionists and cooperative producers during the 1980s. As a rule, the planning and direction of production in the cooperatives was carried out in a top-down manner. Thus, according to one former cooperative member, the extensionist from the State coffee enterprise would pass by twice a month to supervise the state and management of the coffee plants and to advise on application of chemicals on the basis of *recetas* (literally recipes), indicating the type and amount of fertilizer and pesticides to be applied (CaseStudy1 2000).

Several of the *parceleros* interviewed commented on the design of the coffee plantations when CONARCA undertook replanting with coffee monoculture and virtually no shade trees. For instance, Alejandro said that he had disagreed with the extensionist's recommendations to concentrate solely on coffee. For Alejandro, it was important to have alternatives and not to depend on one single product. In the cooperative, the extensionist had not allowed fruit trees and plantains to be grown in the coffee plantation and, according to Alejandro, in those times you had to obey the extensionist, or else you would not receive your credit. However, Alejandro had negotiated with the extensionist for permission to grow some citrus trees on the borders of the coffee plot along the path, a location that he now lamented, because much of the fruit from his trees was stolen. When technical assistance to the cooperatives was abolished in the early 1990s, Alejandro started to include a variety of trees in the coffee-production system (CaseStudy2 2000). As the current shade structure in the *parceleros'* coffee farms shows, many cooperative producers, like Alejandro, decided to plant trees in their coffee plots as soon as they were free to take their own management decisions.

"One needs the firewood, one needs the money"

In a focus-group discussion, one of the participating producers, Andrés, very clearly expressed how resource availability and the different needs of different kinds of producer households required different designs of the coffee-production systems:

I have noticed that Miguel Gonzalez [one of the large-scale coffee producers in San José] produces 60 fanegas per manzana, equivalent to 30 quintales. But now we are talking about technified [production], right? He has all the resources. There he has no citrus, no banana, no nothing, right? But I prefer a farm that has . . . citrus, plantain, etc. Then you will have three harvests a year, or rather, two harvests and you maintain your plantain throughout the year. And yes, it is possible [to have coffee and production of fruit, banana and plantain in the same plot], although the coffee may not yield the same. But what we are talking about is that one needs the firewood, one needs the money, and of course you need the citrus. (Andrés, FocusGroup 2000)

The quotation highlights two aspects of how the design and management of the coffee-production system is linked to the socio-economic situation of the farm household: One is the necessity of financial resources to successfully maintain a modernized, or "technified," coffee-production system without shade trees. The typically larger cash reserves held by larger producers also mean that risk minimization is less crucial, as household subsistence is less vulnerable to variability in coffee income due to a bad rainy season or a temporary fall in price. The other aspect concerns the importance of tree products for poorer farm households, as they contribute to consumption needs (e.g. for firewood or food crops) and allow income smoothing throughout the year.

Differences in Diversity
Although, at first glance, the coffee agroforestry systems investigated seemed quite similar in terms of shade density and composition, the tree diversity encountered on different farms reflected, at least in part, different management and considerations. Tree composition in some cases was the result of carefully planned design and investment decisions based on consideration of household needs and market demands. In other cases, the association of trees appeared to be more the consequence of a laissez-faire approach, with corresponding differences in system output. However, even coffee agroforestry systems that seem similar in terms of output levels, can be the result of different design and investment patterns. For example, a comparison of two case studies shows that system diversification can involve quite different strategies, such as a focus on coffee and a few selected fruit trees or, alternatively, diversification into a broad range of niche products. The two producers, César and Alejandro, earned about the same amount of cash income from the tree products of their coffee agroforestry system, but, as table 8.3 indicates, a lot more tree species and trees were counted in Alejandro's plot.

In the sample plot on César's farm, there were only six tree species, and these all had a very well-defined purpose: fruit for sale, plantain for home consumption and the *madero negro* for shade. His priorities had lead to a substantial income from tree products, mainly concentrated on the sale of avocado and citrus fruit.

In Alejandro's case, 15 species and also a lot more trees were counted in the sample plot. His income from tree products was comparable to César's, but reflected

Table 8.3
Comparison of case studies (1999–2000 season).

	César	Alejandro
Coffee AFS area	3.3 mz	3 mz
Tree product income	5,950C$	5,120C$
Tree species/1,000 m²	6	15
Trees (individuals/1,000 m²)	68	86

a much more complex composition, including plantains, oranges, avocados, other fruit trees, honey, and bamboo. Moreover, he was producing timber for the family's future use and was experimenting with niche cash crops such as cinnamon and pepper. During a walk through his coffee grove undertaken with a group of producers, I asked Alejandro about the *cedro*[15] trees he had. Pondering a little on the issue, he told me that it made him think how he had been struggling for 15–20 years with his plot and that soon his children would be needing to build a house of their own and would need timber: "Therefore the cedro,–well, that is my opinion, maintain it as much as you can." (FocusGroup 2000)

During the same field walk, César expressed a divergent view, arguing that timber trees had too long a time horizon:

You have to grow things that are more rapid, that get there quicker! Well the citrus, the citrus is the most rapid. Sure, let us talk about your "patio."[16] . . . [The timber trees] will give you a little cash, but when? It will take 20 more years, but then our friend here [Alejandro] won't be around anymore! [Everybody laughs.] Well, okay, okay, he might still be around, that could be. But 20 years until you can make some money . . . whereas the orange yields in 3 years! You have to grow things that earn you more money, earn you money more rapidly—at least we who are poor, we who live of the farm. (César, FocusGroup 2000)

Pedro, another producer accompanying us, agreed on the advantages of fast growing trees compared to timber trees. He especially preferred orange trees as shade because they created favorable conditions for the coffee plants and generated income. Although he assigned a certain value to timber trees, such as the laurel,[17] Pedro preferred the fast-growing *madero negro* because it gave him a quicker benefit in the form of poles that he could sell. Moreover, he did not have to sacrifice the shade, which he would if a whole tree was cut for timber (CaseStudy1 2000).

As the discussion shows, when working with trees, the time horizon is a crucial factor in decision making. In contrast to most agricultural crops, trees are a medium-term or a long-term investment. For producers with limited availability of land and cash, it can be difficult to set aside land for tree growing and wait for the returns of long-term investments. That timber did not appear a too attractive product compared to other options, moreover, may be attributed to the fact that felling trees for timber was subject to somewhat bureaucratic procedures and taxation. Finally, the two latter producers' preference for fast growing fruit trees and low-value trees for poles to valuable timber is also interesting in the way that it highlights the importance of local markets and petty commerce in the area.

The Importance of Capital Availability

César had designed his agroforestry systems in a very structured way. His family had established the farm from scratch after they had bought the plot in the late 1960s. The couple had been able to accumulate some capital from the savings of years of work on haciendas during the agro-export boom. Milagro, César's wife, had sold food to the workers. César had worked as foreman and later as farm manager, and he continued in this job until 1994, after he had acquired the farm in San José. When the coffee-production system was established, the family had sufficient means to invest in the planting of coffee and fruit trees all at the same time. Some years later, the surplus from their production allowed them to purchase an additional piece of land, which they planted with coffee and fruit trees in the same manner as the original one.

Unlike César, Alejandro had not had the means to undertake a larger investment in the farming plot he received in the late 1980s when his cooperative was split up into individual holdings. When asked how he had gone about starting to grow trees in his coffee field, he told me that some of the first ones he had planted were *madero negro* trees. At the time of the interview, more than ten years later, he still remembered with gratitude what a stroke of luck it had been that his father-in-law had given him 100 cuttings of *madero negro* for free. Otherwise it would have been more difficult for him to establish proper shade for his coffee at the time. So the fact that there was a market for almost everything in Carazo and Masaya was not an advantage for the producers in all situations. On the one hand, they could sell many different products but, on the other, almost everything they did not produce themselves had to be bought. Everything had a price, even cuttings of *madero negro*.

The example shows that in addition to differences in the producers' preferences and strategies, availability of capital also played a role for the way they could design their coffee agroforestry systems. César had had the means to plant what he wanted when he wanted, while producers with more limited amounts of capital planted what they had, or what they could get hold of cheaply or for free, and they tended to carry out improvements only little by little. Hence, although the capital demands of low-input, diversified coffee agroforestry systems can be considered to be relatively modest, the access to capital to some extent conditioned the possibilities to invest in coffee productivity, product diversification and timely adaptation of the agroforestry system to shifting risks and market opportunities.

Seven Good Reasons to Diversify

The case-study examples indicate how the producers took into account a range of different considerations in their production strategies, including natural and economic production conditions, socio-economic characteristics and needs of the households, and the opportunities and constraints of local, national and international markets. Based on the survey data, the different factors that contributed to promote diversification of the coffee-production systems are examined under the following headings:

agro-ecological adaptation
economic production conditions
risk spreading
land-use optimization
products, income, and market opportunities
income smoothing
household consumption needs.

Coffee Agroforestry as Agro-Ecological Adaptation

In the study area, the natural conditions required for coffee growing are met in some aspects, while in others they are suboptimal. Thus the soils are of volcanic origin with good fertility. They are deep, well-drained soils of medium to coarse texture and neutral to slightly acid pH values (Rice 1990, p. 173). The study area has a pre-mountainous humid climate, but with averages of 1,000–1,400 millimeters per year and erratic patterns of distribution, rainfall is a limiting factor. In the regional experimental center of Unicafé located near San José average precipitation was measured at 1,480 mm/year between 1993 and 1999, varying between 1,005 and 1,992 mm/year (CATIE/MIP 1993–1999). Moreover, the average temperatures of 23–27°C in the study area are somewhat higher than the ideal conditions for coffee growing (Rice 1990, p. 1; INIFOM/AMUNIC 1997, pp. 3–4; Dauner 1998, p. 17).

The natural growing conditions for coffee regarding light, water, and nutrients can be modified to a certain extent by the management practices applied. In the study area, the use of shade had special importance, as shade trees mitigate the impact of extreme temperatures, contribute to conservation of humidity and serve as windbreaks. Moreover litter fall from the tree strata is recycled within the coffee agroforestry system and supplies the soil with important biomass and nutrients, which is highly relevant for low-input coffee cultivation methods such as those practiced in the study area. The foliage and roots of trees provide protection against soil erosion. And finally, shade cover is an effective protection against weed growth reducing the need for mechanical or chemical measures.

The principal disadvantage associated with shade trees in coffee plantations, especially if shade cover is too dense, is lower yields because photosynthesis is slowed down. On the other hand, the slowing down of the photosynthetic processes also contributes to enhance plant longevity. Further potentially negative effects are competition for nutrients and water and damage to the coffee plants from falling branches during winds. A further problem that is sometimes associated with dense shade cover is the risk of certain diseases developing in the more humid conditions. The behavior of pests and diseases in shaded environments, however, is a complicated issue as different pests and diseases have different preferences with regard to micro-climatic conditions (Cambrony 1992, pp. 66–68; Fernández and Muschler 1999, pp. 75–76; Guharay et al. 1999, pp. 80–83)

In recent years, the complex interrelationships determining the impact of shade on coffee productivity have been investigated in different studies, including issues such as climate related stress and conditions for different pest, disease and weed development. Research from Central America has illustrated the relation as a curve topping at a locally specific optimum density, depending on the given natural conditions (Fernández and Muschler 1999; Guharay et al. 1999). Assuming a relation between the variables of shade percentages and coffee yields, the sample data of the present study indicated an optimum density between 40 and 50 percent in the local climatic and management conditions, which lies within the range of 35–60 percent suggested for low, dry zones by Muschler (1997).

The data suggest that adaptation of the shade cover to local growing conditions played a role in the use of shade in the coffee farms in the study area. Thus, the shade demand of coffee plants grown in the lower altitude and drier conditions found here is quite different from higher altitude areas with more rainfall.

Economic Production Conditions

Supply of nutrients and biomass was another factor highlighting the advantage of shade trees for the coffee producers. During the agrarian reform, chemical inputs were cheap and credit also was available to small producers, which stimulated a general increase in their use of fertilizers. In the 1990s, as was noted above, these conditions were transformed by the agricultural policies of the new liberal government. Most small-scale producers lost access to credit and input prices increased significantly, resulting in a precipitous fall in their fertilizer use. At the time of the study, the coffee producers in the sample had reduced input levels further due to climatic adversities and low world market prices. The survey data showed that only

15 (39 percent) producers had used chemical fertilizers (nitrogen or NPK) in the 1999–2000 season. Average application was 148 kilograms per mz, only about one-third of what had been recommended for semi-technified coffee production under CONARCA (Clemens and Simán 1993, p. 10). About one-third of the producers had applied chicken manure, a locally available organic fertilizer, and the remaining group had used neither chemical nor organic fertilizer. In coffee production with low or no use of external inputs, as had become the case in the study area, tree leaves and cuttings are an important contribution of biomass and nutrients produced within the system, which can be considered a factor promoting the use of shade trees.

Risk Spreading

Considerations regarding risk were clearly expressed in the producers' strategies. During the 1990s, the coffee producers in the study area had experienced unpredictable variations in coffee yields and income due to erratic precipitation patterns and fluctuating coffee prices. In view of these risk factors, many producers expressed an interest in planting more fruit trees in their coffee agroforestry systems. For precisely the same reasons, variability of crop yields and prices, it appeared that priority was given to crop diversification over the maximization of one specific crop in the system. It was found that among the 13 producers who had prepared fruit tree seedlings the majority (11) had also produced coffee seedlings, which rather points toward a double strategy of investing in more fruit production while maintaining or improving coffee productivity. This is also supported by the strong concern voiced by the producers to identify the citrus species and varieties that are most appropriate for association with coffee. The interest in this issue was expressed frequently in interviews and discussions and in producers' experiments, in which properties with importance for the association with coffee plants, such as root extension, tree height and foliage cycles, were of special concern, in addition to the more general criteria of yields and the time it took for the fruit trees to become productive. Crop diversification within the agroforestry system can thus be interpreted as a way of spreading risk, and producers' interest in identifying citrus trees that are appropriate for combination with coffee reflects an aim to optimize total system productivity.

Land-Use Optimization

In addition to the trend toward more intercropping with tree products, plantains and bananas a tendency toward increased coffee plant density was identified among the producers. Thus 20 (33 percent) producers stated that they had increased coffee plant density within the five years preceding the study. Better utilization of their limited production areas was mentioned as the most frequent reason. In view of the high pressure on agricultural land and a tendency toward fragmentation of family farms by inheritance, intercropping and increasing plant density are ways of intensifying the use of the horizontal and vertical space available in the mostly very small farming plots. The small land holdings in the study area meant that producers could not just carry on producing in the same way as their parents and grandparents had done, but had to seek new ways to increase the outputs of the agricultural land available to them.

Products, Income, and Market Opportunities

In the 1999–2000 season, average income from coffee production was calculated at 15,174 Córdobas[18] (SD 17,223) for the sample group (N = 39).[19] The standard deviation value shows that this aggregate number conceals considerable variation among the producers. Moreover, it should be noted that the figure only gives a snapshot of the income that the producer households could expect from their coffee production. In the study area the climatic conditions, and therefore also yields, were very variable. Furthermore, coffee prices fluctuated, which meant that even if the harvest was good, incomes could still decline if prices were low. The producers of the sample sold their coffee to export companies or to intermediaries. It was generally sold unprocessed at undifferentiated prices without a quality bonus, and no examples of alternative marketing were encountered.

For the 1999–2000 season, the value of tree products for sale and home consumption added up to approximately 5,000 Córdobas on average, corresponding to more than 30 percent of coffee income. Moreover, 1999–2000 was a relatively good season for coffee incomes, both in yields and in prices. The relative importance of secondary crops from the agroforestry system compared to coffee could thus be of even greater importance in seasons with lower coffee yields or prices.

Average income generated by tree crops was 3,485 Córdobas (SD 10,780, N = 39). As the standard deviation indicates, the figures on income from tree products concealed even greater variation among the sample farms than coffee income. In addition to variation in quantities, there were also differences in the composition of the tree products sold, comprising regular sale of citrus and avocado, occasional sale of a whole timber tree, and/or minor income from other fruit, plantains, poles or firewood. Figure 8.1 gives an overview of the shares of income generated by different kinds of tree products.

As the figure shows, fruits were the tree products generating most income for the farm households. Some producers are increasingly specializing in the production of citrus and avocado, but a wide range of other fruits were sold and consumed at

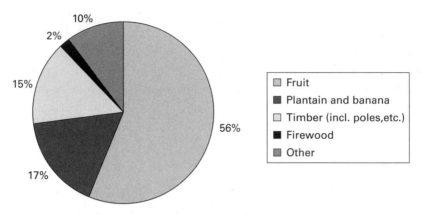

Figure 8.1
Composition of tree product income, 1999–2000 (*N* = 39).

different times of the year. Fruit was mainly sold to nearby markets in Masaya or Managua, frequently via an intermediary, but direct selling was not uncommon. In this case, it was often women who took the fruit produced on the farm to the market, and many also engaged in reselling.[20]

Sale of wood was another income source, be it as poles or firewood or, if a whole tree was felled, also as timber for construction, boards or furniture. In contrast to coffee and fruit, large timber was not produced on an annual basis but when sold supplied the family with a lump sum. While markets in many other regions are limited to timber of large dimensions, in the study area timber of both large and small sizes could be sold as there was a flourishing production of crafts and furniture around Masaya and, for instance, fence posts and firewood were marketable products. As a reflection on the existence of markets for such subsistence type products, it can be noted that in more wealthy regions and countries such products would tend to be replaced by commercial materials (e.g. concrete for construction or gas for cooking).

Income Smoothing

Apart from the amounts of income different products generated, the question of *when* incomes were generated was another important consideration for poor producer households. Income smoothing was thus one of the criteria that producers mentioned for product diversification. Coffee income was concentrated in the months of December and January. In households with limited resources, a single annual lump-sum payment for coffee often implied difficulties in making ends meet toward the end of the agricultural season, especially in a year following a bad harvest. Diversification with tree products meant that subsistence products such as plantains and firewood were continuously available, and it was possible to spread income throughout the year. Although limited in absolute terms, the income gained from the sale of minor amounts of fruit, plantains or firewood could be crucial at times when cash was scarce.

Household Consumption Needs

Besides generating cash income the agroforestry systems supplied several products for household consumption. Thus, in 36 out of 38 cases the annual household demand for plantains, an important element in the local diet, was covered. Moreover, unquantified amounts of fruit of different types contributed to family nutrition.

With respect to wood products, 35 out of 38 respondents stated that wood from the coffee agroforestry system was sufficient to satisfy the household's annual demand for firewood, which was the principal fuel used for cooking. A more irregular demand for timber for construction or repair of houses was also partly met by the trees in the coffee agroforestry system. Thus 6 respondents had extracted timber for domestic use during the year 1999–2000. In practically all cases unquantified amounts of wood had been used for fence posts, house repair, support poles, etc.

Calculating the monetary equivalent of the families' annual consumption of plantain and firewood by reference to local market prices, the average sample household's annual costs were estimated to be approximately 1,200 Córdobas for plantains and 400 Córdobas for firewood. The share devoted to subsistence therefore had a considerably higher economic value than the share that was sold.

Summing up the Argument: From Modern to Traditional Coffee Production?

Contrary to the reductionist dichotomy of "traditional" vs. "modern" farming systems, much more complex dynamics of technology change are revealed in the analysis of coffee producers' practices. Specifically, the transition in the *parcelero* group's production strategies from modern shadeless coffee plantations to diverse coffee agroforestry systems should not be understood as simply a return to an earlier state of "traditional" production. Rather, the changes observed are the result of flexible adaptation to a dynamic social and natural context, where agro-ecological considerations and the need for product diversification each played a role. Producers' decisions regarding the design of their production systems and selection of tree species show how they must face constant challenges. These include how to utilize the resources available to them, not least their farming plots, in the most efficient way, and how best to safeguard themselves and respond to the variable contextual conditions of climate and market.

Diversified Production in Diversified Livelihood Strategies

We are poor and cultivate our farms, but if we study the lives of the different families, they are all seeking, and everyone takes a different path, seeking here, seeking there. . . .
—César, FocusGroup 2000

Diversification was not only an important feature of the production systems, but also of the livelihood strategies of the coffee producer households. Two-thirds of the respondents stated that the household had between one and four other income sources, in addition to farm production. The economic activities mentioned most frequently were farm wage labor in 17 cases, petty commerce in 14 (e.g. a *pulpería*,[21] sale of bread, ornamental plants or seedlings), commercialization of fruit in the local market in 10, and wage labor outside the agricultural sector in 7 cases. In addition to the more or less permanent jobs, it was common for household members to engage in a range of occasional economic activities to supplement incomes.[22]

Livelihood Diversification and the Family Life Cycle

The reasons for, and implications of, livelihood diversification, were investigated in the larger research project[23] on which the present chapter is based. The changes that had taken place in the households' livelihood strategies were studied from two different angles: in-depth case studies in an historical perspective, and a typology comparing socio-economic characteristics of producer households with different livelihood strategy types across the entire sample. The findings show how engagement in off-farm and non-farm work was not only a way of surviving for households, but that livelihood diversification could also contribute to investment in land and farm productivity and, eventually, even offer the opportunity to rely entirely on farming.

Although the percentage of the population depending on farm production has declined at the sectoral level in Nicaragua, livelihood diversification among the sample group is not necessarily to be understood as a step away from farming. Most of the small-scale coffee producers saw a farm-based livelihood as the ideal, but many of them had not inherited land, or at least not enough to maintain a household. Differences between the livelihood strategies of households of different ages can usefully be understood conceptually in terms of the family life cycle dynamic. Thus households trace different stages in a trajectory from dependence on off-farm or non-farm income, followed by a phase with both farming and other economic activities while establishing the production system and, finally, an entirely farm-based livelihood.

For a household to be able to follow the trajectory outlined above in the course of its life cycle, however, it is essential that the dynamics between off-farm and non-farm work and farm production permit some degree of accumulation. This was not the case in all of the producer households included in the sample. Thus the findings revealed that livelihood diversification could form part of either positive or negative dynamics, depending on the ways in which the diversification of economic activities at the household level and management and production of the coffee agroforestry system influenced each other. In cases in which incomes from off-farm or non-farm work were complementary to farm production (e.g. petty commerce or other non-agricultural work), cross investment was possible and this gave the household a wider and more secure range of economic options. In other cases, involvement in off-farm and non-farm work did not generate an investment surplus and led to labor constraints on the household farm, hindering the improvement of farm productivity and thereby resulting in a generally stagnant economic situation for the household. The findings suggest that the characteristics of farm-wage labor made the occurrence of adverse trade-off situations more likely, especially in households where the principal manager of the agroforestry system engaged in farm-wage labor. (For further discussion, see Westphal 2002.)

Influence of Broader Political and Economic Tendencies
The research findings reveal how the possibilities of achieving and maintaining a farm-based livelihood are conditioned by larger structural conditions, such as markets, politics and demographic trends in different historical periods. The redistribution of land through agrarian reform is a case in point. Thus successful small-scale producers, such as Alejandro, might never have left wage laboring had it not been for the Sandinista agrarian reform. More generally, it can be argued that the favorable conditions for coffee production in the early 1980s, in terms of national agricultural policies, access to credit, and good world market prices, allowed those coffee producers of the sample who owned land at the time to establish and invest in their coffee-production system. With the reduced access to agricultural credits in the 1990s, many of the coffee producer households had to engage in off-farm and non-farm work as an alternative way of generating capital.

The importance of these broader structural conditions is reflected in the fact that, during the agrarian reform years, many of the sample households were able to forego farm wage labor and live on their farm income. Conversely, in the late 1990s, at the time of the present study, many of these same households depended heavily on farm wage labor. This pattern is reported in the literature, where the agrarian reform is identified as a factor contributing to a shortage of agricultural wage labor during the 1980s (Rice 1990, p. 236).

The historical approach taken in this study highlights the influence of broader political and economic tendencies in determining whether or not producer households are able live off their coffee farms at different points in time. In short, the conditions enabling rural households to remain or become coffee producers, depended on the timing of the different phases in the life cycle of individual households vis-à-vis changing configurations of economic, political, and social forces and tendencies. At times these countered, and at other times reinforced, their efforts and possibilities to live entirely on their earnings from the farm.

Concluding Remarks

The study shows how the strategies of coffee producer households were constantly changed and adapted in response to the ways in which broader political economic tendencies in the regional, national and international context were articulated with agro-ecological and socio-economic conditions at the farm and household level at different points in time.

Taking the methodological perspective adopted by Netting and much of the small-farm literature, the present study suggests that explanations based on agro-ecological adaptation and producer households' socio-economic characteristics and consumption needs contribute usefully to an understanding of the changes in the agro-ecological structures and management practices of the coffee farms analyzed here.

The adaptation of the coffee-production systems to the local natural conditions was an important explanation for the widespread use of shade trees in the study region, which was warmer and dryer and the rainfall more variable than in more optimal coffee-growing areas of Central America. The characteristics of coffee as a crop that responds positively to cultivation under shaded conditions are important in this context.

The socio-economic characteristics and internal demographic dynamics within the producer households were important in the design and management of the coffee agroforestry systems, both with regard to products for consumption and income generation. Moreover, the diversified coffee agroforestry systems provided the producer households' livelihood strategies with a certain robustness in terms of income smoothing, risk spreading and relatively low maintenance costs. In such a volatile context as Nicaragua, such strategies in many ways are better adapted to poor producer households' economy than specialized, capital-intensive production systems, which involve high risks on the investments made and to the livelihoods that they maintain. The recent drastic drop of international coffee prices is just one example of this.

The analysis of production strategies also showed, however, that an exclusive focus on adaptation to local natural conditions and internal household dynamics could not sufficiently explain the technological changes that had taken place, neither the re-emergence of diverse agroforestry systems in the coffee farms of the *parceleros* specifically, nor the ways in which the small-scale coffee producers in general constantly improved and modified the structures and management of their production systems. Rather, the conceptualization of the technological changes analyzed here needs to be based on a more integral analysis of farm and household-level processes and the broader social, political, and economic dynamics to which coffee producers respond in their production and livelihood strategies. Complex but limited product and labor markets at regional, national and international levels, as well as shifts in agricultural policies, access to credit and coffee world market prices, all played a role in the diversity characterizing both production and livelihood strategies.

Seen from a regional perspective, the natural, social, and economic conditions of the Meseta region promoted the use of diverse shade canopies in coffee cultivation in several ways. First, as already mentioned, shade trees mitigated the possible negative impacts of climatic variations that at times created too hot and dry conditions with regard to the requirements for coffee growing. Second, population density and the patterns of land distribution created reduced land sizes in the small-scale farming sector. For small scale producers, intercropping was a way of optimizing the use of the horizontal and vertical space available. Third, in the Meseta region, the closeness of urban markets and their infrastructural accessibility and the existence of local markets for subsistence type products gave positive incentives for diversification with tree products. In contrast to less densely populated regions, where access to natural resources in uncultivated areas is easier, the need for products for household use, such as firewood and poles, was another consideration taken into account by the producers. Interestingly, the conclusion that the diversity of the farming systems increased due to integration into different types of markets and the characteristics of coffee as the cash crop diverges somewhat from the conventional wisdom of market integration and export crop production being associated with specialization and monoculture production patterns.

In the national context, changing agricultural policies affected small-scale producers' possibilities to pursue capital- intensive production strategies and indirectly contributed to promote the reintroduction of shade trees into the coffee-production systems in the 1990s. The agricultural policies during the Sandinista government had lead to considerable increases in external input use in coffee production in the study area, especially in the production units of the reformed sector, but also among private small-scale producers. With the strongly reduced access to credit and rise in input prices as a consequence of deregulation of the agricultural sector and foreign exchange rates in the 1990s, however, most small-scale producers had to limit or give up the use of such inputs. The advantages of agroforestry practices for low-input coffee growing made these a useful alternative for the producers in this situation.

At the international level, the breakdown of the international coffee price agreement in 1989 was another factor spurring change in household production strategies. Adding to the risk of variable climatic conditions, fluctuating coffee prices motivated many producers to pursue a strategy of risk diversification by intercropping more citrus and other fruit for the national markets in their coffee agroforestry systems.

Livelihood diversification was a conspicuous tendency in the study area, and producer households' different opportunities and strategies to follow this trajectory influenced the availability of household labor and capital for farming, and thereby their possibilities to improve farm productivity.

The present study was designed to achieve an integrated understanding of production practices and strategies in their concrete expression and socio-economic processes at the household level within the context of broader structural change. With this integrative perspective, it represents a contribution to the wider debate on inter-disciplinary methodologies within the fields of small-scale agriculture, political ecology and related areas of research.

Notes

1. The names of respondents of in-depth interviews, case studies, and focus groups have been changed to fictive names by the author.

2. *Contras*: Militant counter-revolutionary movement against the Sandinista government

3. 1 mz = 0.7 hectare.

4. Areas less than 10 mz.

5. Comisión Nacional para la Renovación del Café.

6. International Coffee Organization.

7. The underlying objective in the modernization approach is basically to achieve rural development through increasing agricultural productivity and market integration. Modern farming technology, specialization and increased division of labor are seen as the means to raise the efficiency of agricultural production (see e.g. Johnston and Peter Kilby 1975; Mellor 1985; Tomich, Kilby, and Johnston 1995).

8. The fact that the social organization of cooperatives and state-farms departed from the structures of the large-scale coffee plantations confiscated from the dictator Somoza and his affiliates was another reason.

9. An early contribution to this line of thinking was provided by A.V. Chayanov, who put forward his theories in the 1920s in the Soviet Union (Chayanov 1966; Kerblay 1971).

10. Although botanically they are not trees, banana and plantain plants are dealt with in the same way as trees in the coffee agroforestry systems because they fulfill the same agro-ecological and socio-economic functions.

11. *Gliricidia sepium.*

12. The calculation of shade percentages was based on measurement with a densiometer at four points of the 1,000-m² sample plot, each consisting of four readings.

13. On the basis of a *t* test, the difference was found significant at $P = 0.001$.

14. *Enterolobium cyclocarpum*.

15. *Cedrela odorata*.

16. *Patio*: literally courtyard; used for small farms and home gardens with mixed cultivation of plants.

17. *Cordia alliodora*.

18. 100 Córdobas = 8 US dollars (1999).

19. To calculate the income levels an average sales price of 1,000 Córdobas/mz was used and production costs were subtracted. In reality, however, prices could vary considerably within one harvest period and moreover depended on the individual agreements made between the producer and the buyer.

20. Women have historically had an important role in these economic activities in the study area. For an analysis, see Dore 1997.

21. A *pulpería* or *venta* is a small home-based shop or kiosk with a limited supply of articles for everyday needs.

22. Since such minor and occasional economic activities can be difficult to take into account, the survey data may understate the range and number of activities engaged in by the sample households.

23. The research work carried out for the present chapter formed part of a larger research project documented in the author's Ph.D. thesis (When Change Is the Only Constant: Coffee Agroforestry and Household Livelihood Strategies in the Meseta de los Pueblos, Nicaragua, Roskilde University, Denmark, 2002).

References

Altieri, M. A., F. J. Trujillo, et al. 1987. Plant-insect interactions and soil fertility relations in agroforestry systems: Implications for the design of sustainable agroecosystems. In *Agroforestry: Realities, Possibilities and Potentials*, ed. H. Gholz. Martinus Nijhoff.

Beer, J., R. Muschler, et al. 1998. Shade management in coffee and cacao plantations. *Agroforestry Systems* 38: 139–164.

Bonilla, G. 1999. Tipologías Cafetaleras en el Pacífico de Nicaragua. Turrialba, Centro Agronómico Tropical de Investigación y Enseñanza.

Bunch, R. 1985. *Two Ears of Corn*. World Neighbours.

Cambrony, H. R. 1992. *Coffee Growing*. Macmillan.

CATIE/MIP. 1993–1999. Datos de Precipitación en el Centro Experimental de Café, Jardin Botánico, Masatepe. Masatepe, Programa CATIE-MIP/AF Nicaragua (NORAD).

Chambers, R., A. Pacey, and L. Thrupp, eds. 1989. Farmer First: Farmer Innovation and Agricultural Research. Institute of Development Studies, University of Sussex.

Chayanov, A. V. 1966. *The Theory of Peasant Economy*. Irwin.

Clemens, H., and J. Simán. 1993. Tecnología y Desarrollo del Sector Cafetalero en Nicaragua. Managua, Universidad Nacional Autonoma de Nicaragua, CIES/ESECA and Proyecto MIP, CATIE.

Dauner, I. 1998. Mercados Financieros Rurales en Nicaragua. Managua, Nitlapán-UCA.

Dore, E. 1997. Property, households and public regulation of domestic life: Diriomo, Nicaragua 1840–1900. *Journal of Latin American Studies* 29, no. 3: 591–611.

Escalante, E. E. 1999. Species diversity in small farmers multi-strata agroforestry systems in Venezuela. Presented at International Symposium on Multi-Strata Agroforestry Systems with Perennial Crops, CATIE, Turrialba, Costa Rica.

Fernández, C. E., and R. G. Muschler. 1999. Aspectos de la sostenibilidad de los sistemas de cultivo de café en América Central. In *Desafíos de la Caficultura en Centroamérica*, ed. B. Bertrand and B. Rapidel. Editorial AgroAmerIICA.

Guharay, F., J. Monterrey, et al. 2000. Manejo Integrado de Plagas en el Cultivo del Café. Managua, CATIE.

Guharay, F., D. Monterroso, et al. 1999. Designing Pest-Suppressive Multi-Strata Perennial Crop Systems. Multi-Strata Agroforestry Systems with Perennial Crops. IUFRO, Turrialba, Costa Rica.

INIFOM/AMUNIC. 1997. La Concepción. Managua, Instituto Nicaragüense de Fomento Municipal/Asociación de Municipios de Nicaragua.

INIFOM/AMUNIC. 1997. Masatepe. Managua, Instituto Nicaragüense de Fomento Municipal/Asociación de Municipios de Nicaragua.

INIFOM/AMUNIC. 1997. San Marcos. Managua, Instituto Nicaragüense de Fomento Municipal/Asociación de Municipios de Nicaragua.

Johnston, B. F., and P. Kilby. 1975. *Agriculture and Structural Transformation*. Oxford University Press.

Kearny, M. 1996. *Reconceptualizing the Peasantry*. Westview.

Kerblay, B. 1971. Chayanov and the theory of peasantry as a specific type of economy. In *Peasants and Peasant Societies*, ed. T. Shanin. Penguin.

Lipton, M. 1989. *Why Poor People Stay Poor: Urban Bias in World Development*. Avebury.

Llanderal, T. 1998. Diversidad del Dosel de Sombra en Cafetales de Turrialba, Costa Rica. Escuela de Posgrado. Turrialba, CATIE.

Maldidier, C., and S. J. P. Marchetti 1996. El Campesino-Finquero y el potencial económico del campesinado nicaragüense. Managua, Nitlapán-UCA.

Mellor, J. W. a. G. M. D. 1985. Agricultural change and rural poverty: A synthesis. In *Agricultural Change and Rural Poverty*, ed. J. Mellor. Johns Hopkins University Press.

Muschler, R. 1997. Efecto de sombra de Erythrina Poeppigiana sobre Coffea arábica vars Caturra y Catimor. 180 Simposio Latinamericano de Caficultura, San José, Costa Rica.

Nair, P. K. R. 1993. *An Introduction to Agroforestry*. Kluwer.

Netting, R. M. 1993. *Smallholders, Householders: Farm Families and the Ecology of Intensive, Sustainable Agriculture*. Stanford University Press.

Nielsen, F. 1998. Issues in the Utilization of Indigenous Knowledge in Agroforestry Research. Institute of Geography. University of Copenhagen.

OIM. 1999. Características Socio-Demográficas de la Población Rural de Nicaragua. Managua, Organización Internacional para las Migraciones (OIM), Instituto Nacional de Estadísticas y Censos, Agencia Suiza para el Desarollo y la Cooperación.

Ponte, S. 2001. Coffee Markets in East Africa: Local Responses to Global Challenges or Global Responses to Local Challenges? Centre for Development Research, Copenhagen.

Pretty, J. N., ed. 1995. *Regenerating Agriculture*. Earthscan.

Rice, R. A. 1990. Transforming Agriculture: The Case of Leaf Rust and Coffee Renovation in Southern Nicaragua. Department of Geography, University of California, Berkeley.

Röling, N., and J. Brouwers. 1999. Living local knowledge for sustainable development. In *Biological and Cultural Diversity*, ed. G. Prain et al. Intermediate Technology Publications.

Romero, W., and F. Hansen. 1992. Café Amargo: Pequeños productores de Centroamérica y crisis cafetalera. Managua, CRIES /Latino Editores.

Rosset, P. M. 1999. The Multiple Functions and Benefits of Small Farm Agriculture. Food First/Institute for Food and Development Policy.

Sahlins, M., and E. R. Service 1960. *Evolution and Culture*. University of Michigan Press.

Scott, J. C. 1976. *The Moral Economy of the Peasant: Rebellion and Subsistence in Southeast Asia*. Yale University Press.

Tomich, T. P., P. Kilby, and B., Johnston. 1995. *Transforming Agrarian Economies: Opportunities Seized, Opportunities Missed*. Cornell University Press.

UNICAFE, V. G. d. E. y. P. 1997. Perfil de proyecto, Transferencia de tecnologías en el manejo de cafetales en pequeña y mediana producción. Managua, Unicafé.

UNICAFE, V. G. d. E. y. P. 1998. Numero de productores y áreas de café. Unicafe.1998. Managua, Unicafe.

Vaast, P., and D. Snoeck. 1999. Hacia un Manejo Sostenible de la Materia Orgánica y de la Fertilidad Biológica de los Suelos Cafetaleros. In *Desafíos de la caficultura en Centroamérica*, ed. B. Bertrand and B. Rapidel. San José, Editorial AgroAmerIICA.

Westphal, S. M. 2002. When Change Is the Only Constant: Coffee Agroforestry and Household Livelihood Strategies in the Meseta de los Pueblos, Nicaragua. Department of International Development Studies, Roskilde University.)

Interviews

CaseStudy1 2000. "Pedro," S. M. Westphal. Fátima, San Marcos, Nicaragua.

CaseStudy2 2000. "Alejandro & Magdalena," S. M. Westphal. Fátima, San Marcos, Nicaragua.

FocusGroup 2000. Focus group discussion with case study participants. S. M. Westphal. Fátima, San Marcos, Nicaragua.

9

Farmers' Livelihoods and Biodiversity Conservation in a Coffee Landscape of El Salvador

V. Ernesto Méndez

The production and commercialization of coffee is important to both the environment and the economic development of producing countries (Rice and Ward 1996). Recently, the coffee commodity has suffered from an international price crisis, which has called attention to the high level of environmental and economic vulnerability affecting farmers and landscapes (Ponte 2002; Rice 2001). This chapter examines a particular interface between agriculture, rural development and biodiversity conservation in a coffee landscape of El Salvador during the time of the coffee crisis. The general objective was to gain a better understanding of the socio-ecological relationships associated with "biodiversity friendly" coffee plantation management in three different small-scale farmer cooperatives. The research also analyzed the interactions between different types of cooperatives, shade-tree management and household livelihoods. Parallel to the objectives exclusively relating to research, an action-research process was initiated between the three cooperatives and the researchers involved. This resulted in a solid partnership and the formation of a permanent, non-profit foundation that conducts research and supports local development at the site. This chapter presents both the results of research conducted between 1999 and 2004, as well as reflections on the outcomes of the action-research process to date.

Recent Ecological and Social Research on Coffee

The shaded coffee (*Coffea* spp.) agroforestry system[1] has been shown to maintain production with minimal environmental degradation (Beer et al. 1998). A significant number of studies have also demonstrated its importance in biodiversity conservation of animal, plant and insect species (Gallina et al. 1996; Greenberg et al. 1997a; Greenberg et al. 1997b; Johnson 2000; Llanderal and Somarriba 1999; Moguel and Toledo 1999; Monro et al. 2001; Nestel et al. 1993; Perfecto 1994; Perfecto and Snelling 1995; Perfecto and Vandermeer 1996; Perfecto and Vandermeer 2002; Perfecto et al. 1997; Roberts et al. 2000; Roth et al. 1994; Soto-Pinto et al. 2000;

Williams-Guillen et al. 2001; Wunderle 1999; Wunderle and Latta 1998). This research has adopted primarily an ecological focus, aiming to document the diversity and abundance of species in different types of shade coffee plantations, and discussing how to insert shade coffee into conservation programs.

Another recent area of research examines the benefits and limitations of alternative marketing networks for the sale of specialty coffee, including Fair Trade, organic, and shade-grown. (Bray et al. 2002; Raynolds 2002; Renard 1999; Rice 2001). This work has concentrated on the characteristics, benefits and limitations of these alternative trade channels, and on the requirements expected of farmers in order to participate in them. In addition, and as the price crisis has become more severe, several researchers have focused on further analyzing the characteristics of coffee commodity chains or networks (Bacon 2005; Ponte 2002; Ponte 2004).

These different types of studies have contributed valuable information to the analysis of shade coffee ecology and on the social and marketing networks associated with this type of agroecosystem. However, there is still much to learn about the interactions between the social and ecological dynamics that are present in shade coffee landscapes. Furthermore, understanding the diversity of livelihoods that are emerging in the coffee regions of Mesoamerica requires a particular focus that also documents and links these strategies to broader networks and environmental factors.

Conceptual Framework

In order to better understand the intricate contexts where environmental degradation and rural development meet, a number of scholars from different backgrounds have emphasized the importance of developing and applying interdisciplinary research approaches that are able to link social and ecological processes within and across different geographical scales (i.e. local to global, plot to landscape) (Ewel 2001; Izac and Sanchez 2001; Peet and Watts 1996; Redman et al. 2004; Scoones 1999; Walker et al. 2002). This chapter applies an interdisciplinary research framework, which integrates approaches from the social and the natural sciences. It draws mostly from the fields of political ecology and agroecology, and to a lesser extent from landscape ecology, human geography, and work on agro-food networks. The following discussion briefly notes contributions in these fields, which have influenced my research framework.

Agroecological research uses a combination of ecological, agronomic and social science methods, such as plant inventories and diagrams, studies of soil, geographic information systems and mapping, surveys and interviews (Altieri 1995; Altieri and Nicholls 2002; Ellis and Wang 1997; Gliessman 1998; Gliessman 2000). Agroecology has increasingly sought to integrate both social and ecological factors, in order to achieve interdisciplinary research and applications. One of the most recent conceptualizations of agroecology shows the evolution of the field by defining it as "the integrative study of the ecology of the entire food system, encompassing ecological, economic and social dimensions." (Francis et al. 2003, p. 100).

A recent contribution by a group of researchers led by the Spanish rural sociologist Eduardo Sevilla-Guzmán discusses the need for agroecology to integrate social science approaches in order to develop interdisciplinary frameworks and applications. Their book provides a thorough discussion of the epistemological roots and characteristics of agroecology, and explains why these attributes make it the most adequate foundation for "participatory and endogenous" rural development. According to Guzmán-Casado et al. (1999), these characteristics include: its interdisciplinary nature (integrating agronomy with ecology); its concern for the ecological integrity of the agricultural landscapes; its concern for the social and cultural factors that affect agriculture; and its integration of both local and scientific knowledge, as a way to understand and design more sustainable natural resource management systems. This description points to agroecology as an appropriate conceptual and methodological field to analyze the ecological components of rural landscapes, and as part of interdisciplinary studies (Méndez and Gliessman 2002). In my research, I integrated agroecology with the other approaches discussed in this section to construct an interdisciplinary framework.

Recent approaches analyzing the environment, food, and rural development place significant importance on examining potential causalities and consequences at different geographical and political scales (Marsden 1997). As an internationally traded and consumed commodity, coffee is affected by processes that occur both locally (i.e. production) and globally (i.e. international sales). The production, processing, commercialization and consumption of coffee traverses a great number of social and environmental spheres, starting with farms in tropical countries, and ending in stores and coffee shops in the United States and Europe. My research follows the networks associated with coffee, including the interactions between human actors and their networks, as well as the ecological landscapes where coffee is produced (Whatmore and Thorne 1997).

These interactions take center stage in political ecology, which argues for the importance of a multi-scale analytical approach that integrates ecological settings with the social and political conditions of power that direct access to and management of natural resources (Bryant and Bailey 1997). The contributions of political ecological approaches to the analysis of complex human environment situations can be illustrated by the work of several authors. Thus, a recent study by Rocheleau et al. (2001) uses a political ecology approach to examine the interactions between native tree biodiversity conservation, international development projects and household dynamics in the Zambrana-Chacuey region of the Dominican Republic.

The findings show that the incidence of native trees is the result of both the influence of global actors, as well as the characteristics of households. These results are important because they caution against attributing the changing tree composition of this rural landscape solely to global influences. Furthermore, it underscores the challenge of being able to capture subtle and, many times sensitive, local socioecological interactions.

A second illustration is provided by Bebbington's long-term research in the South American Andes, which draws on four case studies to show the more positive influences of what he terms as "transnational landscapes," which are conditioned by forces operating at both the local and global spheres (Bebbington 2001). The paper compares four rural settings in the countries of Bolivia, Ecuador, and Peru, where rural people have achieved a certain level of local organization, as well as being connected to various types of national and international networks. Bebbington uses these case studies to raise three critical issues. The first is that the interactions between local livelihoods and global actors need not always pose a threat to local people, and may even provide opportunities to enhance their livelihood strategies. This is an especially important consideration when dealing with supposed "crisis" scenarios, which might also present opportunities. Secondly, the results of the interactions between local livelihoods and global actors are highly differentiated and context-dependent processes, and thus cannot be analyzed or explained in generalized terms. This calls attention to the third issue, which deals with the way we undertake research and try to explain interfaces where local, national and global actors meet. In Bebbington's words (2001, p. 432): "Any programme of political ecological inquiry into globalization and livelihoods must therefore revolve around a constant interplay between case-specific depth and comparative breadth." Even when research is focused on a particular locality, and not on several comparative cases, the investigator has to decide upon where to focus his or her resources. Should it be locally, or should it follow the entirety of the "transnational network" more superficially? The fact is that there will probably always be a compromise to be made as to where it is necessary to invest more time and resources. However, the work of Bebbington and his colleagues, which spans more than a decade, provides an example of a particular way of carrying out these types of investigations (Bebbington 1992, 1993, 1996a, 1996b, 1997, 1998, 1999, 2000, 2001; Bebbington et al. 1996; Bebbington and Batterbury 2001). Their research first concentrated on local realities, and then followed networks and interactions progressively "upward" into the national and global scales.

Action-Research Processes

Greenwood and Levin (1998, p. 4) define action-research (AR) as "social research carried out by a team encompassing a professional action researcher and members of an organization or community seeking to improve their situation. AR promotes broad participation in the research process and supports action leading to a more just or satisfying situation for the stakeholders." Different approaches to action-research have since been adopted in a variety of disciplines, including education, psychology, community health sciences, and rural development (Greenwood and Levin 1998; Selener 1997). Participatory action research (PAR) is specifically associated with rural and agricultural development in developing countries (Kenton 2004; Selener 1997). PAR seeks to engage parallel processes of research and local development or social change, through more horizontal relationships between local actors and researchers (Bacon et al. 2005). I used this framework as a guiding principle to work with farmers in a mutually beneficial relationship that led to the formation of Advising and Interdisciplinary Research for Local Development and Conservation (ASINDEC), a local, non-profit foundation that aims to institutionalize this approach and engage in a long-term partnership.

Coffee and Environment in El Salvador

El Salvador is one of the smallest and most densely populated countries of Latin America. With average population densities exceeding 200 inhabitants per square kilometer and with only 2 percent of its original forest cover remaining, the country's natural-resource base has reached a critical level of degradation (CCAD 1998). In El Salvador, coffee holds considerable economic importance as the country's leading export crop (PROCAFE 1998). However, due to the loss of forest cover shaded coffee farms have also acquired particular significance, as they are now the main providers of ecosystem services attributable to the natural forest (Cuéllar et al. 1999; Herrador and Dimas 2000; Rosa et al. 1999). For this reason, several state institutions and non-government organizations have initiated projects that integrate shaded coffee into the country's environmental management and conservation efforts.[2]

· Approximately 74 percent of El Salvador's coffee farms are smaller than 7 hectares, covering an estimated 40 percent of the total area under coffee cultivation (table 9.1). In contrast to larger holdings, most of these small farms utilize traditional shade management, characterized by a diverse canopy of shade trees and limited use of potentially contaminating agricultural inputs (Galloway and Beer 1997). These characteristics increase the potential of these farms to act as providers of environmental services, such as water provision and conservation, soil protection, and conservation of flora and fauna (Dominguez and Komar 2001; Herrador and Dimas 2000; Moguel and Toledo 1999; Monro et al. 2001; PRISMA 1995). For these reasons, small shaded coffee farms constitute a potentially important component of conservation efforts. However, in many cases, these farmers lack the capital to finance costs associated with certification, such as organic or Rainforest

Table 9.1
Coffee farmers and plantations in El Salvador, classified by farm size. Source: PROCAFE 2001.

Size of farm (ha)[a]	Number of farmers	Number of farms
0–7	13,128	13,653
7–14	1,355	1,725
14–35	1,193	1,806
35–70	578	736
≥ 70	364	432
Total	16,618	18,352

a. Most cooperatives that are collectively managed consist of plantations with areas ≥ 35 hectares.

Alliance. Experiences with organic coffee have also shown that successful integration of small farms into alternative markets requires external support. Specifically, assistance is essential in order to improve management practices and develop local organizational structures (Bray et al. 2002).

Despite the great number of small coffee farms in El Salvador, detailed information on their agroecological and socioeconomic characteristics is scarce. Very little is known about the management rationale behind traditional shade and the problems or advantages associated with this practice. Attempts to incorporate small farmers into conservation-oriented strategies need to be based on a clearer understanding of the social and ecological dynamics of their farming communities and organizations.

The Coffee Landscapes of Tacuba

The research was carried out in the coffee landscapes of the municipality of Tacuba in western El Salvador. Tacuba is 188 kilometers from San Salvador, the capital city, and 18 km from Ahuachapán, the nearest large urban center (figure 9.1).

The municipality of Tacuba has an area of approximately 130 km², with a total population estimated at 29,176 inhabitants. Altitudes range between 600 and 1,400 meters above sea level, and annual precipitation ranges between 1,650 and 2,100 millimeters. Tacuba is part of the Apaneca mountain range and is characterized by an accentuated topography (Cienfuegos 1999; CNR 1990). The site is adjacent to the Parque Nacional El Imposible (PNEI), one of the most important conservation reserves in the country, and also within the projected area for the Mesoamerican Biological Corridor (MBC).[3]

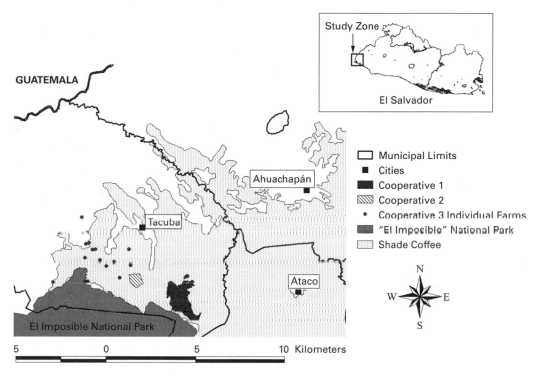

Figure 9.1
Locations of the three cooperatives in the municipality of Tacuba in western El Salvador.

The Coffee Cooperatives

The work was carried out with three distinct small-scale farmer cooperatives. (See figure 9.1 for their locations.) Each cooperative represents a different type of organization with particular origins and development (table 9.2).

Cooperative 1 was established in the first phase of the agrarian reform of 1980 (decree 154).[4] These cooperatives have faced many challenges since their formation, including their indebtedness, lack of social infrastructure, inadequate training for its members, and corruption both at the cooperative and higher institutional levels (PRISMA 1996).

Cooperative 2 was formed through two stages, which stemmed from the initiative of eight farmers. The first step was to negotiate 36 hectares through decree 207, the third phase of the agrarian reform. This land was divided into individual plots for residences and subsistence crops. Afterwards, these leaders negotiated to acquire 31.5 hectares cultivated with coffee, and were able to obtain a loan to purchase the

Table 9.2
Characteristics, total area, and sample sizes for the selected organizational strata. Source: Méndez 2004b.

Cooperative type	Sample size per stratum	Total population per stratum	Total area (ha)	Management
Coop 1: agrarian reform cooperative	25 households, 20 plots	99 members and households	195	Shade coffee managed collectively under supervision of board of directors
Coop 2: traditional cooperative (formed through decree 207)	10 households, 14 plots	22 members and households	31.5 ha under collective management; individual plots range between 0.7 and 3.5 ha	Shade coffee area managed collectively under supervision of board of directors; individual plots managed independently
Coop 3: Farmer Association	17 households and plots	28 members and households	Individual farms range between 0.7 and 3.5 ha	Farms managed independently
Total	51 plots, 52 households	149 members and households		

land. Membership opportunities were opened to other farmers in the community until they formed the cooperative with an initial membership of 22.

Individual farmers formed cooperative 3 in the hopes of improving coffee commercialization through alternative markets, tapping into development projects and acquiring land for collective management. Full legal representation took members at least three years (2000–2003) to achieve under the rubric of an "association of agricultural production," as established by the Ministry of Agriculture.

Biophysical characteristics of the three cooperatives, including soils (Shapiro, forthcoming), elevations, rainfall, and land uses are very similar (MARN 2003). The main differences between the cooperatives were the size of their holdings and their management strategies (collective vs. individual).

Methodology

Social Networks and Rural Livelihoods
In order to assess rural livelihoods and the relevance of social networks in relation to shade management and tree conservation, research was done at three levels: (1) households, (2) cooperative board of directors, and (3) relevant second- and third-

level organizations. Information was collected through five methods: (a) participant observation, (b) a combination of semi-structured interviews and a survey with 52 households, (c) 15 focus groups with different combinations of groups, including the boards of directors of each cooperative, local NGO staff, and other informants, (d) interviews with staff from second- and third-level farmer organizations and NGOs, and (e) review of secondary information and records.

Shade-Tree Biodiversity

Shade-tree composition and abundance were evaluated through inventories. For this purpose, 51 1,000-m² plots were systematically laid out in the cooperatives, as follows: 20 and 14 plots in cooperatives 1 and 2, respectively, and 17 plots in cooperative 3. In each of the plots, all trees with a height ≥ 2 m. were measured, sampled, and taken to the herbarium at La Laguna Botanical Garden at the end of each week for identification and storage.

Importance of Conserving Native Trees

In order to assess the native-tree-conservation potential of the coffee plantations, I compared species found in the cooperatives with those of the nearby PNEI forest. Although I was unable to personally sample in the forest, I was able to compare data from a recent study of tree biodiversity in the PNEI (Ramírez-Sosa 2001), which used similar sampling techniques and the same plot size as my study. In this section, I present an analysis of incidence-based tree composition similarity between the cooperatives and the forest sites using the Jaccard coefficient of similarity.[5] Since the Jaccard index is sensitive to sample size (Magurran 1988), and the samples in the cooperatives and the forest are different, species richness figures were transformed using sample-based rarefaction with the software EstimateS, version 6.0 (Colwell 1997).

Results

Household Livelihoods

Basic Services Eighty-four percent of the sampled households in cooperative 1 reported having access to electricity and more than 96 percent has access to a health clinic, running water, and schooling until the sixth grade. Access to these resources has been made possible by the direct action of the cooperative. Cooperatives 2 and 3 both lack electricity and only recently acquired access to running water. In all locations, existing schools only teach up to the 6th grade.

Income On average, there were seven members per household in the three cooperatives (n = 52), generating an average monthly income of US$93 per household in

Table 9.3
Frequency of benefits other than shade reported for shade trees by 52 households in three coffee cooperatives in Tacuba, El Salvador. Source: Méndez 2004b.

	Frequency		
	Coop. 1 (N = 25)	Coop. 2 (N = 10)	Coop. 3 (N = 17)
Firewood	96%	90%	77%
Fruit	84%	40%	100%
Timber	20%	70%	44%
Organic matter for coffee	4%	10%	0%

2001. Average monthly income for cooperative 1 and cooperative 2 households was US$142 and US$66 respectively. In these two cases incomes were relatively fixed and associated to agricultural work related to coffee work and harvesting. Other economic activities mentioned were paid domestic work (exclusively women) and work as laborers in Ahuachapán or San Salvador. These jobs were mostly held by younger members of the family. Average monthly income reported for cooperative 3 households was US$71 per month. These figures represent estimations of fixed income, since all farmers engage in a variety of activities that produce income for their families, and which usually vary year to year.

Subsistence Crops   All families farm plots of maize (*Zea mays*) and beans (*Phaseolus vulgaris*), which are separate from shaded coffee. These crops are the basis for the yearly subsistence of all households. In cooperative 1 there are assigned areas within the cooperative, while households in cooperatives 2 and 3 use their own plots, or in some cases rent land to grow their subsistence crops.

Products and Benefits from Shade Trees   For cooperatives 1 and 2, which manage collective coffee plantations, the most frequent benefit reported from shade trees was firewood (table 9.3). Households use an estimated 5 *pantes* (volumetric measure) of firewood, with an approximate cost of US$14.30 per *pante*. Given that most families obtain their firewood for free from the shade canopy, this represents savings equivalent to at least US$71.50 per year. For members of cooperative 3, who manage individual plots only, the most important benefit was fruit production, followed by firewood.

Favorite products for the market were bananas and the inflorescence of *pacaya* (*Chamaedorea tepejilote*), a palm that grows in middle and higher elevations. When asked about his reasons for maintaining plant diversity in his coffee farm, Don

Pedro Rosales said: "A couple of years ago a technician from an NGO came and told me to take out all my bananas. She said it was affecting my coffee production. I didn't do it, and she did not like it. But, today, I am getting more money, more regularly, from selling my bananas, than from the coffee. I am glad I did not heed her advice."

In cooperative 3, farms are small and thus most families need to obtain some of their firewood from other areas, or to purchase it. At the same time, these farms contain a large number of fruit trees in comparison with the other cooperatives. Even in cooperative 2, where members own and manage independent residential plots, the incidence of fruit trees in these plots is relatively low.

For cooperative 3 members, the coffee farms represent a source of a great diversity of products and benefits. Half of the farms in cooperative 3 grow and sell a diversity of bananas and other fruit . Bananas and plantains are entirely absent from collective plots in cooperatives 1 and 2, and fruit trees are rare.

Other uses reported for shade trees are medicinal and the consumption of edible parts besides fruit. Further study of local ethnobotanical knowledge is necessary to gain a better understanding on the habits and uses of the different species.

Diversification as a Way to Face the Coffee Crisis Two focus groups held during the month of March 2002 discussed farmers' ideas on what measures to take to face the coffee price crisis. Two main activities were proposed as potential production alternatives: (1) tapping into alternative coffee markets (Fair Trade, organic, shade-grown) and (2) diversifying the coffee plantation with marketable crop or tree species.

In relation to plantation diversification, farmers identified short-term and long-term species. For the short term, the two species that show the greatest potential were different varieties of bananas and plantains, and *pacaya*. Especially in the case of cooperatives 1 and 2, there is much room for improvement, since their plantations do not contain these crops. Members of cooperative 3 were interested in finding a way to plant *pacaya*, which now only grows through natural regeneration (it is dispersed by bats), and wanted to introduce more banana varieties that are in demand at the markets (especially the local "red" and "manzano" varieties). Longer-term diversification strategies focused on the introduction of timber and fruit species.

Perceptions of Cooperatives and Networks

The main functions reported by farmers and the existing infrastructure of the three cooperatives is presented in table 9.4.

Members of cooperative 1 consider ensured employment as the main benefit provided by the cooperative. Cooperative statutes stipulate that members have

Table 9.4
Perceived and observed benefits of three coffee cooperatives in Tacuba, El Salvador. Source: Méndez 2004b.

Cooperative	Benefits of organization (% of interviewees that expressed this as the most important benefit)	Infrastructure
Coop 1: agrarian reform cooperative (N = 25)	Employment (72%) Negotitations with external actors, including markets (20%) Land for subsistence crops (8%)	Complete coffee processing plant. Office, storage and other buildings. Roads are in relatively good condition. Most members have access to electricity and water. Local school.
Coop 2: traditional cooperative (N = 10)	Organization is necessary to achieve better livelihoods (70%) Employment (20%) Negotiations with external actors, including markets (10%)	Office, storage building, and school, which are located in the communal plot.
Coop 3: farmer association (N = 17)	Organization is necessary to achieve better livelihoods (82%). Negotiations with external actors, including markets (18%).	None

priority for employment positions on all agricultural work related to coffee management. Although cooperatives 1 and 2 have similar systems of collective management, members of cooperative 2 perceive that the main benefit of their cooperative is to act as an instrument to improve their livelihoods, not just as a source of employment. This sentiment is also shared by members of cooperative 3, for whom the main benefits are related to the cooperative's organizational capacity to improve their economic situation and negotiate with external actors.

Origins and Development of the Cooperatives   The differences in perceived cooperative benefits can be traced to the varied origins of these organizations and their subsequent evolution. Cooperative 1 was formed by the agrarian reform of 1980. Although peasant leaders assumed control of the boards of directors, the process was instituted without the active mobilization and motivation of the people that enrolled as members of the cooperatives. In these cases, the formation of the organizations was not a conscious decision taken by the farmers, but rather an opportunity provided by the state. This resulted in much confusion and corruption among the different organizations and actors that were involved. In addition, it prevented many members from achieving a real sense of ownership over the land. Furthermore, at present, members are suspicious about the dealings of the board of directors with outside markets and actors. This leads to a perception that benefits that could potentially be negotiated with external actors might only benefit the board of directors, and not the rest of the membership. Thus, the only benefit that seems certain is employment, which is guaranteed by the cooperative statutes.

In contrast, cooperatives 2 and 3 were formed through the initiative and effort of their membership. The sense of ownership of their land and shared organization in these cooperatives is high. The household interviews show that most of the members of these cooperatives consider collective organization to be essential for their socio-economic development. During the interviews and focus groups, members repeatedly expressed the view that being part of a cooperative is considered a proactive effort to improve their individual livelihoods, as well as those of the broader community. An example of this sense of identification is illustrated by how members of cooperative 2 have consistently worked their collective coffee plantation, with or without payment. As expressed in the focus groups, members of cooperative 2 were only partially paid for the agricultural work they invested in the coffee plantation in 2000 and 2001. When asked about this situation, they responded that it was what they had to do in order to keep their land in the best shape possible. As a result, many of the cooperative's members had to seek employment on other coffee farms, something that is highly disliked by most farmers. Nonetheless, there have been very few conflicts over working without payment. In contrast, one of the greatest sources of conflict in cooperative 1 is having to work for delayed payments or no payment. In part, this is justified by the members' need to receive their income in a timely manner, in order to meet their immediate necessities. However, it also shows a lack of perception that working for the cooperative represents an investment directly related to their livelihoods.

For members of cooperative 3, forming a cooperative was perceived as the only way to improve their condition as coffee farmers. In the words of the cooperative's president: "We are small coffee producers. If we don't unite, if we don't form our cooperative, or are not organized, we have to sell to the *coyote* (intermediary), who pays what he wants and sells to a higher price. That is why we seek to get organized, to find better markets, with better prices, to seek projects, and ultimately to improve our lives." (Escalante 2001)

Structure and Governance   In theory, all cooperatives function through democratic decision making. To reach any decisions the board meets with the general assembly, consisting of all cooperative members, and motions are decided by majority vote. Agrarian reform and most traditional cooperatives (cooperatives 1 and 2) follow rules set out in the General Law of Cooperative Associations of El Salvador. The Salvadoran Federation of Agrarian Reform Cooperatives (FESACORA), provides support on legal matters, and mediates in governance conflicts. Alternatively, farmer associations like cooperative 3 follow a different set of rules established by the Ministry of Agriculture, which also follow the General Law.

In the three cooperatives, the board of directors consists of up to 15 members with the usual postings of president, vice-president, treasurer, secretary, etc. Board members are elected by majority vote, and they hold the charge for 4-year periods. At cooperative 1, it is common practice to review the board every 2 years. If justified, the cooperative members are able to modify the board at this time. Any changes that conflict with established regulations need to be mediated by the regulating agencies. If representatives from FESACORA or the Ministry of Agriculture do not approve them they are considered illegal.

In practice, there is a tendency for the board of directors to make decisions without consulting the general membership. Some of these are justified, since it becomes unmanageable to call for general assemblies on every occasion. However, many conflicts arise due to what is perceived as a lack of communication and transparency.

Among other things, some of the conflicts dealing with governance have to do with the scale of each cooperative. It is a lot harder to get 99 people (cooperative 1) to participate than 22 or 32 (cooperatives 2 and 3, respectively). However, this situation is also the result of the way in which the organizations were formed (as discussed in the previous section). In general, decisions and actions are much easier to implement in cooperatives 2 and 3.

Relationships with External Actors and Markets   Most relationships with external actors and markets are formed only if these individuals and institutions approach the cooperatives. Cooperative 1, with its larger size and volume of production, has established links with a diversity of institutions and individuals over the years. These include banks, development NGOs, second- and third-level cooperatives, and national associations. It is through the Association of Small Coffee Producers of El Salvador (APECAFE) that cooperative 1 is now selling up to 100 percent of its production at a stable Fair Trade price of US$1.26 per pound. In addition, through the Fair Trade networks, cooperative 1 has been the recipient of several state and international cooperation projects. Cooperatives 2 and 3 initiated the process of becoming members of APECAFE in 2004, and hope to become Fair Trade certified in 2005. Moreover, all three cooperatives are seeking organic certification, also mediated through APECAFE.

Relationships with local and external actors and networks have been complicated by the degree of geographical isolation of the three cooperatives. Cooperatives 2 and 3 are very isolated and can be reached only with a four-wheel-drive vehicle. This makes for difficult communications, even with communities and organizations relatively nearby. When it comes to external actors and institutions, it becomes a challenge, both for outsiders to come to the cooperatives and for the members to travel to other locations. On the other hand, cooperative 1 has easier access by vehicle. It is also closer to the town of Tacuba. This is part of the reason why this cooperative has seen a greater number of organizations approach it for different types of projects.

The current coffee crisis has motivated a large number of non-government organizations and cooperative federations to seek projects for coffee-growing areas, including Tacuba. The networks that the cooperative leadership developed in the last 5 years (personal observation) have paid off by linking them to several opportunities. The most significant support received in this period was a project sponsored by the Spanish International Cooperation Agency (AECI), which provided financial resources for coffee management and certification, beginning in August 2003. The project has a duration of 2 or 3 years and has provided direct economic alleviation of the price crisis for the three cooperatives. In addition, the Salvadoran Foundation for Coffee Research (PROCAFE) promoted an exotic timber tree species (*Acrocarpus fraxinifolius*) to diversify the coffee plantation of cooperative 1. However, the cooperative was unable to successfully integrate this tree into its plantation, and had very high mortality rates the first year.

In their dealings with research and development actors, the cooperatives still exhibit weak negotiating power and lack of administrative skills. This was specifically expressed by the boards of directors and leadership in several workshops we undertook between January and March 2004. In general, cooperative members feel they are obliged to do exactly what project organizers and agencies tell them to do, even when they are not fully in agreement. However, they choose to follow these demands in order to have access to financial and technical resources. The administrative problem represents a structural deficiency, further complicated by the low educational levels of cooperative members (average level of education of the board of directors is 6th grade). Although many members have received training in this topic, very little progress has been made in achieving more efficient administration.

Shade-Tree Biodiversity

Tree Composition of the Shade Canopy   A total of 2,743 trees were identified and measured in the three cooperatives. The sample contained 169 tree species, representing at least 46 families. Of these, 123 have been identified, and include 109 native, and 14 exotic species. Tree species richness and abundance[6] were compared

Table 9.5
Summary of ecological characteristics of tree communities in three coffee cooperatives of Tacuba and three forest sites of PNEI, El Salvador. Adapted from Méndez 2004b.

	Coop. 1	Coop. 2	Coop. 3	Forest site 1[a]: selectively cut	Forest site 2: abandoned coffee plantation	Forest site 3: selectively cut
No. of plots	20 (2 ha)	14 (1.4 ha)	17 (1.7 ha)	10 (1 ha)	8 (0.8 ha)	10 (1 ha)
Total species richness per site[b]	69	48	93	103	110	67
Mean richness per plot (minimum per ha)	12	12	22	31	34	26
Mean stem density (trees per plot)	39	35	89	106	162	147

a. Data for forest sites from Ramirez 2001.
b. Total species richness figures are not fully comparable between the cooperatives and the forest, because data necessary to perform rarefaction of forest plots were not available.

to a recent study on the vegetation of the nearby PNEI, which found 174 tree species in 28 plots of the same size (1,000 m²), distributed in three different sites (Ramírez-Sosa 2001) (table 9.5).

The individual farms of cooperative 3 showed significantly higher levels of tree species richness and abundance than the collective plots of the other two cooperatives.[7] Both seem to be related to the livelihood strategies of the households. Since these farmers are unable to count on the relatively secure employment provided by the other two cooperatives, they are much more dependent on what they can produce on-farm. Maintaining a high level of tree species richness and abundance seems to be a generalized strategy of these types of farmers. The characteristics of the shade canopy found in the three cooperatives were very similar to other shade coffee agroecosystems in other areas of Mesoamerica (Méndez et al. 2001; Moguel and Toledo 1999; Soto-Pinto et al. 2000).

Forest plots showed much higher average levels of tree density and species richness than cooperatives 1 and 2. On the other hand, cooperative 3 contains similar levels of total richness, but lower levels of diversity per unit area, when compared to forest plots. Tree density in all cooperatives is much lower than that in forest plots. Cooperative 3's abundances were the closest to those of the forest.

Comparison of Tree Species Composition in the Cooperatives and the Forest
Comparative analysis between the cooperatives and the forest was limited by a lack of data from forest sites. Therefore, the variables presented in table 9.5 should be used only as a preliminary visualization, and considered with caution. A total of 227 tree species were identified to the genus level including the three cooperatives and the three forest sites. The total of identified species richness does not differ considerably (141 vs. 123 species in forest sites and cooperatives, respectively).

Jaccard coefficients (CC_J) of community similarity were calculated using the total 227 species identified at least to the genus level in the three cooperatives and the three forest sites. Cooperatives 1 and 2 were the most similar sharing 47 percent of their tree species, while cooperatives 2 and 3 are the most different, only sharing 38 percent of the tree species present in each. Overall, the three cooperatives only shared 35 out of the 123 species identified.

Jaccard coefficients were lower in comparisons between cooperative and forest sites (≤ 0.12), than in comparisons within cooperatives (≥ 0.38) or within forest sites (≥ 0.30). In general, forest site 2, which was an abandoned coffee plantation, showed the most similarity to all cooperative sites. Forest and cooperatives are very different, the highest similarity being 0.12 between cooperative 1 and forest site 2. In total, 36 species were shared between cooperatives (pooled) and the forest (pooled), which represents only 16 percent of the total of 227 species used in the analysis (Jaccard coefficient = 0.163).[8]

Farmers' Perceptions on Tree Conservation
Farmers consider themselves providers of environmental services (Méndez et al. 2002). Some of the services mentioned were water and soil conservation, and improvement of the environment by maintaining trees in their coffee plantations. All of the farmers that were interviewed showed a disposition to participate in efforts to conserve native trees. However, an equally unanimous response was that they were willing to contribute, if and only if, they could keep control of and access to their shade coffee plantations. Specifically, they mentioned the need to continue to prune and selectively harvest trees for firewood and timber.

Action-Research Results and Direct Support to Farmers' Livelihoods
Researchers began supporting farmers in Tacuba initially by organizing networks and training on organic coffee farming, alternative markets, computers and livelihood diversification in 2001. These activities were separate from the original research and were intended to improve farmers' knowledge and capacities for decision making.

The other important dimension of action research in El Salvador was to invest in concerted activities to support farmer livelihoods. These have taken several forms, ranging from contributing bananas and plantains for plantation diversification and making direct market links with the Community Agroecology Network (CAN) to providing access to infrastructure through the office in Tacuba (phone, fax, space for meetings, computers, etc.). The formalization of ASINDEC, as a local foundation made these initial relationships permanent and consolidated the work directed to livelihood support (Bacon et al. 2005). Some of the major achievements from this 5-year process can be summarized as follows:

• ASINDEC is supporting the three cooperatives in the establishment of the Association of Organic Coffee Farmers of Western El Salvador (ACOES), following the Nicaraguan model they observed at the Central of Coffee Cooperatives of the North (CECOCAFEN) through a farmer-to-farmer exchange in Matagalpa (Méndez 2004a).
• Both researchers and farmers have committed to a long-term Participatory Action Research (PAR) process.
• Through the partnership with CAN, farmers are establishing a direct marketing channel for their coffee at prices equal to or higher than Fair Trade. In addition, Tacuba is part of an internship network that supports their research and development efforts, both financially through internship fees, and through the work of interns.
• Farmers have received up to 15 trainings in topics ranging from tree biodiversity management to alternative marketing.
• Farmers continue to use the office infrastructure, including phone, fax, space for meetings, and computers.
• This case has been very successful in facilitating interdisciplinary research, including one concluded doctoral dissertation in environmental studies (University of California, Santa Cruz), one doctoral dissertation in anthropology (University of Texas, Austin), two completed master's theses in forestry and environmental science (Yale University), and two ongoing master's theses in sustainable agriculture (University of El Salvador).

Discussion

Organizations and Networks for Conservation and Livelihoods

Different types of organizations and networks (e.g. social, research, marketing) are very important for both conservation and rural development efforts focusing on improving rural livelihoods. The three cooperatives are organizations that are still representative of their membership. The comparative analysis shows that the smaller models, constructed by farmers, have less internal conflicts, but these are integrated in fewer networks with actors related to coffee and development. In addition, these organizations are perceived by their members as a means to improve their livelihoods. The smaller size makes it easier for members to be better informed and for boards of directors to be more accountable. On the other hand, after 22 years, the larger, agrarian reform model is still affected by internal conflict and the low motivation of its members. Most members still perceive their cooperative as a source of employment, and question its ability to have a direct impact on their livelihood situation. However, this larger cooperative has been able to develop more sophisticated networks that have resulted in significant benefits, notably access to Fair Trade markets. The other two cooperatives are largely lacking these contacts and remain much more isolated.

Notwithstanding the concerns expressed above, the farmer cooperatives remain the most appropriate organizations to promote both conservation and rural development initiatives. For example, all members, even if critical, respect the authority of the majority vote to decide on management strategies and other important issues. In addition, it is a familiar form of organization, seen as the institutional arrangement that remains close to their daily realities. In general, most members attend the assemblies and participate in decision making.

The household surveys confirmed that cooperative members are almost entirely dependent on coffee to generate incomes. The severity of the coffee price crisis, greatly affected all cooperative members' financial capital, and called for a re-evaluation of their dependence on coffee. Tacuba offers several possibilities for livelihood diversification, including eco-tourism, and proper conditions for other crops (e.g. bananas and *pacaya*). However, the cooperatives need training and support to start exploring these alternatives. This represents an opportunity to combine work with improving the organizational structure and capacities of cooperatives, with specific themes that could directly improve the livelihoods of cooperative members.

Improving the efficiency and reputation of the cooperatives requires intensive work on accountability, more transparent communications, and effective capacity building for better administration and management. Given the long history of inefficient trainings from outside actors, this calls for a re-evaluation of external support in order to focus more directly on empowerment and capacity building adapted to local realities. Working with the cooperative members to resolve the observed problems should be a first step for any environmental and development efforts. Other models in the region, such as the regional unions of cooperatives that exist in Nicaragua, could be the basis for restructuring and innovation to improve the cooperative models (Méndez 2004a). In addition, a concerted effort needs to be made to develop direct networks, as opposed to intermediated relationships, with actors that can support these efforts. These actors include donors, researchers, coffee buyers, and alternative trade organizations, among others.

Tree Biodiversity Conservation and Rural Livelihoods: Using Biodiversity

The levels of native tree richness found in the cooperatives are lower, but similar, to those observed in the nearby forest. This shows promise for conserving high numbers of tree species within shaded coffee plantations. However, the comparisons also show that tree compositions in the forest and the coffee plantations are very different. This means that some of the forest species that may be threatened or endangered are not necessarily found in the coffee plantations. In addition, analyses by Méndez et al. (2007) show that tree species of global conservation importance[9] are very different from those appreciated and preferred by the Tacuba farmers. Similar results have been found in other areas of Mesoamerica (Gordon et al. 2003). This calls for a need to work with farmers to develop coffee plantation management that serves both their livelihood needs, as well as conservation goals.

Similarity coefficients for the three cooperatives show that the composition of shade trees in their plantations is very different. This affords an opportunity for cooperative members to learn and share knowledge on the habit and use of at least 88 tree species that are being grown in only one or other of the cooperatives . This seems especially promising at this time since the cooperatives are currently seeking to diversify their shade stratum as a strategy to cope with the coffee price crisis.

Another important result of this research is the confirmation that trees are important for household livelihoods. Considering the low incomes of the majority of households, it can be expected that the products derived from shade trees represent considerable savings for these families. In this context, researchers and planners need to address the issue of how valuable to conservation is the maintenance of tree species that are also of importance to household livelihood strategies.

Small, individual farms showed the highest levels of both tree biodiversity and dependence on this diversity for livelihood strategies. This points to the importance of including these farmers in conservation activities, contrary to current practice under most initiatives, as was largely the case with the recent Coffee and Biodiversity Project, funded by the Global Environmental Facility (GEF).

The evidence of my research supports the proposition that the tree biodiversity maintained and used by shade coffee farmers is important to conservation efforts. Farmers have adopted these strategies of their own accord, without direction or incentives from technical or development organizations. It seems logical that if support and technical assistance were provided to increase "used" biodiversity, farmers would be willing to collaborate. This is further confirmed by the trends in alternative coffee markets that support the use of diversified shade with price premiums. However, any attempts to integrate coffee farmers into conservation strategies need to ensure that these initiatives will not negatively affect the socioeconomic situation of their households. Restricting the use of shade-tree resources by low-income families, without compensation, will no doubt further impoverish these households.

Reflections on the Action-Research Process

The Tacuba process illustrates a case where research preceded the concrete actions supporting livelihoods, but included approaches that made it easy to transform it into a PAR process. The experience demonstrates that it is possible to conduct rigorous research in combination with action objectives, but that this will probably require more time and resources. Furthermore, the generation of reliable information, as well as the facilitation of linkages for farmers, has been very useful for the cooperatives in their efforts to create and expand development and support networks. Basic support, such as providing access to infrastructure through the local ASINDEC office has also contributed to farmer's capacity-building and networking activities.

As research has increasingly given way to more direct support to farmers, ASINDEC has embraced the following process-oriented principles, outlined in Gómez and Méndez (2004) as characteristic of "pro-community" organizations[10]:

• It has a commitment to strengthen the political empowerment of its local partners.
• It focuses on the institutional development of community organizations and human capital, through the development of local capacities.
• It is committed to the learning process of local actors, allowing them to assume the leadership, even if they make mistakes. Learning from mistakes is an important part of the process.
• It avoids paternalism and the creation of external dependency.
• It is long-term and dynamic, and it utilizes a complexity of networks and contacts.
• It prioritizes long-term processes, not short-term projects.
• It invests in relationships of mutual trust with local actors.

These principles present the reflections of individuals and organizations seeking to improve models of monitoring and support for poor rural communities. It is also important to include self-critical reflection as an important part of this process-oriented approach, in order to ensure that its evolution is in step with local needs and development (Kenton 2004). As the PAR process evolves in Tacuba, these principles have demonstrated their value as essential guidelines to balance the action and research work successfully.

Notes

1. Agroforestry is a land-use system incorporating trees with agricultural crops and/or animals, in which their ecological interactions are managed to obtain multiple social, economic, and/or environmental products and benefits. (This definition was adapted from Nair 1993 and Somarriba 1998.)

2. These efforts resulted in the first conference on Biodiversity Friendly and Sustainable Coffee Production, held in El Salvador in December 1997, which launched a 3 year project on Coffee and Biodiversity funded by the Global Environmental Facility (GEF) of the World Bank.

3. The MBC is the most ambitious conservation initiative to be undertaken in the region in the past several decades. It aims to develop linked conservation areas from Mexico to Panama, including land under human management.

4. In 1980, the Salvadoran government implemented an agrarian reform. The first phase expropriated farms larger than 500 hectares (decree 154). The second phase was supposed to expropriate land larger than 250 hectares, but it was never implemented. In 1983, the third phase of the reform, decree 207, transferred titles for plots of up to 7 hectares to tenants who had been renting for more than 10 years. As in phase 1, this land was bought from owners and resold to farmers.

5. The Jaccard index is calculated through the formula $CC_J = j/(a + b - j)$, where j is the number of species common to both sites, a is the number of species at site A, and b is the number of species at site B. A CC_J value of 1 indicates that two sites contain the same species; a value of 0 denotes that the sites share no species (Magurran 1988; Harvey and Haber 1999).

6. Richness refers to the number of species present in a site, while abundance refers to the number of individuals of each species present at a site (Magurran 1988).

7. For statistical analyses see Méndez 2004b and Méndez et al. 2007.

8. For more in-depth ecological analyses, including statistical comparisons of richness and abundance of tree species between cooperatives, and an examination of the importance of the species found for global conservation efforts, see Méndez 2004b and Méndez et al. 2007.

9. Méndez et al. (2007) established global conservation importance of tree species through the lists of species of conservation concern developed by the World Conservation Union (IUCN) and the United Nations Environment Program (UNEP) (IUCN. 2002; UNEP-WCMC 2003).

10. These principles draw from my work with the Salvadoran Research Program on Development and Environment (PRISMA) and specifically from discussions with Ruben Pasos.

References

Altieri, M. A. 1995. *Agroecology: The Science of Sustainable Agriculture.* Westview.

Altieri, M. A., and C. I. Nicholls. 2002. Un metodo agroecologico rapido para la evaluacion de la sostenibilidad de cafetales. *Manejo Integrado de Plagas y Agroecologia* 64: 17–24.

Bacon, C. 2005. Confronting the coffee crisis: Can fair trade, organic and specialty coffees reduce small-scale farmer vulnerability in northern Nicaragua? *World Development* 33, no. 3: 497–511.

Bacon, C., V. Méndez, and M. Brown. 2005. Participatory Action-Research and Support for Community Development and Conservation: Examples from Shade Coffee Landscapes of El Salvador and Nicaragua. Research Brief 6, Center for Agroecology and Sustainable Food Systems, University of California, Santa Cruz.

Bebbington, A. 1992. Searching for an "Indigenous" Agricultural Development: Indian organizations and NGOs in the Central Andes of Ecuador. Centre of Latin American Studies, University of Cambridge.

Bebbington, A. 1993. Fragile lands, fragile organizations: Indian organizations and the politics of sustainability in Ecuador. *Transactions of the Institute of British Geographers* 181: 179–196.

Bebbington, A. 1996a. Indigenous organizations and agrarian strategies in Ecuador. In *Liberation Ecologies: Environment, Development, Social Movements*, ed. R. Peet and M. Watts. Routledge.

Bebbington, A. 1996b. Organizations and intensifications—campesino federations, rural livelihoods and agricultural technology in the Andes and Amazonia. *World Development* 24: 1161–1177.

Bebbington, A. 1997. New states, new NGOs? Crises and transitions among rural development NGOs in the Andean region. *World Development* 25: 1755–1765.

Bebbington, A. 1998. Seeking common ground in Ecuador. *Environment* 40: 42–43.

Bebbington, A. 1999. Capitals and capabilities: A framework for analyzing peasant viability, rural livelihoods and poverty. *World Development* 27: 2021–2044.

Bebbington, A. 2000. Reencountering development: Livelihood transitions and place transformations in the Andes. *Annals of the Association of American Geographers* 90: 495–520.

Bebbington, A. 2001. Globalized Andes? Livelihoods, landscapes and development. *Ecumene* 8: 414–435.

Bebbington, A., and S. Batterbury. 2001. Transnational livelihoods and landscapes: political ecologies of globalization. *Ecumene* 8: 369–380.

Bebbington, A., J. Quisbert, and G. Trujillo. 1996. Technology and rural development strategies in a small farmer organization: Lessons from Bolivia for rural policy and practice. *Public Administration and Development* 16: 195–213.

Beer, J., R. Muschler, D. Kass, and E. Somarriba. 1998. Shade management in coffee and cacao plantations. *Agroforestry Systems* 38: 139–164.

Bray, D. B., J. L. Plaza-Sanchez, and E. Contreras-Murphy. 2002. Social dimensions of organic coffee production in Mexico: Lessons for eco-labeling initiatives. *Society and Natural Resources* 15: 429–446.

Bryant, R. L., and S. Bailey. 1997. *Third World Political Ecology*. Routledge.

CCAD. 1998. Estado del ambiente y los recursos naturales en Centro America. CCAD, Guatemala.

Cienfuegos, R. E., ed. 1999. Fomento de la participacion ciudadana para el desarrollo sostenible en el municipio de Tacuba, Departamento de Ahuachapan, Fase Piloto FUNDESYRAM-IIZ. IIZ-El Salvador, San Salvador, El Salvador.

CNR. 1990. Monografias del departamento y municipios de Ahuachapan. Centro Nacional de Registros (CNR), San Salvador.

Colwell, R. K. 1997. Estimates: statistical estimation of species richness and shared species from samples. http://purl.oclc.org.

Cuéllar, N., H. Rosa, and M. E. Gonzalez. 1999. Los servicios ambientales del agro: el caso del cafe de sombra en El Salvador. PRISMA, San Salvador, El Salvador.

Dominguez, J. P., and O. Komar. 2001. Investigacion aplicada de los criterios ecologicos para las plantaciones de cafe amigables con la biodiversidad. Proyecto Cafe y Biodiversidad-SIMBIOSIS/IRG, Nueva San Salvador, El Salvador.

Ellis, E. C., and S. M. Wang. 1997. Sustainable traditional agriculture in the Tai Lake Region of China. *Agriculture, Ecosystems & Environment* 61: 177–193.

Escalante, M. Y. 2001. Transcription of the Workshop on Alternative Markets. Shade, Livelihoods and Conservation Project-UCSC/PROCAFE/CFHF, San Salvador, El Salvador.

Ewel, K. C. 2001. Natural resource management: The need for interdisciplinary collaboration. *Ecosystems* 4: 716–722.

Francis, C., G. Lieblein, S. Gliessman, T. A. Breland, N. Creamer, R. Harwood, L. Salomonsson, J. Helenius, D. Rickerl, R. Salvador, M. Wiedenhoeft, S. Simmons, P. Allen, M. Altieri, C. Flora, and R. Poincelot. 2003. Agroecology: The ecology of food systems. *Journal of Sustainable Agriculture* 22: 99–118.

Gallina, S., S. Mandujano, and A. Gonzalez-Romero. 1996. Conservation of mammalian biodiversity in coffee plantations of Central Veracruz. *Agroforestry Systems* 33: 13–27.

Galloway, G., and J. W. Beer. 1997. Oportunidades para fomentar la silvicultura en cafetales en America Central. CATIE, Turrialba, Costa Rica.

Gliessman, S. R. 1998. *Agroecology: Ecological Processes in Sustainable Agriculture*. Ann Arbor Press.

Gliessman, S. R. 2000. *Field and Laboratory Investigations in Agroecology*. Lewis.

Gómez, I., and V. E. Méndez. 2004. Análisis de contexto: el caso de la Asociación de Comunidades Forestales de Petén (ACOFOP). Proyecto Innovativo FORD-CIFOR/PRISMA, San Salvador, El Salvador.

Gordon, J. E., A. J. Barrance, and K. Schrekenberg. 2003. Are rare species useful species? Obstacles to the conservation of tree diversity in the dry forest zone agro-ecosystems of Mesoamerica. *Global Ecology and Biogeography* 12: 13–19.

Greenberg, J. B., and T. K. Park. 1994. Political ecology. *Journal of Political Ecology* 1: 1–12.

Greenberg, R., P. Bichier, A. C. Angon, and R. Reitsma. 1997a. Bird populations in shade and sun coffee plantations in Central Guatemala. *Conservation Biology* 11: 448–459.

Greenberg, R., P. Bichier, and J. Sterling. 1997b. Bird populations in rustic and planted shade coffee plantations of Eastern Chiapas, Mexico. *Biotropica* 29: 501–514.

Greenwood, D. J., and M. Levin. 1998. *Introduction to Action Research: Social Research for Social Change.* Sage.

Guzmán-Casado, G., M. González de Molina, and E. Sevilla-Guzmán. 1999. *Introducción a la agroecología como desarrollo rural sostenible.* Ediciones Mundi-Prensa, Madrid.

Harvey, C. A., and W. A. Haber. 1999. Remnant trees and the conservation of biodiversity in Costa Rican pastures. *Agroforestry Systems* 44: 37–68.

Herrador, D., and L. Dimas. 2000. Payment for environmental services in El Salvador. *Mountain Research and Development* 20: 306–309.

IUCN. 2002. IUCN Red List of Threatened Species. http://www.redlist.org.

Izac, A. M. N., and P. A. Sanchez. 2001. Towards a natural resource management paradigm for international agriculture: the example of agroforestry research. *Agricultural Systems* 69: 5–25.

Johnson, M. D. 2000. Effects of shade-tree species and crop structure on the winter arthropod and bird communities in a Jamaican shade coffee plantation. *Biotropica* 32: 133–145.

Kenton, N., ed. 2004. *Participatory Learning and Action: Critical Reflections, Future Directions.* IIED.

Llanderal, T., and E. Somarriba. 1999. Tipologias de cafetales en Turrialba, Costa Rica. *Agroforesteria en las Americas* 23: 30–32.

Magurran, A. E. 1988. *Ecological Diversity and Its Measurement.* Princeton University Press.

MARN. 2003. Mapas del Sistema de Información Ambiental. Ministerio de Medio Ambiente y Recursos Naturales (MARN), El Salvador.

Marsden, T. 1997. Creating space for food: The distinctiveness of recent agrarian development. In *Globalising Food: Agrarian Questions and Global Restructuring*, ed. D. Goodmand and M. Watts. Routledge.

Méndez, V. E. 2004a. Intercambio de "agricultor a agricultor" entre cuatro cooperativas cafetaleras de El Salvador y la Central de Cooperativas Cafetaleras del Norte (CECOCAFEN), Nicaragua. ASINDEC, San Salvador, El Salvador.

Méndez, V. E. 2004b. Traditional Shade, Rural Livelihoods, and Conservation in Small Coffee Farms and Cooperatives of Western El Salvador. Ph.D. thesis, University of California, Santa Cruz.

Méndez, V. E., and S. R. Gliessman. 2002. Un enfoque interdisciplinario para la investigación en agroecología y desarrollo rural en el trópico Latinoamericano. *Manejo Integrado de Plagas y Agroecologia* 64: 5–16.

Méndez, V. E., S. R. Gliessman, and G. S. Gilbert. 2007. Tree biodiversity in farmer cooperatives of a shade coffee landscape of western El Salvador. *Agriculture, Ecosystems & Environment* 119: 145–159.

Méndez, V. E., D. Herrador, L. Dimas, M. Escalante, O. Diaz, and M. Garcia. 2002. Cafe con sombra y pago por servicios ambientales: riesgos y oportunidades para impulsar mecanismos con pequenos agricultores de El Salvador. PRISMA/Fundacion FORD, San Salvador, El Salvador.

Méndez, V. E., R. Lok, and E. Somarriba. 2001. Interdisciplinary analysis of homegardens in Nicaragua: Micro-zonation, plant use and socioeconomic importance. *Agroforestry Systems* 51: 85–96.

Moguel, P., and V. M. Toledo. 1999. Biodiversity conservation in traditional coffee systems of Mexico. *Conservation Biology* 13: 11–21.

Monro, A., D. Alexander, J. Reyes, M. Renderos, and N. Ventura. 2001. Arboles de los cafetales de El Salvador. Natural History Museum of London and University of El Salvador.

Nair, P. K. R. 1993. *An Introduction to Agroforestry*. Kluwer/ICRAF.

Nestel, D., F. Dickschen, and M. A. Altieri. 1993. Diversity patterns of soil macro-Coleoptera in Mexican shaded and un-shaded coffee agroecosystems: An indication of habitat perturbation. *Biodiversity and Conservation* 2: 70–78.

Peet, R., and M. Watts, eds. 1996. *Liberation Ecologies: Environment, Development, Social Movements*. Routledge.

Perfecto, I. 1994. Foraging behavior as a determinant of asymmetric competitive interaction between two ant species in a tropical agroecosystem. *Oecologia* (Berlin) 98: 184–192.

Perfecto, I., and R. Snelling. 1995. Biodiversity and the transformation of a tropical agroecosystem—ants in coffee plantations. *Ecological Applications* 5: 1084–1097.

Perfecto, I., and J. Vandermeer. 1996. Microclimatic changes and the indirect loss of ant diversity in a tropical agroecosystem. *Oecologia* (Berlin) 108: 577–582.

Perfecto, I., and J. Vandermeer. 2002. Quality of agroecological matrix in a tropical montane landscape: Ants in coffee plantations in Southern Mexico. *Conservation Biology* 16: 174–182.

Perfecto, I., J. Vandermeer, P. Hanson, and V. Cartin. 1997. Arthropod biodiversity loss and the transformation of a tropical agro-ecosystem. *Biodiversity and Conservation* 6: 935–945.

Ponte, S. 2002. The "latte revolution"? Regulation, markets and consumption in the global coffee chain. *World Development* 30: 1099–1122.

Ponte, S. 2004. Standards and Sustainability in the Coffee Sector: A Global Value Chain Approach. IISD/UNCTAD, Winnipeg.

PRISMA. 1995. El Salvador: dinamica de la degradacion ambiental. PRISMA, San Salvador, El Salvador.

PRISMA. 1996. La deuda del sector agropecuario: implicaciones de la condonacion parcial. PRISMA, San Salvador, El Salvador.

PROCAFE. 1998. Boletin estadistico de la caficultura Salvadorena. PROCAFE, San Salvador, El Salvador.

PROCAFE. 2001. Boletin estadistico de la caficultura Salvadoreña. PROCAFE, San Salvador, El Salvador.

Ramírez-Sosa, C. R. 2001. Vegetation of a subtropical pre-montane moist forest in Central America. Ph.D. dissertation, City University of New York.

Raynolds, L. T. 2002. Consumer/producer links in fair trade coffee networks. *Sociologia Ruralis* 42: 404–424.

Redman, C. L., J. M. Grove, and L. H. Kuby. 2004. Integrating social science into the long-term ecological research (LTER) network: Social dimensions of ecological change and ecological dimensions of social change. *Ecosystems* 7: 161–171.

Renard, M. C. 1999. The interstices of globalization: The example of fair coffee. *Sociologia Ruralis* 39: 484–500.

Rice, R. A. 2001. Noble goals and challenging terrain: Organic and fair trade coffee movements in the global marketplace. *Journal of Agricultural and Environmental Ethics* 14: 39–66.

Rice, R. A., and J. R. Ward. 1996. Coffee, conservation, and commerce in the western hemisphere. Smithsonian Migratory Bird Center/Natural Resources Defence Council, Washington.

Roberts, D. L., R. J. Cooper, and L. J. Petit. 2000. Use of premontane moist forest and shade coffee agroecosystems by army ants in western Panama. *Conservation Biology* 14: 192–199.

Rocheleau, D., L. Ross, J. Morrobel, L. Malaret, R. Hernandez, and T. Kominiak. 2001. Complex communities and emergent ecologies in the regional agroforest of Zambrana-Chacuey, Dominican Republic. *Ecumene* 8: 465–492.

Rosa, H., D. Herrador, M. Gonzalez, and N. Cuellar. 1999. El agro Salvadoreno y su potencial como productor de servicios ambientales. PRISMA, San Salvador, El Salvador.

Roth, D. S., I. Perfecto, and B. Rathcke. 1994. The effects of management systems on ground-foraging ant diversity in Costa Rica. *Ecological Applications* 4: 423–436.

Scoones, I. 1999. The new ecology and the social sciences: What prospects for a fruitful engagement? *Annual Review of Anthropology* 28: 479–507.

Selener, D. 1997. *Participatory Action Research and Social Change.* Cornell University Press.

Shapiro, E. N. Forthcoming. Interdisciplinary Agroecological Research as a Tool for Environmental Restoration. School of Forestry and Environmental Science, Yale University.

Somarriba, E. 1998. Que es agroforesteria? In *Apuntes de clase del curso corto: Sistemas agroforestales,* ed. F. Jimenez and A. Vargas. CATIE, Turrialba, Costa Rica.

Soto-Pinto, L., I. Perfecto, J. Castillo-Hernandez, and J. Caballero-Nieto. 2000. Shade effect on coffee production at the northern Tzeltal zone of the state of Chiapas, Mexico. *Agriculture, Ecosystems & Environment* 80: 61–69.

UNEP-WCMC. 2003. UNEP-WCMC species database: trees. http://www.unep-wcmc.org.

Walker, B., S. Carpenter, J. Anderies, N. Abel, G. Cumming, M. Janssen, L. Lebel, J. Norberg, G. D. Peterson, and R. Pritchard. 2002. Resilience management in social-ecological systems: A working hypothesis for a participatory approach. *Conservation Ecology* 6: 14.

Whatmore, S., and L. Thorne. 1997. Nourishing networks: Alternative geographies of food. In *Globalising Food: Agrarian Questions and Global Restructuring,* ed. D. Goodman and M. Watts. Routledge.

Williams-Guillen, K., C. M. McCann, A. Roque, D. Osornio, and C. Gomez. 2001. Composicion y fenologia de la comunidad de arboles en el cafetal con sombra de la hacienda La Luz, Volcan Mombacho, Nicaragua. In Libro de Resumendes del V Congreso de la

Sociedad Mesoamericana para la Biologia y la Conservacion (SMBC). SMBC, San Salvador, El Salvador.

Wunderle, J. M. 1999. Avian distribution in Dominican shade coffee plantations: Area and habitat relationships. *Journal of Field Ornithology* 70: 58–70.

Wunderle, J. M., and S. C. Latta. 1998. Avian resource use in Dominican shade coffee plantations. *Wilson Bulletin* 110: 271–281.

III

Alternative South-North Networks and Markets

10

Social Dimensions of Organic Coffee Production in Mexico: Lessons for Eco-Labeling Initiatives

David B. Bray, José Luis Plaza Sanchez, and Ellen Contreras Murphy

In recent years, a burst of academic and conservation attention has been accorded to traditional coffee farms as havens for biodiversity and shade-tree or Bird-Friendly coffee as a new conservation oriented marketing strategy (Perfecto et al. 1996; Rice et al. 1997; Moguel and Toledo 1999).[1] This has driven an unusually rapid conversion of scientific findings into the marketing of a biodiversity-based product, shade-tree coffee. Until the rise in interest in shade-tree coffee, almost all efforts to find products that can fuel market-based conservation initiatives have been focused on natural forests, mostly with reference to timber logging and non-timber forest products (Clay 1992; Johns 1997; Crook and Clapp 1998). Shade-tree coffee, however, is the first conservation-oriented market product that focuses both on a broader agricultural landscape and on a major agricultural commodity. It focuses research and conservation on biodiversity in managed systems in general and agroecosystems in particular, not only natural ecosystems (Power and Flecker 1996, Vandermeer and Perfecto 1997). It effectively shows that community-based conservation can also be "commodity-based conservation," with potential impacts on managed landscapes beyond protected areas.

Thus, a major world agricultural commodity already being produced by small farmers could be the vehicle for habitat preservation and conservation. This is quite different from the slow, painful creation of products and markets that most efforts to link biodiversity to marketing have had to face (Millard 1996). Marketing partnerships between coffee roasters, non-government conservation organizations, and public research organizations have rushed to place eco-labeled shade-tree and Bird-Friendly coffees on the market, trying to capture the consumer interest of millions of declared birdwatchers (Greenberg 1997a; Bray 1999a 1999b). These efforts were designed, in part, to draw a line in the sand against the advance of "sun coffee," highly technified monocrop coffee plantations that create biological deserts (Rice and Ward 1996).

Thus far, efforts to study and market shade-tree coffee have been dominated by biologists, conservation organizations, and coffee marketers. This has led to an emphasis on developing bio-physical criteria for certifying shade-tree coffee and consumer consciousness campaigns (Greenberg 1997b; Smithsonian Migratory Bird Center 1999). However, the sudden interest in shade-tree coffee forces renewed attention on a preexisting agroecological production and marketing strategy, certified organic coffee. Organic coffee stands as a relatively successful model of how niche marketing of a product with high social and environmental content can have a positive impact on small-farmer incomes and ecosystems. A closer study of the organic coffee production also suggests that shade-tree coffee academics and advocates, by focusing on biophysical criteria and marketing, are leaving out a long chain of social factors that heavily influence the marketing of a biodiversity-oriented product. Shade-tree coffee efforts and eco-labeling initiatives in general have much to learn from a deeper understanding of the institutional transformations and social self-organization processes that led to the emergence of organic coffee as a viable option for impoverished farmers living in diverse ecosystems. In Mexico, the world's leading exporter of organic coffee, an estimated 11,590 certified producers cultivate 15,000 hectares and produce more than 100,000 60-kilogram sacks, with virtually all of this production coming from organized producers, not individual farms. This represents about 4 percent of Mexico's coffee producers and 2 percent of the coffee landscape (Hernández 1997; Martínez Torres 1997; see also Poniatowska and Hernandez Navarro 2000), a small but striking advance for an alternative, environmentally friendly product. In US markets, certified organic coffee now constitutes as much as 5 percent of the gourmet coffee market and $150 million in sales in 1998 (Rice and McLean 1999).

Organic coffee did not emerge because of organic certification criteria, and only partly from finding committed consumers. As the following analysis of the emergence of organic coffee in Mexico will attempt to demonstrate, the conversion to organic coffee by small farmers also arose from over a decade of populist agrarian organizing and accompanying organizational innovations. The emergence of organized organic coffee producers in Mexico depended upon the substantial amount of pre-existing "social capital accumulation" in the Mexican countryside (Fox 1996; Ostrom 1999), a process of self-organization and institutional learning (Folke, Berkes, and Colding 1998), and the existence of significant subsidies. The producers themselves see organic coffee as only one of a "basket of benefits" received as a result of being organized, with positive environmental outcomes being of secondary importance. It is argued that what allowed Mexico, of all the coffee-producing countries, to emerge as the leading organic coffee producer, was a shifting political and social matrix that led to relative empowerment of small farmer organizations (Porter 2000). And, as Rice and McLean (1999, p. 23) have noted, "For small farmers, organization is perhaps the key ingredient to obtaining market access. By getting organized into cooperatives or companies, farmers are able to achieve the economies of scale necessary to market their own harvests in a cost effective and competitive manner, thus tapping into potential price incentives offered by the market (for quality, organic Fair Trade, etc.)." In the Mexican case, by being organized, producers learned how to improve quality at every processing stage, learned about their customers and the benefits of forward integration (Fairbanks and Lindsay 1997a,b), and have also learned, as Ronald Nigh has noted, that "organic coffee is a typical postmodern economic product with high symbolic and aesthetic content in which 'organic production as an entire way of life' is part of what is being marketed" (Nigh 1997; see also Hernández Castillo and Nigh 1998). The case of organic coffee producers in Mexico also indicates that organic production does not necessarily exclude marginal producers, as has been argued elsewhere (Raynolds 2000). The production of organic coffee is also intertwined with the domestic economy of small coffee producers, and this study will also describe some of the labor allocation issues and social and economic benefits of organized organic coffee production.

Organic coffee is a recent outcome of a long history of interaction between coffee-related institutions, social capital, and the environment in Mexico. This interaction between social systems and ecosystems requires a conceptual framework that sees social systems and ecosystems not in competition, but as mutually dependent and mutually constraining systems (Berkes and Folke 1998). The coffee landscape of Mexico and its varied outputs were created by particular social institutions, but institutions constrained by the nature of the landscape ecology on which it has depended. In order to understand how the relatively positive environmental and social outcome of organic coffee emerged, we will first provide an overview of the institutional and ecological system of coffee in twentieth-century Mexico in relationship to the coffee landscape, and then proceed to a closer case study of the emergence of organic coffee farming in the La Selva Ejido Union (Unión de Ejidos La Selva) in Chiapas. We then draw out the lessons from organic coffee production for conservation and development in general, for a Bird-Friendly coffee marketing strategy in particular, and for collaborations between social and natural scientists on these issues.

Social Systems and Ecological Systems in Mexican Coffee Production

Organic coffee has been the most recent outcome of a series of institutional transformations among small coffee producers in Mexico, including processes of migration and colonization of tropical forests, shifts in Mexican public policy toward coffee production and marketing, fluctuations in international coffee markets, and development interventions that encouraged an emerging niche for small-scale organic production in global markets.

Coffee was introduced into Mexico in the late nineteenth century, but expansion was slow, and by 1970 it occupied only 356,253 hectares. But the growth over the next 22 years was explosive. By 1992, the land occupied by coffee had grown to 761,899 hectares, a nearly 114 percent increase. This expansion was driven by a period of relatively high and stable coffee prices, which resulted from the International Coffee Agreements (ICA),[2] and high rural population growth, some of which was channeled into tropical colonization. This expansion had a major effect on the environment in the states where coffee expanded. Scattered evidence permits a preliminary sketch of the landscape ecology and sociology of this expansion. In 1970, there were 97,716 coffee producers farming 346,531 hectares, an average of 3.55 hectares of coffee per producer. By 1992, there were 282,593 producers farming 761,165 hectares for an average of 2.69 hectares per farmer, a decline in average farm size of some 25 percent. Sixty-nine percent of all coffee producers have less than 2 hectares and 60 percent are indigenous peoples (compared to around 12 percent indigenous peoples nationally) (Nolasco et al. 1985; Regalado Ortíz 1996). Thus, small, poor, indigenous producers dominate coffee production in Mexico.

The first part of the 1970–1992 expansion period partially coincides with the expansion of tropical colonization in Mexico (Revel-Mouroz 1980), and it is known that many indigenous small farmers expanded into lowland and montane tropical forests, particularly in Oaxaca and Chiapas, during this period. In the 1970s, only 15 percent of Mexican coffee production was found to be below an altitude of 500 meters. By the 1990s, 37 percent of all coffee production was estimated to be below 600 m (the studies used slightly different categories) (SAGAR 1996; Nolasco et al. 1985). Thus, it appears that a significant amount of the coffee expansion was into lower montane tropical regions inappropriate for the production of good-quality coffee. One study of increases in coffee production in the 1970–1982 period in Veracruz also suggested that this was due to the incorporation of new land into coffee production, not improvements in yields, with coffee substituting for both other agricultural systems and undisturbed forests (Nestel 1995).[3]

But small coffee farmers do not normally produce only coffee. In fact, the principal product of their farming systems is corn, with coffee forming the smallest land-use component. A study by the parastatal Instituto Mexicano del Café (INMECAFE) of coffee-producing municipios in the early 1980s found that in 44 percent of the coffee municipios, only up to 15 percent of total farm agricultural area was in coffee, but up to 35 percent was in corn (Nolasco et al. 1985, p. 124).

Research on the ecology of coffee production has thus far focused mostly on the significant biodiversity in the coffee plot itself. In the early 1970s, researchers found that heavily shaded traditional coffee farms in Veracruz harbored high rates of biodiversity in birds, epiphytes, and mosses (Aguilar-Ortíz 1982; Gomez-Pompa 1997). The INMECAFE study also noted that small farmer coffee plantations "imitate the structure of the natural forest" (Nolasco et al. 1985, p. 112). In the 1990s new research in Mexico and elsewhere built more evidence of the relationship between traditional shade-tree coffee farms and biodiversity (Gallina et al. 1992; Nestel and Altieri 1993; Perfecto and Vandermeer 1994).

Moguel and Toledo (1999) have classified the coffee landscape, constructing a continuum that includes "rustic," where coffee bushes are inserted into the understory of a natural forest, "traditional polyculture" or "coffee garden," where coffee and other food crops are grown under an intact original canopy, "commercial polyculture" where the overstory is significantly reduced, "shaded monoculture," with a sparse canopy over monoculture coffee, and unshaded monoculture or "full sun." Thus, at one end of the continuum, coffee can almost become a kind of introduced "non-timber forest product" in natural forests, and at the other end it creates a "biological desert." They also conclude that two-thirds of the coffee landscape of Mexico is either in rustic or "coffee garden" production regimes or under traditional shade. The coffee landscape was produced by particular institutions. In the 1970s and the 1980s coffee production was heavily controlled by INMECAFE, which in the early 1980s purchased 47 percent of all coffee and provided subsidies for small coffee farmers, thus regulating rural coffee prices. In a contribution to social capital in the sector, and in order to deliver both services and political control, it also organized the farmers, creating at least 2,671 local organizations including nearly 120,000 producers, some two-thirds of all producers at the time (Nolasco et al. 1985, pp. 186–187).[4] Although delivering a measure of social justice, for a period, to Mexico's small farmers, INMECAFE was a disaster from a quality and ecological point of view. It unsuccessfully attempted to promote sun coffee, emphasized volume over quality, mixing coffee from all different altitudes, and it lost for Mexico a reputation for quality regional coffees.

In the period 1980–1989, coffee organizations focused solely on getting higher prices from INMECAFE and were not able to form enduring organizations. But the disappearance of INMECAFE in the early 1990s, brought about the neo-liberal policies of President Carlos Salinas de Gortari, forced a dramatic shift in organizational strategies, "competing with larger producers to gain control over financing, processing, and marketing coffee. . . . The shift to market reforms in Mexico contributed to more efficient kinds of small farmer organizations by changing the incentive structures of these organizations." (Porter 1997) A major step forward in the creation of social capital and self-organization was marked by the emergence of the Confederación Nacional de Organizaciones Cafeteleras (CNOC) in the late 1980s, including over 50,000 small-scale coffee producers in most coffee-producing regions of Mexico (Hernández 1991; Hernández and Bray 1991; Bray 1995).

Organic production, later adopted by CNOC affiliates, first emerged from a separate Catholic Church-based effort. In the early 1980s a Dutch Catholic priest linked up small coffee farmers in southern Oaxaca with European Fair Trade markets, making the Union of Indigenous Peasants of the Isthmus Region (UCIRI) the first small farmer organic producers in Mexico.[5] The emergence of organic farmer organizations built on the "dense forms of social capital already existing in various indigenous communities" (Porter 1997). In the mid 1980s, a second church-based organization, the Indigenous Peoples of the Sierra Madre of Motozintla (ISMAM), also began experimenting with growing and exporting organic coffee (Nigh 1997; Bray 1991).[6]

The convergence of church-based organizations and agrarian populist ones on an organic strategy occurred with a devastating collapse of world coffee prices in July 1989, stimulated by the rupture of the quota-based International Coffee Agreement. In 1988 the average international price for 100 pounds of coffee was $103; by 1991 it was $50. In addition, the end of the export quota system meant that transnational corporations were able to command bargain prices for green beans while maintaining high prices for retail, dramatically reducing the percentage of share of commodity chain profits that went to producers (Talbot 1997).

The desperate need to find new niche markets drove major transformations in small farmer production and organization. Most Mexican coffee producers had been tagged as coffee "gatherers" rather than farmers; that is, no attention was paid to the crop until harvesting time. The organizational innovations carried out by the producers included a new focus on quality control. The effort required farmers to introduce quality control into every stage of the complex process of coffee production and marketing was enormous, as the Mexican agronomist Eduardo Martínez Torres noted:

Next comes choosing the right time for harvesting; harvesting only mature berries; not allowing harvested berries to heat up; sorting berries on intake; making sure the beans don't crack during the depulping process; double sorting after depulping; making sure fermentation lasts the right length of time, i.e. between 24 to 48 hours, depending on the altitude and average temperature; thoroughly washing the berries; grading; properly drying, preferably both in the sun, as well as in a drier in order to avoid mildewing. The drying temperature should be moderate. The temperature should never be turned up to speed the process and save time, since an uneven drying process can significantly damage bean quality. When drying is done on patios, layers should not be too thick and beans should be constantly stirred. Never mix together beans of different grade of quality, beans at different stages of dryness, or beans from different altitudes. Selection, patience and care are the operative words during processing, since all these things make for the best bean quality and, consequently the best price for (the) product. (1997, p. 216)[7]

With the disappearance of INMECAFE, the only source for the training and technical support for this quality conversion was the small farmer organizations themselves. Although training was carried out using inexpensive farmer-to-farmer methodologies, it still required trained professional staff and substantial overhead expenses, subsidies largely met by the church, international organizations, and the Mexican government. But it was the accumulated and growing social capital in the small farmer organizations that gave them the capacity to orchestrate and deliver the training at every step of the process. These organizations developed an empowering vision of appropriating the commodity chain even to the point of selling coffee by the cup to the ultimate consumer, where the highest profits are concentrated. Thus, organic emerged as only one component of a broader strategy of strengthening the capacity of small farmer organizations to manage their own production and marketing processes in the face of dramatic shifts in the policy and commercial environment.

However, the ability to respond to this vision appears to involve some equity issues. Initial evidence suggests that the organic strategy reached only a particular segment of small coffee producers. Studies of organic coffee producers in three organizations indicates that they are predominately from the 2–5-hectare stratum, and not from the two-thirds of Mexican coffee producers who have less than 2 hectares, as table 10.1 indicates. This suggests, as will be analyzed further in the La Selva case study (discussed next), that there are important barriers to entry into organic production in Mexico.

However, according to Pérezgrovas Garza et al. (1997), organic coffee producers organized in the peasant organization Majomut in the highlands of Chiapas average only around 1.5 hectare per producer, probably reflecting the higher degree of land fragmentation in highland Chiapas.

Table 10.1
Average size of total coffee plot in three organic coffee organizations. Sources: Porter 1997; Heinegg and Ferroggiario 1996; Plaza Sánchez 1997.

Organization	No. of members	Average size of plot (ha)
UCIRI	1,825	3.32
ISMAM	953	3.3
Unión de Ejidos La Selva	535[a]	2.7

a. at beginning of 1991

The remainder of this chapter presents a case study of one organic coffee organization, the Unión de Ejidos La Selva ("La Selva" for short), and what it can tell us about the limitations and possibilities of environmentally friendly coffee production as a conservation and development strategy.[8]

The Unión de Ejidos La Selva

The Unión de Ejidos La Selva (based in Comitán, Chiapas) was one of the founding members of CNOC in 1988. Beginning in 1988, La Selva was inspired by the experiences of UCIRI and ISMAM to begin exploring the organic option, with an initial grant from the Inter-American Foundation (IAF).[9] Its area of operation is in the southwestern Lacandon Rainforest, in the buffer zone of the Montes Azules Biosphere Reserve, established in 1978 with an area of 331,200 hectares. It is a colonization zone, with the most intensive settlement taking place in the 1960s and the 1970s. A significant percentage of the coffee was established in the low-quality regions below 600 m. Survey data showed that most of the farmers are indigenous, 67 percent being descended from Tojolobals (now primarily Spanish speakers), 20 percent Tzotzil speakers, and 13 percent mestizo. Education levels are very low; 44 percent of the members never attended school, and the illiteracy rate is 35 percent, typical of small farmers in the region. Access to health care, education, public utilities, and other public services is extremely limited. By 1992, they had some 200 of their members who had converted up to 1 hectare of coffee to organic production, the base for launching a $300,000 three-year initiative to achieve 1,000 hectares of certified organic coffee.[10] As of 1997, the project had been extended to 1,304 families in 57 communities.

Despite daunting obstacles (including the outbreak of guerrilla war in January 1994, which caused La Selva to lose a substantial portion of its membership to the ranks of the Zapatistas), it achieved 1,287 hectares of organic coffee by late 1996.[11] La Selva had a technical team consisting of six professionals that supervised and trained more than 30 community-selected small farmers. They attended workshops to learn the organic technological package, which they then taught to their peers in a "farmer-to-farmer" methodology. The technical team also visited the communities to support and advise the community promoters and to record the activities carried out by each producer. The decision to pursue an organic strategy was taken by a small number of agronomist advisors who had studied the markets and the experiences of UCIRI and ISMAM, and who set out to convince the small farmers of the wisdom of the strategy. The strategy was not chosen for environmental reasons, but because of the high premium that organic coffee marketed by solidarity groups in

Europe was commanding. It was understood, however, that reaching these markets required assiduous attention to the requirements for organic certification, which included components of environmental protection, particularly with respect to soil conservation, and required building both human capital through training and social capital through organizational strengthening.

In this subsection, study data will be used to indicate the impact of the transition to organic coffee production on the natural resource base, economic well-being, and general welfare of the small coffee farmers. The data are derived from a baseline survey instrument applied to a sample of 320 producers from 12 locations prior to the initiation of the project and at its conclusion, providing detailed data at the household level. Survey conclusions will only be indicated here, and a more detailed economic analysis is available (AICA Consultores 1997).

The Environmental Setting and the Agroecosystem

The mountainous terrain where La Selva members farm includes elevational variations from 200 to 2,500 meters, eight climate areas, a complex mosaic of eleven types of vegetation, and areas of very high biodiversity (Toledo et al. 1991). Until recent years, continued colonization, population growth, cattle ranching, and fires have driven a continuous expansion of the agricultural frontier. Forest has been cleared for corn fields (called *milpa* in Mexico), and population pressure has shortened the rotation of plots by reducing the fallow period, creating new pressures on the forest. Milpa, not coffee, is the major proximate driver of land-use change, with an estimated 77 percent of cultivated land in the region used for milpa, and only about 10 percent devoted to coffee.[12] Land clearing for pasture, a major factor in forest loss elsewhere in the Lacandon, is less significant in this region. But, although coffee preserves forest cover and biodiversity in the agricultural plot, the overall farming system has occasioned extensive deforestation. A study of deforestation in various Lacandon subregions using satellite imagery reveals that part of the Las Margaritas region has been very substantially deforested, well above the 26 percent deforestation found for the overall Lacandon, with forest cover, and thus coffee, being increasingly restricted to mountain ridges (O'Brien 1998). Since the 1980s, federal and state authorities have attempted to slow the clearing of land, with a 1989 Chiapas state logging ban, partially lifted in recent years, placing strict prohibitions on forest clearing for any reason (Bray 1997).

Coffee farms were established by gradually altering the natural forest through clearing out the understory, as well as the establishment of coffee plots in former milpa clearings, over time restoring a modified forest cover. In the Moguel-Toledo (1999) classification, almost all of La Selva's coffee would be considered as either rustic or traditional polyculture.

Costs, Subsidies, and the "Basket of Benefits"

In this section, without entering into a formal economic analysis, we will discuss some of the costs, subsidies, and social and economic benefits of organic coffee production. A formal cost-benefit analysis from private and donor points of view can be found in AICA Consultores (1997). We will then make an exploratory comparison of the benefits of organized organic coffee production compared to unorganized conventional coffee and shade-tree coffee. In the process, we will indicate some of the challenges at the level of the domestic economy that a transition to more intensive forms of coffee production require, and the significance of subsidies and social capital to that transition in the La Selva case.

Any transition to a more intensive form of management will place new demands upon the available land, labor, and capital resources of the small farmer household. In the case of La Selva, the costs in labor were considerable, and frequently exceeded available household resources, forcing the expenditure of cash. As figure 10.1 shows, the peak labor demands for the new economic activities associated with organic production coincide with the traditional labor demands of the milpa. From February to May, much household labor time is spent in preparing and planting the land. In the latter part of this period and beyond, organic coffee demands significant labor time in pruning, shade management, maintenance of compost piles, and construction of terraces. The second half of the year has less significant overlapping of labor demands.

The overlapping of demands on household labor was foreseen from the beginning of the project, and it was feared that it would constitute a serious entry barrier. In the first years of the project, the study discovered that not only did labor demands overlap, but that the labor requirements of the transition were substantially higher than originally projected. The initially estimated labor time was 164 days for conversion of 1 hectare of coffee, while the actual labor time was found to be 268 days, more than 60 percent higher. The costs for additional labor and training in organic production and quality management could not be met out of current incomes of the farmers. These costs were thus subsidized in two ways: (1) by a market price premium both as Fair Trade and as organic and (2) by direct training and organizational support subsidies from international organizations, the latter constituting investments in both human and social capital.

With reference to the price premiums, table 10.2 shows a comparison of the price to the producers that La Selva paid for organic and conventional coffee from 1992 to 1996, and the price paid for conventional by intermediaries in the same period. As the table indicates, members could receive two possible price premiums, one for selling conventional coffee to La Selva, and another for selling organic coffee

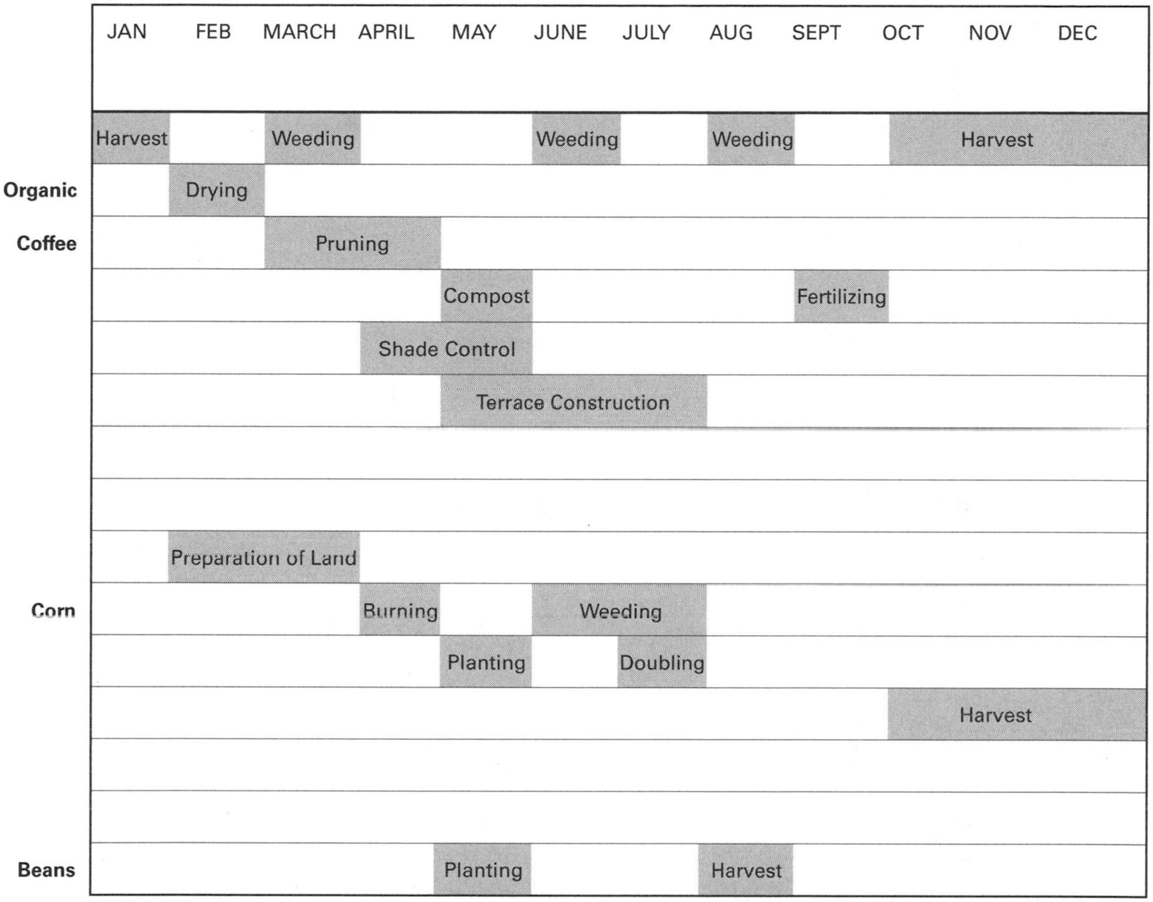

	JAN	FEB	MARCH	APRIL	MAY	JUNE	JULY	AUG	SEPT	OCT	NOV	DEC
Organic Coffee	Harvest		Weeding			Weeding		Weeding			Harvest	
		Drying										
			Pruning									
					Compost				Fertilizing			
				Shade Control								
					Terrace Construction							
Corn		Preparation of Land										
				Burning		Weeding						
					Planting		Doubling					
										Harvest		
Beans					Planting		Harvest					

Figure 10.1
Seasonal overlap of labor in production of organic coffee and milpa. Adapted from AICA Consultores, S.C 1997.

to La Selva, since European solidarity markets gave both Fair Trade and organic premiums. As the table indicates, in the 1992–93 harvest period, La Selva members received 51 cents a pound for conventional coffee and 58 cents a pound for organic, while non-members who sold non-organic coffee to intermediaries earned only 41 cents a pound, a 43 percent advantage for organic. However, in the subsequent years, due to a quota system in the European solidarity markets, La Selva struggled to retain any price differential in conventional coffee, and the organic premium declined to between 14 and 19 percent in the three subsequent years.

Table 10.2
Average coffee prices (per pound) to producers for organic and conventional coffee by La Selva and conventional coffee by intermediaries in program region, 1992–1996. Exchange rate used is average for harvest months (November–April); prices in dollars.

	1992–1993	Increase from conventional	1993–1994	Increase from conventional	1994–1995	Increase from conventional	1995–1996	Increase from conventional
Conventional coffee intermediary price	0.41	—	0.50	—	1.16	—	0.78	—
Conventional coffee La Selva price	0.51	24%	0.51	2%	1.20	3%	0.81	4%
Organic coffee La Selva price	0.58	41%	0.57	14%	1.32	14%	0.93	19%

In general, the gap between the price paid for organic coffee and conventional coffee grows larger when overall coffee prices drop, and has narrowed to as little as 5 percent when coffee prices have gone way up. When the premium for organic coffee goes down, as it always will, farmers are less likely to undertake the transition. Although there may be a variety of environmental and societal benefits, a private farmer may not view these benefits as his own. This is the classical problem of the "privately provided public good" (Field 1997). Most initiatives propose to pay for this privately provided public good through a market premium, as did the European Fair Trade markets and as do most biodiversity marketing efforts. However, in the case of La Selva and most other organic coffee producers in Mexico, the market premium was essential but not sufficient. Since most biodiversity-based marketing efforts focus exclusively on the search for markets, this is a crucial point. The transition to a higher-quality environmental product also required substantial training subsidies, drew upon the accumulated social capital of cohesive small farmer organizations, and are beyond what can be supported by any market premiums.

Extensive use of the farmer-to-farmer methodology for building human capital kept training costs lower than they otherwise would have been, but the costs of making the organization more effective were significant. For example, the IAF contributed $439,130 for training, marketing, and general organizational expenses over approximately a five-year period. This represented a subsidy of $336.76 to each of the 1,304 organic producers registered in 1995 over the entire period. In addition to this subsidy, there were also other subsidies from the Mexican government and other international organizations that have not been systematically quantified.

These subsidies to human capital, social capital, and organic production produced a series of private and public goods, which can be summarized as environmental (primarily public), social (public and private), and economic (public and private), discussed next.

Environmental Benefits The environmental benefits are suggested by field observations and the known environmental consequences of organic production. It is hoped that this study will stimulate further research by ecologists on these questions. The environmental benefits include the elimination of agrochemicals that enter organisms and watersheds, use of locally available materials for fertilizers, eliminating pollution from manufacture and transportation of chemical fertilizers, installation of small-scale terraces and other structures to enhance the formation and conservation of soil, and accelerated changes in the richness of soil organic matter. The farmers also note dramatic improvements in plant health, particularly in the abundance of foliage and the size of coffee plants, as well as productivity gains in the number and quality of fruits. Of crucial significance, some farmers also began experimenting with organic techniques in the milpa as well, creating the possibility that this technological package could extend to the whole farming system. This would suggest that conservation attention to intensify milpa farming could be an important strategy for promoting shade-tree coffee.

Social Benefits   The producers perceive a range of both economic and social benefits from organic production, revealed in a series of focus group sessions and surveys carried out with program participants, both men and women in separate groups. Ten focus groups with 147 producers, where discussions were transcribed and a survey administered, showed a high degree of satisfaction with the results of the organic coffee program. Seventy-three percent of those surveyed as part of the focus group felt that the higher price for organic coffee they received compensated for the additional work and labor needs. A resounding 97 percent thought that the program had benefited them in general. Most recognized the high entry costs, but are interested in continuing because they believe they will receive a permanent flow of higher income for less effort and expense, now that they are repositioned as high-value niche producers. Separate focus groups with women indicate that the additional income generated by the project has gone for food and basic articles such as soap, clothing, and shoes. Finally, the producers do not see the benefits of organic coffee in isolation, but as one of a "basket of benefits" they receive from the organization that includes credit, housing, emergency food, health programs, more production infrastructure, and trips to Mexico City for their children.

Economic Benefits   The subsidies and the fact that their organization was working on every step of the production and marketing chain brought a series of economic benefits to the producers which will only be indicated here. The benefits fall into three categories:

(1) Farm-level benefits, including increased return to producers, increases in yields (which averaged around 15 percent), and steady increases in the percent of the coffee plot dedicated to organic (producers added ¼ hectare annually, having a full hectare by the fourth year). The transfer of the technology to the milpa is also an economic benefit, as well as an environmental one, if it leads to productivity gains in corn production.

(2) Processing benefits. The forward integration opportunities created by the organization and the privatization of INMECAFE processing facilities meant that farmers, for the first time, were able to become owners of the *beneficio seco* (dry processing) phase of operations. This allowed them to reduce losses and increase efficiency and quality at the processing plant. They also established a tasting laboratory on the premises of the dry processing plant to increase quality control.

(3) Marketing benefits. These included increases in export sales prices due to better timing, increase in quantity of coffee placed on export markets, and new sales in the national markets due to the establishment of a small chain of coffee shops (*cafeterias*) owned by La Selva. As of 2001 there were six such coffee shops, four of them in Mexico City and one in Chiapas. The *cafeterias* have not yet resulted in much direct profit for the producers but these poor coffee farmers are the proud owners of their own chain of coffee shops. Thus, the original vision, conceived in 1988, of breaking through the chains of intermediaries to the final consumer has been realized, although the long-range economic viability of this strategy is yet to be determined. Technical problems in the collection of the data do not permit a calculation of the exact increase in income per producer because of organic production, but the indications are that modest real increases in average income were obtained and are projected for the future.

Almost none of these benefits would have accrued to small farmers acting individually. They came about because of collective action through an organization, a major contribution from social capital. The greater social and economic benefits for organic coffee come from the virtues of vertical integration and social benefits of being organized. Studies of shade-tree coffee, while they make reference to the importance of working with small producers and mention the cases of successful organizations, focus almost exclusively on the issue of market premiums and biophysical criteria such as structural and floristic diversity (Rice and Ward 1996; Greenberg 1997; Workshop 1999).

Conservation, Development, and Organic Coffee: Lessons for "Bird-Friendly" Coffee and Other Forms of Biodiversity Marketing

We have suggested that biodiversity marketing, in general, and shade-tree coffee marketing, in particular, will benefit from a broader understanding of how a major environmentally friendly agricultural product emerged in Mexico over the last two decades. A deeper analysis that incorporates how social systems and ecosystems have interacted and impacted the coffee landscape reveals both the potential and problems with using coffee as a vehicle for both social justice and biodiversity conservation, and suggests that conservation and development strategies that go far beyond certification and consumer consciousness campaigns are necessary. The crucial role of "social capital" provided by several decades of organizing efforts (government-induced in the early period, but much more autonomous in the latter period) is emphasized. The producers themselves perceive that they are receiving a "basket of benefits" from organizing that go beyond the price of organic coffee. The relative success of the program was not simply due to market signals, but to the fact that La Selva had achieved a significant degree of organizational consolidation before they began implementing the program. The organization also increases the producer's negotiating capacity within Mexican society and in the international economy, and contributes to the democratization of rural Mexico. The project made good use of traditional community structures, such as producers committees, as a basis for training paraprofessional agronomists from the ranks of community members. It is these organizational innovations, an accumulation of social capital, along with the impact of neo-liberal market reforms in Mexico,[13] which allowed Mexico to emerge, of all the coffee-producing countries in the world, as the leading organic producer, with almost all of that production coming from organized small producers, not individual farms (Martínez Torres 1997). This affirms that for any market-based transition to more sustainable agricultural landscapes to take place that "scaling up from one community to several, from community level to higher levels, is imperative" (Strum 1994). Likewise, the development of shade-tree coffee will depend upon the development of sustainable institutions at the community, regional and national levels (Murphree 1994).

The potential of sustainable coffee production to be one component of a diversified strategy for small, poor farmers to increase their well-being while conserving resources is clear. Our study suggests that organic coffee production is an important production alternative for those who are able to take advantage of it. Despite high entry costs, it permanently repositions small farmers to receive among the highest prices available in the market. The study also argues that coffee must not be isolated from the larger agroecosystem of which it is a part. It is fruitless to celebrate shade-tree coffee if the larger farming system of which it is an integral part continues to destroy forests. Organic can drive an entire agroecosystem toward a more sustainable and biodiverse path, while shade-tree alone only focuses on one component of the agroecosystem.

Thus, this chapter indicates the crucial importance of intensification of sustainable agricultural production in the milpa as a major priority for habitat conservation in the region, just as important as buying organic or Bird-Friendly coffee. Slash and burn agriculture is deeply rooted in tradition, and further training efforts are necessary to quicken the spread of organic methods into the milpa, as well as studies of the economic and social implications for the household of this activity. La Selva members continue to clear land for corn, even while declaring on questionnaires that they had met the conservation demands of the program.

While the general conclusion of this study is that organic coffee cultivation is an economically viable and sustainable development alternative, it is not a panacea and its limitations must be recognized. As we saw earlier, organic coffee appears to be primarily reaching the "middle" strata of small producers. The entry costs to organic production, even subsidized, appear to be too high for the smallest producers, but is a significant option for the slightly larger producers. This does not mean that "marginal producers are . . . excluded" from the benefits of organic production, as Raynolds (2000) has argued, since this "middle stratum" is still quite poor by any measure. But it does suggest the need for greater subsidies or greater economies of scale in trying to reach the below 2-hectare stratum with an organic strategy. And despite the environmental advantages of organic coffee and its potential positive impact on the larger agricultural landscape, there are still questions. The article pointed out that the interaction between social systems and ecosystems in Mexico resulted in a significant percentage of Mexican coffee being at lower altitudes not appropriate for high-quality coffee. If these low-altitude areas are also providing supportive habitat for biodiversity, then it raises difficult issues. Should farmers be encouraged to continue to cultivate an inappropriate and poor-quality crop because it provides habitat? Will the market subsidize this? Should it? Are there economically viable and environmentally friendly alternatives at these lower altitudes? Other environmental damage from coffee production is only beginning to be addressed, such as water contamination from the wet coffee processing (although there is an increasing tendency to use the pulp in compost).

Ecologists, social scientists, and green business experts will need to form research and action teams to understand how the social systems and ecosystems have interacted to produce the outcome of organic coffee, and what is required of those systems to provide both environmental protection and economic opportunity. Eco-labeling initiatives like Bird-Friendly coffee call for agricultural practices that incorporate most aspects of organic and Fair Trade, but go "beyond" them to incorporate far more systematic aspects of farm management conducive to a higher degree of biodiversity, such as buffer zones, hedgerows around farms, and greater use of pollution control technologies (Greenberg 1997b). The transition to the more intensive ecological management of coffee farms called for by Bird-Friendly coffee advocates and other biodiversity-based products will require intensive farmer-to-farmer training, subsidies (Gobbi 2000), and the organization of producers into more or less cohesive associations that can deliver a broad variety of benefits to their members. It must also be understood that few consumers will buy a product or pay a premium only because of its environmental or social values, it must also be of high quality (Millard 1996; Commission for Environmental Cooperation 1999; Rice and McLean 1999). Achieving higher quality will also require subsidies of various kinds and the social capital of organizations to make effective use of the subsidies.

Understanding and encouraging the social processes, from the household economies of small farmers to the values of middle-class consumers, that lead to successful sustainable agriculture and biodiversity marketing must be joined with our appreciation of ecological processes and a focus on eco-labeling criteria if we are to achieve more sustainable landscapes. In Latin America and elsewhere, organizations of small farmers are crucial vehicles for rural sustainability in coffee landscapes or any other kind. Greater attention must be given to building and strengthening them if eco-labeling initiatives are to have products to label. Small-farmer organizations are important to both biodiversity conservation and green coffee marketing.

Acknowledgments

We thank Robert Rice, Mahadev Bhat, and the anonymous reviewers for *Society and Natural Resources* for saving us from many errors of fact and interpretation. Thanks also to Christine Cairns for editorial assistance.

This chapter is reprinted, with permission, from the journal *Society and Natural Resources*.

Notes

1. The most significant distinction in coffee classification has traditionally been that between arabica (higher-quality) and robusta (lower-quality, used in blends) varieties, with numerous subspecies adapted to different microenvironments or with different qualities. However, in recent years the terms 'organic', 'shade-tree', 'Bird-Friendly', 'small-farmer' or 'Fair Trade', and 'sustainable' have emerged, both as categories salient for marketing purposes and as reflective of variations in production. 'Small-farmer' and 'Fair Trade' refer to solidarity marketing strategies, where niche consumers who are willing to pay a premium for coffee if some or all of their premium will go directly to benefit the producers (although elimination of intermediaries can also lower the cost of Fair Trade coffee). 'Organic' refers to coffee certified by third parties as produced without chemical inputs of any kind, as well as other associated production conditions. 'Shade-tree' and 'Bird-Friendly' (synonymous as currently used) refer to coffee produced in a particular kind of habitat, one characterized by a high number of shade trees and other forms of biodiversity. 'Sustainable' is more of an umbrella term; it refers to organic, shade-tree/Bird-Friendly, and some forms of small farmer production.

2. The International Coffee Agreements (ICA) were administered by the International Coffee Organization, and were agreements between producing and consuming countries to fix quotas on the exports of producing countries in order to stabilize prices (Talbot 1997).

3. It has been documented that Chatino coffee producers in Oaxaca sought out distant forests when they first began producing coffee in the 1950s and the 1960s (Hernández Diaz 1987).

4. The local organizations were called Economic Unions for Production and Marketing (UEPC-Uniones Económicas de Producción y Comercialización).

5. Organic coffee was produced on a private finca in Chiapas in Mexico, Finca Irlanda, going back to the 1960s.

6. See also Cadaval 1994; Hernández Castillo 1997; Hernández Castillo and Nigh 1998.

7. For a description of the many quality improvements in coffee production and processing made as part of the La Selva organic project, see AICA Consultores, p. 33.

8. For other reports on this study, see Murphy 1995; Plaza Sánchez 1997, 1998; Bray 1999a,b.

9. All three authors of this chapter worked for the Inter-American Foundation at the time this grant was made, two full-time and one as a contracted researcher.

10. The La Selva effort also included members of two additional ejido unions, Juan Sabines and Maravilla Tenejápa in the municipalities of Las Margaritas and La Independencia, also in the state of Chiapas.

11. Most coverage of the Zapatista uprising in Mexico has painted a picture of a polarized conflict between guerilla sympathizers and PRI traditionalists. Among other things, La Selva attempted a "third way": self-organized alternative production strategies. Many of the Zapatistas were farmers who quite literally chose the gun over organic coffee. See Bray 1999b.

12. These proportions clearly vary substantially even in coffee-producing regions of Chiapas. For example coffee farmers from the organization Indigenas de la Sierra Madre de Motozintla (ISMAM) in southwestern Chiapas have 62 percent of their land in coffee and 30 percent in corn (Heinegg and Ferroggiaro 1996).

13. Organic coffee cultivation and globalization is an implicit theme in this chapter. Much has been written about the impact of globalization on small farmers in Mexico and elsewhere, but one of the points of this chapter is that globalization is uneven in its impacts and can actually benefit some small farmers. The organic coffee producers described in this chapter, with great effort, have been able to position themselves to take advantage of new market opportunities. Most small farmers in Mexico have not been so well placed.

References

Aguilar-Ortiz, Félix. 1982. Estudio Ecológico de las aves de cafetal. In *Estudios Ecológicos en el agroecosistema cafetelera*, ed. E. Ávila-Jiménez. Xalapa. INIREB.

AICA Consultores, S.C. 1997. Evaluación del Programa de Producción de Café Orgánico en Las Margaritas, Chiapas, Mexico. Informe Final.

Berkes, Fikret, and Carl Folke. 1998. Linking social and ecological systems for resilience and sustainability. In *Linking Social and Ecological Systems*, ed. F. Berkes and C. Folke. Cambridge University Press.

Bray, David. 1991. Where markets and ecology meet: Organic coffee from the Sierra Madre of Chiapas. In *1991 in Review*. Inter-American Foundation.

Bray, David Barton. 1995. Peasant organizations and "the permanent reconstruction of nature": Grassroots sustainable development in rural Mexico. *Journal of Environment and Development* 4, no. 2: 185–204.

Bray, David Barton. 1997. Forest and protected areas policies in the Lacandon Rainforest, Chiapas. Paper presented at XX International Congress of Latin American Studies Association, Guadalajara.

Bray, David Barton. 1999a. Coffee that eases the conscience. *New York Times* editorial page, July 5.

Bray, David Barton. 1999b. A bird in the cup: Grinding towards social and environmental justice in the coffee world. *Orion Afield* 4, no. 1: 30–35.

Cadaval, Olivia. 1994. Coffee farmers and the politics of representation: What's in a good cup of coffee? *New York Folklore* 20, no. 3–4: 43–53.

Clay, Jason. 1992. Some general principles and strategies for developing markets in North America and Europe for nontimber forest products. In *Sustainable Harvest and Marketing of Rain Forest Products*, ed. M. Plotkin and L. Famolare. Island.

Commission for Environmental Cooperation. 1999. Measuring Consumer Interest in Mexican Shade-Grown Coffee: An Assessment of the Canadian, Mexican and US Markets. Commission for Environmental Cooperation, Montreal.

Crook, Carolyn, and Roger Alex Clapp. 1998. Is market-oriented conservation a contradiction in terms? *Environmental Conservation* 25 , no. 2: 131–145.

Fairbanks, Michael, and Stace Lindsay. 1997a. Changing Latin America's economic mind-set. *Miami Herald*, September 21

Fairbanks, Michael, and Stace Lindsay. 1997b. *Plowing the Sea: Nurturing the Hidden Sources of Growth in the Developing World*. Harvard Business School Press.

Field, B. 1997. *Introduction to Environmental Economics*. McGraw-Hill.

Folke, Carl, Fikret Berkes, and Johan Colding. 1998. Ecological practices and social mechanisms for building resilience and sustainability. In *Linking Social and Ecological Systems*, ed. F. Berkes and C. Folke. Cambridge University Press.

Fox, Jonathan. 1996. How does civil society thicken? The political construction of social capital in rural Mexico. *World Development* 24, no. 6: 1089–1103.

Gallina, S., S. Mandujaro, D. Gonzalez, and A. Romero 1992. Importancia de los cafetales mixtos para la conservación de la biodiversidad de mamíferos. *Boletín Veracruzana de Zoología* 2: 11–17

Gobbi, José A. 2000. Is biodiversity-friendly coffee financially viable? An analysis of five different coffee production systems in western El Salvador. *Ecological Economics* 33: 267–281.

Gómez-Pompa, A. 1997. Biodiversity and agriculture: Friends or foes? In *Proceedings of the First Sustainable Coffee Congress*, ed. R. Rice, A. Harris, and J. McLean. Smithsonian Migratory Bird Center.

Greenberg, Russell. 1997a. Why birds like traditionally grown coffee and why you should care. In *Proceedings of the First Sustainable Coffee Congress*, ed. R. Rice, A. Harris, and J. McLean. Smithsonian Migratory Bird Center.

Greenberg, Russell. 1997b. Criteria Working Group Thought Paper. In *Proceedings of the First Sustainable Coffee Congress*, ed. R. Rice, A. Harris, and J. McLean. Smithsonian Migratory Bird Center.

Heinegg, Ayo, and Karen M. Ferroggiaro. 1996. Inter-American Foundation Strategy in the Mexican Coffee Sector: A Case Study of ISMAM. Manuscript.

Hernández, Luís. 1991. Nadando con los tiburones: La Coordinadora Nacional de Organizaciones Cafeteleras. In *Cafeteleros: La Construcción de la Autonomía*. CNOC/IAF.

Hernández, Luis. 1997. Café: La Pobreza de la Riqueza/Riqueza de la Pobreza. In *Proceedings of the First Sustainable Coffee Congress*, ed. R. Rice, A. Harris, and J. McLean. Smithsonian Migratory Bird Center.

Hernández, Luis, and David Bray. 1991. Mexico: Campesinos and Coffee. *Hemisphere* 3, no. 3: 8–10.

Hernández Castillo, Rosalva Aída. 1997. Nuevos espacios organizativos y nuevos discursos culturales en el sureste mexicano: El catolocismo agroecológico entre los mames de chiapas. *Journal of Latin American Anthropology* 2, no. 2: 76–105.

Hernández Castillo, Rosalva Aída, and Ronald Nigh. 1998. Global processes and local identity among Mayan coffee growers in Chiapas, Mexico. *American Anthropologist* 100, no. 1: 136–147.

Hernández Díaz, J. 1987. El café amargo: diferenciación y cambio social entre los Chatinos. Instituto de Investigaciónes Sociológicas/UABJO: Oaxaca.

Johns, Andrew Grieser. 1997. *Timber Production and Biodiversity Conservation in Tropical Rainforests*. Cambridge University Press.

Martínez Torres, Eduardo. 1997. Outlook on ecological coffee farming: A new production alternative in Mexico. In *Proceedings of the First Sustainable Coffee Congress*, ed. R. Rice, A. Harris, and J. McLean. Smithsonian Migratory Bird Center.

Millard, Edward. 1996. Appropriate strategies to support small community enterprises in export markets. *Small Enterprise Development* 7, no. 1: 4–16.

Moguel, Patricia, and Victor M. Toledo. 1999. Biodiversity conservation in traditional coffee systems of Mexico. *Conservation Biology* 13, no. 1: 11–21.

Murphy, Ellen Contreras. 1995 La Selva and the magnetic attraction of markets: The cultivation of organic coffee in Mexico. *Grassroots Development* 19, no. 1: 27–34.

Murphree, Marshall W. 1994. The role of institutions in community-based conservation. In *Natural Connections: Perspectives in Community-Based Conservation*, ed. D. Western and R. Wright. Island.

Nestel, David. 1995. Coffee in Mexico: International market, agricultural landscape, and ecology. *Ecological Economics* 15: 165–178.

Nestel, D., and M. Altieri 1993. Diversity patterns of soil macrocoleoptera in Mexican shaded and unshaded agroecosystems: An indication of habitat perturbation. *Biodiversity and Conservation* 2: 70–78.

Nigh, Ronald. 1997. Organic agriculture and globalization: A Maya associative corporation in Chiapas, Mexico. *Human Organization* 56, no. 4: 427–435.

Nolasco, Margarita, et al. 1985. Café y Sociedad en Mexico. Mexico, D.F.: Centro de Ecodesarrollo.

O'Brien, Karen L. 1998. *Sacrificing the Forest: Environmental and Social Struggles in Chiapas*. Westview.

Ostrom, Elinor. 1999. Social capital: A fad or a fundamental concept? In *Social Capital: A Multifaceted Perspective*, ed. P. Dasgupta and I. Seragedlin. World Bank.

Pérezgrovas Garza, V., et al. 1997. El Cultivo de Cafe Organico en La Union Majomut. Red de Gestión de Recursos Naturales, Fundación Rockefeller.

Perfecto, I., and J. Vandermeer 1994. Understanding biodiversity loss in agroecosystems: Reduction of ant diversity from transformation of the coffee ecosystems in Costa Rica. *Entomology (Trends in Agricultural Sciences)* 2: 7–13.

Perfecto, Ivette, et al. 1996. Shade coffee: A disappearing refuge for biodiversity. *BioScience* 46, no. 8: 598–608.

Plaza Sánchez, José Luis. 1997. Conservación y Desarrollo Sostenido: La producción de café orgánico en Las Margaritas, Chiapas. In *Semillas para el Cambio en el Campo: Medio Ambiente, Mercados y Organización Campesina*, ed. L. Paré et al. Instituto de Investigaciones Sociales, Universidad Autónoma de Mexico.

Plaza Sánchez, José Luís. 1998. Organic coffee production and the conservation of natural resources in Las Margaritas, Chiapas. In *Timber, Tourists and Temples: Conservation and Development in the Maya Forest of Belize, Guatemala, and Mexico*, ed. R. Primack et al. Island.

Porter, Robert. 1997. Mexico's new coffee producer movement: A case study of the emergence of the Union of Indigenous Communities of the Isthmus Region (UCIRI) from Oaxaca. Paper presented at meeting of Latin American Studies Association Meeting, Guadalajara.

Porter, Robert. 2000. Politico-economic restructuring and Mexico's small coffee farmers. In *Poverty or Development: Global Restructuring and Regional Transformations in the US South and the Mexican South*, ed. R. Tardanico and M. Rosenberg. Routledge.

Power, Alison, G., and Alexander S. Flecker. 1996. The role of biodiversity in tropical managed ecosystems. In *Biodiversity and Ecosystem Processes in Tropical Forests*, ed. G. Orians et al. Springer-Verlag.

Raynolds, Laura T. 2000. Re-embedding global agriculture: The international Organic and Fair Trade movements. *Agriculture and Human Values* 17: 297–309.

Regalado Ortiz, Alfonso. 1996. *Manual para la Cafeticultura Mexicana*. INCA Rural, Mexico City.

Revel-Mouroz, Jean. 1980. Mexican colonization experience in the humid tropics. In *Environment, Society and Rural Change in Latin America*, ed. D. Preston. Wiley.

Rice, Paul D., and Jennifer McLean. 1999. Sustainable Coffee at the Crossroads. White paper prepared for Consumer's Choice Council.

Rice, Robert A., Ashley M. Harris, and Jennifer McLean, eds. 1997. *Proceedings of the First Sustainable Coffee Congress*. Smithsonian Migratory Bird Center.

Rice, Robert A., and Justin R. Ward. 1996. *Coffee, Conservation, and Commerce in the Western Hemisphere*. Smithsonian Migratory Bird Center and Natural Resources Defense Council.

SAGAR. 1996. Programa para el Café 1995–2000. Alianza para el Campo/Consejo Mexicano del Café.

Smithsonian Migratory Bird Center. 1999. Defining Shade Coffee with Biophysical Criteria. Workshop, Commission for Environmental Cooperation, Montreal.

Strum, Shirley C. 1994. Lessons learned. In *Natural Connections: Perspectives in Community-Based Conservation*, ed. D. Western and R. Wright. Island.

Talbot, John M. 1997. Where does your coffee dollar go? The division of income and surplus along the coffee commodity chain. *Studies in Comparative International Development* 32, no. 1: 56–91.

Toledo, Victor. 1991. Conservación y Desarrollo Sostenido en la Selva Lacandona: El Caso de las Canadas, Chiapas. Centro de Investigación sobre Energía y Desarrollo A.C.

Vandermeer, John, and Ivette Perfecto. 1997. The agroecosystem: A need for the conservation biologists's lens. *Conservation Biology* 11, no. 3: 591–592.

11

Serve and Certify: Paradoxes of Service Work in Organic Coffee Certification

Tad Mutersbaugh

In this chapter I will examine the paradoxical effects that transnational product certification standards have on the service workers who certify organic coffee in Oaxaca, Mexico.[1] These front-line service workers must bridge between two service modalities, standards-based certifier practices arising in transnational certification norms and *cargo* practices rooted in the indigenous service culture of Oaxacan coffee producer villages. Yet as these service workers seek to speak across this divide and make standards-based certification intelligible to certified parties, they risk making the results of certification work unacceptable to certifying agencies. "Certified" organic agriculture is a monitored food production system that organizes the movement of organic products from farm fields to consumers. Within certified organic food markets, worldwide production currently totals over $20 billion in sales and encompasses 10.5 million hectares (Willer and Yussefi 2000). Organic coffee holds a 2–3 percent US/EU market share (Rice 2001) and has $US75–125 million in US sales (Griswold 2000). At least 200 certifying agencies audit farmsteads and post-harvest processing, storage, and transport facilities across a global span, with 88 operating in the United States (OFRF 2000) and eight in Mexico (Gomez Tovar et al. 1999).

In this study I focus on monitoring aspects of certified agriculture. I analyze interactive service work activities (Leidner 1993) such as on-farm inspections and document reviews via the theories and methods of labor process studies (Burawoy 2000; Pred 1990). These monitoring activities—audits of producer organization documents and warehouse receipts, farmsteads, transport, production, and other sites along the certified product custody chain—document the integrity of certified organic products.[2] Certification-monitoring tasks do not, however, include organic production activities such as applications of organic soil amendments. Although these activities are important, their social effects are quite distinct from the service work activities studied here.

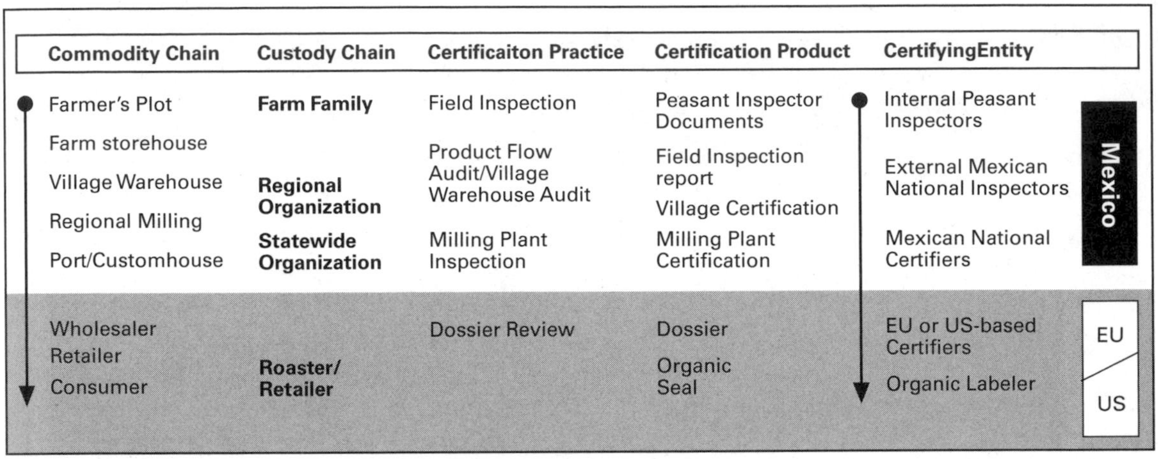

Commodity Chain	Custody Chain	Certificaiton Practice	Certification Product	CertifyingEntity
Farmer's Plot	**Farm Family**	Field Inspection	Peasant Inspector Documents	Internal Peasant Inspectors
Farm storehouse		Product Flow Audit/Village Warehouse Audit	Field Inspection report	External Mexican National Inspectors
Village Warehouse	**Regional Organization**		Village Certification	
Regional Milling	**Statewide Organization**	Milling Plant Inspection	Milling Plant Certification	Mexican National Certifiers
Port/Customhouse				
Wholesaler		Dossier Review	Dossier	EU or US-based Certifiers
Retailer	**Roaster/ Retailer**		Organic Seal	Organic Labeler
Consumer				

(Right margin labels: Mexico — upper band; EU / US — lower band)

Figure 11.1
Certified agricultural commodities and custody chains: the case of Mexican coffee.

Figure 11.1 shows how certification services are arrayed along a transnational "chain of custody" and documented by an audit trail. Oaxacan certified organic products are handled by separate EU and US networks. A Certimex/IMO-Control/Naturland link is regulated by EU 2092/91 rules, and an OCIA-Mexico/OCIA-International/OCIA-USA link by USDA National Organic Program (NOP) rules. Certifying agencies inspect adherence to organic norms but do not set norms; labelers and national bodies set norms but may not certify compliance. Rule-making is thus at once decentralized, with production standards produced and operations monitored by independent agencies set within competing certified organic commodity networks, and also unified, with certification standards increasingly "harmonized" (brought into agreement) under the ISO/IEC guides 65 (published in 1996) and 68 (published in 2002).

Before examining the paradoxes that attend certified agricultural service work, I think it important to signal its positive effects. First, certified organic coffee production has enhanced the livelihoods of certified organic coffee farmers—11,590 in Mexico, the world's largest producer of organic coffee (Bray 2002)—through an "organic price premium" that, depending on world market prices, as much as doubles the price paid to small producers (Rice 2001). Second, as this study will show, certified production promotes the development of a skilled work force at the village level and channels resources to educate and train village-based certification workers. A third, less obvious positive effect results from an associated "benefits scope economy." This scope benefit is realized because the tightly networked administrative structures—extension agents, marketing and administrative staff, governmental liaisons, secretaries, village-level inspectors and community technical officers required to sustain certified production—provide access to social and economic benefits unavailable to non-certified producers such as NGO- and government-based development aid, producer support payments, and medical care.

However, as certified agriculture provides benefits, it also provokes change. Financial and bureaucratic costs are introduced (Rice 2001; Mutersbaugh 2002), and new certification roles reshape inter-organizational linkages entities within organic agriculture alternative trade organizations (ATOs) (Whatmore and Thorne 1997; Hernandez-Castillo and Nigh 1998; Rice 2001). To examine the changes wrought by certification, I use the concepts and methods of labor process ethnography (Burawoy 2000; Lee 1998; Watts 1992; Pred 1990; Pringle 1989) to analyze certification as a form of interactive service work (Leidner 1994, Hochschild 1983) shaped by transnational institutions.

I will argue that two paradoxes attend the entry of certified agriculture into village social space, and affect the functioning of certification.

First, transnational certification standards embed a structural contradiction between rules that regulate the communication of information about certification procedures between parties—norms that dictate what may be said and to whom— and economic rules that prohibit the economic exclusion of poorer groups. In practice, although certification standards require that officers who judge norms be separate from those who explain and implement norms, the economic marginality of Oaxacan villages brings transnational strictures against economic discrimination into play, leading to a result in which certification service workers take on dual responsibilities that should otherwise be mutually exclusive.

Second, certification work embeds a paradox of performance. Organic coffee certification requires service workers to perform (transnational) standards-based service work within a local cultural milieu unfamiliar with this service modality: from either side, that is to say from either certifier or *cargo* service standpoints, the opposing service work modality and the actions of its workers often appear illegible. Given this incommensurability, service workers confront the difficult challenge of making their work legible to service recipients (certified parties) who must participate if workers are to accomplish their service duties. When, however, service workers modify standards-based service work to make it intelligible to service recipients, they run the risk of making their certification results unacceptable to certifying agencies. This difficulty is further compounded by structural conditions within Oaxacan organic coffee smallholder communities. In particular, the geography of remote agricultural fields limits surveillance and increases the cost of certification work. This makes it necessary to enroll local certification workers from Oaxacan indigenous coffee-growing communities to act as certification service workers in peasant inspector (PI) and community technical officer (CTO) capacities. Since Oaxacan village-based workers look to *cargo* work to inform their service work norms—and indeed certified agriculture makes use of *cargos* to appoint local certification service workers, this paradox of performance is internalized within producer communities. In this context, the resolution of structural contradictions and performance paradoxes falls to service workers. By exercising their creative agency within training programs and at multiple sites and administrative layers, workers find a means to bridge between service traditions.

To support these arguments, this research draws on case study material from a coffee producer's confederation in Oaxaca, Mexico. The CEPCO farmer's union (Coordinadora Estatal de Productores de Café de Oaxaca) manages organic sales, inspections, and trainings for about 5,000 smallholder organic coffee producer families of 31 regional organizations in 126 producer villages. They market certified organic coffee through Naturland (EU) and OCIA (Organic Crop Improvement Association–Mexico) certified coffee labelers, and hence make use of two Mexican national organic certifiers, respectively, Certimex and OCIA-Mexico. In the year 2000, I canvassed producer communities and attended inspections in five villages. Since then my ethnographic research has focused on tracing certification work up the chain, attending organic producer meetings in Oaxaca, Mexico (2001–2002), inspector trainings (2000, 2002) and conducting interviews with certifying agency officials. This was augmented in 2002 by a study of archived certification documents covering CEPCO certifications since 1994, and of transnational standards such as the ISO guides 65 and 68 and the USDA National Organic Program.

Service Workers and the Labor Process

From the standpoint of labor process ethnography, the present case is interesting for its focus on how two distinct service traditions—a *certifier* tradition emerging from international standards institutions and a *cargo* (service) tradition common to southern Mexican indigenous communities—overlap at the point of production in field inspections. The tensions arising in this overlap are both normative and practical, felt both as attitudes about how service work should mesh with community social relations and as material, spatio-temporal work qualities. Labor process ethnography is centrally concerned with the study of how social and work relations (class, gender) are materially shaped, contested, and reordered at the point of production, reproduction, and/or consumption, and is undertaken via an analysis of workplace tasks, labor organization, and contestation/cooperation. This approach views work as "a kind of technology itself, involving mastery of human interaction, labor allocation, and organizations as means of production" (Sayer and Walker 1992, p. 17).

Within labor process studies, the present research contributes to debates over how local cultural milieus provide templates for workplace relations such as labor management and for worker resistance. Industrial, gender-development and agrarian studies literatures each point to ways in which culture shapes social relations within factories and fields, with effects ranging from the accordance of economic advantage to important cultural players (Saxenian 1994; Wright 1997; Hernandez-Castillo and Nigh 1998) to providing modes of resistance for workers (Ong 1987; Fantasia 1988) or both (Chari 2000; Lee 1998; Hart 1991). Much of this research has focused on factory labor regimes, yet a growing body of literature has taken the insights of industrial studies and applied them to weakly surveilled agrarian and service sector labor processes that, while needing to conform to workplace rules forged within corporate headquarters and/or government agencies, unfold in agrarian contexts such as contract farming (Watts 1992; Freidberg 1997), international development projects (Schroeder 1999; Carney and Watts 1991), and service work (Leidner 1993; Hochschild 1983; Urry 1987).

Theoretically, product certification intersects questions of workplace surveillance, long a key area of investigation within labor process studies because the workplace forms a critical site within capitalism for the extraction of surplus value from workers bodies (Chari 2000; Wright 1997; Lee 1998; Pred 1990; Burawoy 1979; Marx 1867). Certification itself constitutes a form of surveillance in which the workers are to ascertain whether desired "organic" qualities have indeed been embedded into commodities and maintained during subsequent processing and transport, and so requires production monitoring. Yet from a labor process standpoint, agricultural product certification is doubly weak in that it combines difficult-to-surveil agricultural fields with a form of service work (field inspections) that is weakly tied to certifying agencies via contracts. This structural relation, then, directs attention to the front-line, interactive service workers upon whom certification depends for a reliable determination of whether a commodity possesses advertised qualities.

Within labor process studies, a subliterature examines interactive service work (Leidner 1993; Pringle 1989; Hochschild 1983; Steinberg and Figart 1999; England and Folbre 1999; Kunda and Van Maanen 1999; Leidner 1999). A focus relevant to certification is analysis of "routinization" practices used to get workers to act in the interests of service providers by producing consistent, repeatable results. Leidner's (1993) study identifies practices that include "scripting" of service interactions through texts and instruments that constrain workplace interaction, "industry training" programs that set out expected parameters of service work interaction, and, of particular importance to highly trained workers who must act with a high degree of autonomy (as is the case in field certifications) "transformation" practices that seek to instill a sense of professional identity in service workers so that they will desire to act in ways deemed appropriate by service providers. (See also Hochschild 1983.) In the present study, certifying agencies use each of these tactics in a bid to shape the actions of village coffee certifiers.

However, the control exercised by service providers over employees is far from complete (Leidner 1994; Urry 1987). Workers encounter spaces and moments within the labor process where they must exercise agency in adapting service provision to local conditions. Additionally, skilled workers in "information" services, of which certification employees provide an example, have leverage since service providers (1) need skilled, self-motivated workers that may be in short supply and must allow them leeway even as they require that services be provided in standardized and predicable ways (Busch 2000; Urry 1987), (2) need to have services carried out under varied local conditions which requires that employees adapt protocol to circumstance, and (3) must have services provided in a manner acceptable to clients who may have other provider options, and who bring set of expectations that, while reducing transactions costs (Busch 2000; Leidner 1994), may also constrain the leeway of service providers.

This study, then, extends labor process theory to workplaces decidedly outside of the industrial, intra-firm contexts to examine workplaces linked through social divisions of labor (Sayer and Walker 1992), governed by transnational work rules, and subject to the exigencies of interactive service work. From this standpoint, the study also speaks to network-oriented approaches such as actor-network theory (Whatmore 2002; Whatmore and Thorne 1997; Hernandez-Castillo and Nigh 1998; Latour 1999) by using political economic- political ecological methods to question whether and how service provision cultures operative within networks may shape the character, content and geography of those networks.

Transnational Certification Rules: Structural Contradictions

Since 1999, organic certification has intensified, broadening in scope to include a wider array of activities and requiring more labor inputs from village-based certification workers. To understand intensification and its impact upon front-line organic certification service providers, it is useful to examine the codification of transnational certification standards since the early 1990s. Although a full analysis of transnational certification rules is beyond the scope of this chapter, an overview of standards development will lend support to two contentions of this chapter, namely, that there has been an ongoing harmonization of rules (the latter term refers to the practice of having the many national codes agree in content), and that these rules have a hand in shaping village-level certification work.

At an international scale, recent years have seen global food certification standards become increasingly harmonized under the ISO (International Organization for Standardization) guide 65 and guide 68 rules. Thus, standardization receives its impetus from two factors, the increasing volume of international food sales driven by global appetites and cheap labor (Watts and Goodman 1997), and the food safety crises during the 1990s of which BSE (bovine spongiform encephalopathy, "mad cow disease") has become emblematic. However, the framework of global certification, both in terms of inter-institutional arrangements and legal authority, is shaped by global organizations dedicated to trade promotion. Harmonization of global standards—such as ISO/IEC Guide 65, EN 45011, NOP subpart E, and EU 2092/91—is enforced by a "strategic alliance" between the World Trade Organization and the ISO: WTO signatory nations must comply with trade standards (of which certification is an example) or risk trade sanctions under the Technical Barriers to Trade (TBT) provision (ISO 2000a; ISO 2000b), which provides a standards database and an enforcement structure.

Harmonizing Transnational Certification Standards

In recent years, harmonization of certification standards has resulted in an increasing conformity across national contexts. In particular, initiatives such as the ISO/IEC guide 68 (published in 2002) require recognition by all WTO signatory nations of all products, regardless of national origins, as long as they are certified under procedures harmonized to the ISO/IEC guide 65 norm.[3] The object of this section is not, however, to analyze the why of harmonization, but rather to trace the effects of global standards on field-level service work.

In the following paragraphs I will address standards harmonized from ISO/IEC guide 65 to the US NOP and then to certifying agency rules: these standards include organizational, documentary, and personnel norms and are designed, in the words of the ISO/IEC guide 65: 1996 (E), section 4.2 on organization, to "foster confidence in its certifications." Of particular interest are standards sections that create divisions within and between organizations, for example between commercial and certifying activities [section 4.2(l)] and between inspectors and evaluators [4.2(f)], and that affect communications between certifier and certified such as the [4.2(o)(2)] prohibition that certification bodies shall not "give advice or provide consultancy services to the applicant as to methods of dealing with matters which are barriers to the certification requested," a standard placing constraints on the [4.2(e)] admonition that "this [certification] structure shall enable the participation of all parties significantly concerned in the . . . certification system" and the [8.1.1]

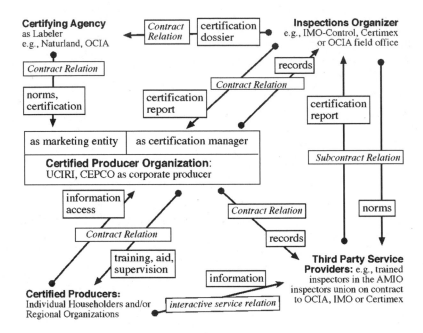

Figure 11.2
Contractual relations between certifying agencies, inspectors, and certified parties. Sources: fieldwork, Quality Guarantee Manual (Naturland n.d.).

rule stating that "the certification body shall provide to applicants an up-to-date detailed description of the evaluation and certification procedures." These conflict-of-interest and transparency rules keep parties from having a hand in certifications that directly affect them and make certification procedures clear to participants, and yet also place limits on conversations between certifier and certified, and between members within certified organizations who take on these differing roles.

US NOP standards incorporate similar language. The NOP website provides guidance for certified organic producers:

Section 205.501(a)(8) [NOP 2003a] requires that certifying agents provide applicants with sufficient information to enable them to comply with the OFPA and regulations. Section 205.501(a)(11)(iv) prohibits certifying agents from giving advice or providing consultancy services to certification applicants or certified operations for the purpose of overcoming barriers to certification. In other words, certifying agents must explain the regulations, but they cannot tell producers or handlers how to correct a noncompliance. . . . (NOP 2003b)

Enacting Standards in Certifier-Certified Contractual Relationships
The standards affect service work when certifying agencies use them to structure inter- and intra-organization contractual linkages.

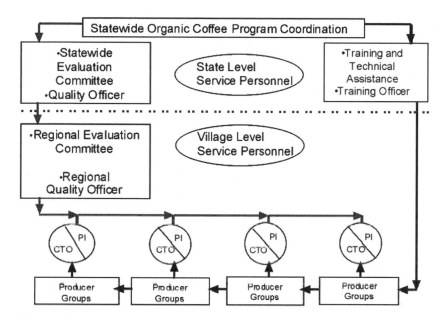

Figure 11.3
Organizational chart of village- and state-level certification roles. Adapted from PI/CTO certification training presentation.

Figure 11.2 illustrates the actual contractual and service relations engendered by transnational certification standards. Several points demonstrate the impact of certification standards on inter-organizational relations. First, certification requires that information exchange be unequal (as also noted in the NOP Q&A): certifying agencies provide norms and feedback including reports and controls; certified parties provide information, access to records, and compliance. Second, the key contract in the diagram is that between the inspections organizer (often a subsidiary of the certifying agency) and the certified producer organization. Certification standards set strictures on communications such that there is not supposed to be any communication outside of approved connections. This prohibition on conversation, however, extends itself through the "audit linkage" connecting certified producers/ certified producer organization/inspections organizer /certifying agency: this is illustrated on the diagram by showing the split between "certification manager" and "marketing entity" capacities within the certified producer organization. As a result (and as followed up in figure 11.3), communication channels within the producer union are disrupted. Third, the expansion of offices and communications channels required by transnational certification standards sharply increases the minimum organizational capacity necessary if organizations are to successfully certify. Producer group costs are increased, and, although the ISO Guide 65 states that "there shall not be undue financial or other conditions" placed upon applicants, producers often find certification standards burdensome because they increase many organizational costs and require expanded labor inputs that rest disproportionately on certified producers with sufficient educational and social capacity to perform these management functions (Mutersbaugh 2002b).

Structural Contradiction in Implementation of Standards

To see how standards *cum* contractual relations are expressed as workplace rules, we must examine the social and economic circumstances confronted by Oaxacan producers. The year 1999 was experienced as somewhat of a watershed by Oaxacan certified organic producers because it was then that ISO-mandated standards, taken to require annual certification of 100 percent of farms, came into full force (Michelsen 2001; Barrett et al. 2001).[4] Within the economic context of mass producer groups, the only means of bringing producers into conformance was to implement village-managed certification structures to survey 100 percent of plots, with these verified by a random sample of 10–20 percent of plots conducted by external, third-party inspectors on contract to the certifying agency.

Figure 11.3 illustrates certification standards-based structural relations implemented within villages: if producers are to earn higher organic premiums, they must work within this structure. Reading these relations, we see several aspects of interest. First, types of service workers increase to include quality officers, coordinators, training officers, evaluation committees, community technical officers, and peasant inspectors. Second, the bifurcated structure envisioned in transnational certification, with training and extension services separate from inspections and certifications is reproduced. At diagram right, organization lines trace training and technical assistance flow from the statewide organic coordinator office to village CTOs (community technical officers) who develop organic work plans for each village producer and provide village-level training. At diagram left, organizational lines trace monitoring and inspections from statewide offices down through regional quality officers to village-level PIs (peasant inspectors) who undertake inspections of certified organic producers in, as arrows indicate, neighboring villages to avoid interest conflicts. Third, the PIs and the CTOs are in fact one and the same PI/CTO person (represented by the split circles at the bottom of figure 11.3); this embedding of a dual role reflects the realities of village life in which workers with skills sufficient to undertake this labor are often in short supply, and the economic burden of supporting two, separate workers is beyond limited village resources. It also, however, embeds the structural contradiction between inspection and extension work.

Paradoxes of Performance in Interactive Service Work

This chapter has thus far traced the structural impacts of multilateral (ISO) and national (US NOP) standards on certifying agency and producer group organization. The chapter will now consider how these affect the performance of front-line service employees operating within a local cultural milieu. To this end, subsequent sections will consider *cargo* and *certifier* village-level service work, and examine ethnographic moments in which tensions between competing service modalities come to the fore. First, however, I would like to prefigure these tensions between service modalities by examining a discussion during a union meeting at which payment of salaries to PI/CTOs was proposed. Confronted with a high attrition rate among PI/CTOs, the agricultural extension staff of the CEPCO Organic Section offered salary support. Although the payments were small ($500.00 pesos/month about USD$55.00), the proposal provoked strong reactions at an Organic Coffee Producer Meeting in July 2001:

Extension agent: [The] technical assistance structure would . . . provide [salary for] community technical officer/peasant inspector. . . . But producers must finance this. . . . To speak in a somewhat bluntly: you'll [have to pay for] a CTO if you want to be in the (organic) program. . . ."

Coffee producer #1: I think that there's a simple solution. If we want to be in the organic program, it is best that each community select their technician, but as a social service, not as personnel working for money or tips. . . .

Extension agent: What is the biggest problem we have? . . . There aren't any [village] PIs/CTOs because they've gone to "el norte" [the United States]. They're in their jobs a month, half a year, a year, and they're gone. When the technical officer goes, the group has difficulty staying in the organic program. So, why don't we help . . . by [providing salaries]?"

Coffee producer #1: [In our] village we're not accustomed to the idea of paid employees. . . . This is the dynamic, the culture we've followed. . . ."

Heated arguments continued for quite some time, yet this selection points to key differences in how the practices and meanings of service work are understood: certifying agencies and extension agents view the wage relation as compatible with PI/CTO work; village-based producers, on the other hand, view this work as something best performed by unpaid *cargo* holders. Although some producers were certainly concerned about the bottom line, my discussions with producers after the meeting brought out two additional sentiments. One widely held view was that the proposed payment structure excluded many of those most active in organizing and managing village-level certifications. Although from a certifying agency vantage point the

PI/CTO office is the only essential village-level office, from a village standpoint this work is seen to require an extensive supporting cast (Mutersbaugh 2002b). A second concern expressed by attendees (and finding resonance in the words of producer #1 above) was that this work should be performed as a social service, that a receiver of "money or tips" could not be a leader. Village producers, I will argue, desire a service provider who will champion village interests above selfish or outside interests and who is accountable to villagers, a person that, in the *cargo* tradition, is "in front" of the community. In other words, PI/CTO duty in which accountability is to outside entities is not legible as a service when seen from within the *cargo* tradition, thereby complicating efforts to stabilize PI/CTO work.

Cargos as a Template for Village-Level Certification Service Work

Oaxacan coffee-producing villages make use of *cargos,* a form of unpaid service work performed by village members, to administer communal property and provide social services in indigenous communities. With the advent of certified organic coffee, this *cargo* service modality has been extended to articulate with transnational markets (Nigh 1997), and has become the principal governance institution used by Oaxacan coffee producer organizations.[5]

The structural importance of *cargos* is mirrored by their social importance, and it is important to consider how *cargo* holders are both constrained and enabled in the exercise of their duties. A case in point is that of Ramon who took on a "major" *cargo* as coordinator of a village development project. The general assembly appointed him to the post after two previous *cargo* holders had failed. As is generally the case with *cargo* holders, he was not able to refuse his post: failure to accept a *cargo* appointment can be sufficient cause for expulsion from the village. On the other hand, he was given the considerable authority accorded a major *cargo* holder, namely, the ability to request village approval for tequios (labor drafts or corvées), levy taxes to cover expenses, and jail any villager who failed to perform assigned tequios. Success, however, required Zu-Bin to make tremendous personal sacrifices that included making dozens of trips to Oaxaca City with little compensation, arguing for the general assembly to approve many, burdensome tequios which earned him the enmity of many villagers, and spending late nights poring over government funding request forms and accounting sheets. He was ultimately successful where others had failed because he had earned *el mando*, a respect relation in which an individual is held in high regard for having demonstrated a willingness to place village interests first and a capability to manage village labor projects in a manner that produces a product of acceptable quality.

Regional and village coffee organizations utilize *cargos* not only as a method to recruit administrators such as PI/CTOs, regional quality officers, and organization leaders but also as a template to inform ideas about service work assessment and accountability norms. Whatever the particulars of the local *cargo* structures, all hold in common the notions that quality norms should be locally determined and adjudicated and that group members are accountable to the social body. *Cargos* thus provide a set of interlocking norms and practices ("our culture" of service, in the words of the coffee producer cited above) that makes work legible and intelligible. Within this context it is difficult to separate work performed as a *cargo* from the associated social labor norms.

Certification Service Work

The next three subsections make use of participant observation data to trace certification service work through, respectively, training of PI/CTOs, field inspections by PI/CTOs, and interpretation of external field inspection results. In each of these instances, document production plays a central role. As one agricultural extension officer remarked, with only minimal irony, "organic coffee is documents." The three sections highlight the tension between, on the one hand, the standards-based structural schema of certification service work, and, on the other, the cultural context within which certification service work is performed. Within this context, prominent dynamics include unequal power relations, particularly in information transmission and work interpretation, and negotiation over how, when, and where standards-based certification allows haggling over the context and meaning of work. These dynamics figure in certification service work for despite codification and routinization under the aegis of transnational standards, it remains a performance subject to the ambiguities of local conditions, the exercise of service worker agency, and expectations of service consumers.

Training PI/CTOs
As with any village *cargo*, PI/CTOs are appointed to their posts by a village assembly. However, in contrast to other *cargo* holders PI/CTOs must attend a certifier-approved PI/CTO training course and then pass an exam in order to receive certifying agency approval. The names of individuals who have passed required courses are held in databases: any organic inspection report signed by an individual not on this list will be rejected. This policy results in the exclusion of villagers who do not have the requisite skills to pass exams even though they may be proficient organic farmers. A case in point is that of Roberto (not his real name), a committed organic farmer who has not only passed each inspection of his organic coffee plots, but has also reclaimed land for shade coffee from that cleared for goat pastures, a difficult process requiring farmers to carry water to shade saplings on the hot, dry

slopes of Oaxaca's Mixteca region. He was proud to be named to a PI/CTO post by his village-level organic producer organization and welcomed the opportunity to be "in front" of his community. Unfortunately, although he attended the trainings he was unable to pass the written exam since his command of Spanish was only at a conversational level, and has since worked as a regional organic production coordinator. This exclusion was doubly felt because it jarred with the sense of community empowerment implicit in having one's *mando* recognized, and also because it deflated his sense of mission, widely felt by farmers who see organic agriculture as part of a "greater indigenous identity, working like our grandparents" in which organic farmers have "the duty to be better informed than others" and "not even think that we are organic until we have taught ecology to our children" (comments made during an organic producer's meeting, February 8, 2000).

During 2000 and 2001 I attended PI/CTO trainings that were repeated at various sites around the state of Oaxaca to minimize attendee travel time and cost. Trainings used a certifier-approved template and focused on certification standards and document production. For instance, producers were (humorously) enjoined from conflicts of interest: "It is incorrect to think 'I can inspect my relatives and they all pass, but if it's my enemy, well . . . I'll tell the external inspector that he/she applied herbicides.'" This injunction was supported by hours of discussion covering norms and sanctions, and their application to common problems such as what to do if the registered producer migrates to "el norte" to work and leaves his/her spouse who is not registered as an organic farmer in charge of the plot. This insistence on standards was balanced, however, against a sense of social solidarity with organic farmers in Oaxaca and across the globe. Oaxacan organic farmers were understood to be united in the performance of tasks agreed to in somewhat difficult negotiations with northern certifiers, the latter characterized somewhat ambiguously in a February 25 training session: "Europeans, like Gringos, are very rigid, but are also flexible at times."

In document production, trainees were taught to fill out forms both during the lecture portion that takes about a day, and in a subsequent field practicum. Each of the eleven basic village-level forms used at the time was covered in detail, the trainer filling in an overhead as the trainees filled in practice forms. Trainees also trained in pairs, practicing the technique of holding forms so that the person in the role of the service recipient (certified party) sat opposite in standard interview fashion. These practical points were emphasized at length. On February 25, 2000, a trainer said: "If we don't have documents, our work in the fields doesn't serve us." This sentiment was expressed even more forcefully at an April 2000 producer meeting on document production: "Organic coffee is documents! Has anyone said that you don't know how to produce organic coffee? Has anyone said that you don't know how to process and warehouse organic coffee? What [certifiers] have said is that you don't know how to fill out forms!" But in June 2001 the trainer sounded a cautionary note, seeking to couch document production in terms of organic producer solidarity: "If we judge strictly, many of our compañeros will be broken. If our goal is that we all advance, and we do just a single evaluation, many will fail. We are going to have to reunite many times."

Trainings invariably concluded with a discussion of PI/CTO salaries. On February 8, it was noted that pay should be "dependent upon the quality and timeliness of work" but that "there is a norm which states that the inspector cannot receive pay directly, but only from a source not connected with the inspected party." In other words, PI/CTO pay is to be based on the quality of the work, but quality is to be judged and compensated by an external agency.

Trainings, then, expressly grapple with the tension between transnational certification standards and their enactment in local service work. Service workers are prepared to follow scripts (prepare documents in a routinized manner) and warned against deviating from the script for friend or foe, and yet are also called upon to make multiple efforts to get producers certified. This calls really upon both sides of the *cargo*/certifier service work paradigm, with standards-based scripting performed by *cargo* holders in a context of social solidarity to pull co-producers through the process. This also, I might add, increases certification labor costs and shifts them to PI/CTOs, who, as we have seen, face financial uncertainties. Further exacerbating this burden, testing requirements exclude many who wish to accept this work for the social good of their organizations as requirements of certifier approval for PI/CTOs shifts work to a relatively small group of skilled workers within villages. At the same time, recalling the experience of Roberto, it generates a feeling of injustice among those who "flunk out" of the program: this negative external (certifying agency) assessment of personal worth runs both against the organic farmer's sense of mission grounded in a concept of indigenous earth stewardship and against a *cargo* experience in which personal worth, embedded in *el mando*, is determined by one's community.

PI/CTOs: Field Inspections and the Negotiation of Service

A second pivotal moment in certification service work concerns the performance of field inspections. These take place in two stages, first as internal inspections performed by PI/CTOs (contract relation #3 in figure 11.2), and second as external inspections performed by independent inspectors on contract to certifying agencies (contract relation #1 in figure 11.2). This section examines the first, internal phase in which PI/CTOs play a key role in adapting certifier/labeler-sponsored norms to the lived realities of village coffee production, documenting organic production plans and making producers aware of international norms such as biodiversity maintenance, organic composting and wastewater control.[6]

One of the striking aspects of my participant observation studies was to see how PI/CTOs work in two-person teams to organize inspections and facilitate information transfers. The PI, following certification standards, visits from another community outside of the inspective village's regional organization. In the village, the PI joins forces with the local CTO who organizes records and more importantly, calls upon certified parties to be present in their plots and serves as a guide during field inspections. This latter point is important, for certified parties treat the inspections as a form of tequio: the CTO selects certification participants from a list and informs them of their duty to be present for inspections. Following transnational standards, PI/CTOs acting in their PI role during cross-inspections of other producer organizations are permitted to explain how the inspections process functions but not what producers may do to correct deficiencies. However, PI/CTOs acting in their CTO role in home villages may explain deficiencies and how to correct them. By operating in local-extralocal teams, PI/CTOs may overcome the injunction against explaining how norms may be met during inspection.

Instances during an internal inspection in 2000 demonstrate these mutually supporting PI/CTO roles. During visits to organic coffee plots, the PI pulled out producer documents from a dossier that he had brought along in an embroidered shoulder bag. Juxtaposing forms including the producer's "organic improvement plan" form filled out in consultation with the local CTO the year before, the blank certification inspection form, and last year's inspection results—and holding these records so that the producer could not see them—the visiting PI would question the producer. On occasion a producer would turn to the local CTO, seated with the producer, and ask why the question was being asked. (Producers considered the requirement that they state how much coffee they had sold the year before when they knew that PI could see the figure recorded on last year's inspection results to be particularly odd.) The CTO would explain that that is how inspectors must act, that they need to perform their tasks in the manner according to their training. The CTO performed other tasks as well, spending a portion of nearly every plot visit performing cultivation tasks such as pruning and brush clearing, tasks performed mirroring points raised during inspections. On one occasion, the PI, referring to the organic improvement plan, asked a plot owner where the 200 *recepas* (a radical pruning in which the trunk is cut through at 40 centimeters above the ground so that it will re-sprout into a productive coffee bush) were to be found. At this, the CTO pulled out a saw and cut through the trunk of a nearby coffee tree.

These instances, I would argue, constitute a significant change in local modalities of service provision. In retrospect, what I have found perhaps most surprising is the degree to which service providers focus on teaching certified parties how to consume standards-based certification services and assert the "organicness" of their plots within this document-based context. Although from a US or EU perspective script-based inspections may seem a normal service modality, they are not experienced as such by certified parties from a radically different service culture. In sum, while certification instills new styles of authoritative practice into village service work, these remain bound to notions of service in which plot owners expect *cargo* holders to negotiate quality with certifiers. Village-based service workers anticipate these expectations and spend substantial time with producers drawing up documents and explaining how inspectors examine documents during limited, random-sample inspections. Given the frequency of documentation failures and consequent need to re-inspect, this joining of PI/CTO forces provides leverage.

Negotiating Results: The Post-Inspection Interpretation of External Inspections

This final empirical section takes up the "external" phase in the certification process. After PI inspections have been undertaken and results reviewed by a producer-union review panel, an external certifying-agency-approved inspector is contracted. The external inspector travels to the village, examines a random sample of coffee plots and undertakes an exhaustive review of documents including coffee warehousing, transport, and marketing receipts. The end product is an inspection report that documents material findings and informs a later determination, by first national and then international certifying agencies, of whether an organization and its producers are to "pass" and receive harvest and/or individual organic producer certifications.

Incidents observed during an external inspection, conducted in three villages within a regional organization (RO) during a three-day period in June 2000 again demonstrate the tensions between standards-based certification and certification service workers. The circumstances of this particular inspection were difficult: the previous year's certification had failed, the previous leaders had resigned, and the new regional organization leaders found themselves in a difficult spot. Perhaps for this reason, the external inspector maintained strict professionalism. Each village inspection began with farmstead visits: our group included the external inspector, an RO officer, a CTO from the village, Mexican government SINDER[7] extension agents assigned to provide organic coffee extension services, this researcher, and the farmers whose plots were to be inspected.

Our group climbed through organic plots on the surrounding slopes, checking coffee trees as we wended our way among mossy boulders strewn under a high shade canopy. At each plot, the owner participated in a lengthy survey of farm operations, tenure status and production quantities, answering questions while the inspector consulted an archive of previous certification results carried by the local CTO. The inspector traversed each coffee plot looking for signs of chemical use, inspecting soil quality, shade intensity and shade-tree diversity, and noting companion planting arrangements. During the three days of inspections (23 producers each with multiple plots inspected), producers passed inspections without failures: there were, however, many potential problems requiring review and possible future correction. As with internal inspections, the external inspector could not explain how faults might be addressed. It was possible, however, for the external inspector to speak with the SINDER agents who had a contractual relation that falls outside of the figure 11.2 schema, since they were employees of the Oaxacan state government and since they were with this researcher. Although careful not to speak within earshot of certified parties, the external inspector did discuss organic norms within the context of local agricultural practice with us.

After spending a day inspecting organic coffee farms, we returned to the village's certified organic warehouse to wait as producers brought in personal copies of paperwork to check against organization records. As we rested on quintal sacks of coffee, the inspector's check found that producers had sold a small quantity of certified coffee as non-certified coffee. The producers were surprised that there was anything wrong with this act: while it is remiss to incorporate non-certified coffee into a certified coffee shipment, what could be wrong with the reverse act of selling higher-quality certified coffee as a non-certified type?

The inspector explained that certifying agencies might take the incident as evidence that villagers were engaging in "parallel production" in which a certified producer also cultivates non-certified plots. Agencies routinely compare vendor lists, organized by village, of certified versus non-certified sales. If a name appears on both lists, the producer is assumed to be engaging in parallel production and certifiers must decertify the producer and the regional organization's harvest, or require clarifications and (expensive) follow-up inspections. In response, producers pointed out that the suspect sales occurred at harvest's end when remaining quantities of certified coffee were insufficient to justify the expense of a separate shipment (certified coffee must be segregated from non-certified coffee during transport). A producer held out her sales slip as evidence, pointing to the sales date and the fact that only 11 kilograms had been sold. "How," the inspector responded, "can you risk your organic certification for only eleven kilos of coffee?" Plainly troubled, he suggested that village organization members produce a memo providing a rationale for these actions.

That night the external inspector worked with RO officials until 3 A.M. cross-checking documents such as coffee warehouse receipts, warehouse log entries and producer dossiers against each other and with union records. Despite these efforts, and despite the fact that no producer lost organic certification, the lack of a clear audit trail documenting the passage of coffee from field to processing plant resulted in a less-than-satisfactory report and a subsequent need for a costly re-inspection.

What is interesting in this example is that, again, service workers are constrained in their actions by transnational standards, yet they find creative ways to provide assistance within these limits. In this instance, the external inspector was acutely aware and respectful of the principles of standards-based certification and the constraints to information exchange that they impose, carefully limiting comments made to certified parties to explanations of process and appeals permitted under standards-based certification. In the example above, the inspector explains what the problem is, and how to appeal it, but does not to indicate what rationale should be used in the memo nor say whether it would be successful. Earlier, confronted with potential issues in organic farming practice, he engaged in "off-stage" conversations with government-employed agricultural extension agents with whom he was able to speak more freely about problems arising in discussions of organic agriculture, thereby recreating at a different contractual level the PI/CTO strategy of information exchange.

Conclusion

I have examined how transnational certification standards are enacted within organic inspections and have found that this process generates paradoxes with respect to the communication of information vital to organic agriculture: standards-based certification requires that certification service employees not communicate with certified parties, and yet the success of certified organic agriculture depends upon precisely this communication. Fortunately, however, certification also provides new spaces for the agency of certified organic service workers who are able to resolve this paradox at each organizational level. In villages, local service workers find ways to bridge between *cargo* and certifier service traditions, thereby making the process legible to service consumers unfamiliar with standards-based service work; at the regional level, external certifiers also find a means to communicate with service consumers (certified parties) without compromising the integrity of standards-based certification.

With respect to labor process theory, this study signals the importance of culture to an understanding of interactive service work. Since the publication of Burawoy's 1979 book *Manufacturing Consent*, a focus of labor process studies has been on the question of why there is so little conflict on the shop floor: however, where Burawoy looked to labor organizational logic for the answer, recent research on workplace conflict, cooperation, and solidarity has looked to culture. The present study affirms this focus on culture, and yet highlights the role of the consumer and her culture as a shaper of workplace social relations. In this study, standards forged in international institutional milieus become codified in service work modalities. What is different about certification workplaces, however, is that work is only successful if and when consumers consent to standards-based interactive service, collaborating in production of documents and surveillance of their formerly independent workspaces. Farmers become, in effect, participants whose labor forms a necessary part of the service dyad: that they are able to integrate effectively, however, is a testament to the creative use of *cargo* modalities to organize and inform farmers.

Service workers nevertheless confront a conundrum. They must rework certification modalities to make them intelligible to local service consumers, yet if service workers alter the transnational template by too great a degree, their work becomes illegible to transnational certifying agencies. In this instance, service work becomes as much a process of transposing standards into local practice and vice versa as of gathering data. Fortunately, service workers are able to construct parallel channels for transmission of necessary information and train consumers in the act of service consumption. Returning to Burawoy, we see that service workers manufacture the consensual relations necessary to service consumption (which is inextricably bound up with production in interactive service work). The trouble with this schema, the contradiction that may undercut it, is that this service work arrangement is both expensive and exclusive. It draws upon *cargos* to mobilize village labor, and yet the requirement of certifier approval limits participation and challenges the underlying concept of merit via *el mando* that legitimates and sustains *cargo* service.

By understanding the diversity of service work cultures, we may better see that the contemporary standards-based certification service paradigm based on professionalization, scripting, and information restrictions is itself a cultural construct of recent origin. And, as this research shows, it has the inadvertent effect of constraining crucial inter-organizational communication. Organic agriculture ATOs have come to rely upon standards-based services, yet by casting network links in these terms they hazard introducing the notion that commonplace negotiations between certifiers, labelers, and producers within ATOs may be unethical. This runs against the social-movement ethic that motivates organic agriculture (Rice 2001), and may undercut its long-term vitality (see Michelsen 2001). No matter how this internal paradox is resolved, in the near term the PI/CTO service jobs crucial to certification are under pressure from historically low coffee prices and the consequent increased migration of trained personnel. This creates a local concern with no easy solution:

village certification services probably cannot survive on *cargo* labor sans incentives, yet changes to *cargo* service modalities run the risk of undercutting the service ideals that sustain it.

Notes

1. The certification service relations associated with certified organic coffee are more complex and greater in scope than those found to date in other certified agricultural systems such as Fair Trade (Raynolds 2000, 2001). Hence, the findings in this chapter must be selectively applied to agricultural certification systems such as sustainable harvest timber (Dudley et al. 1997), Mesoamerican Biological Corridor products (World Bank 2000), Bird-Friendly coffee (Rice 2001), carbon sequestration (Brown 2002), and non-genetically modified organism foods (Roseboro 2002).

2. Certifications include "proof-of-origin" checks to determine that foodstuffs originate on certified farms, "proof-of-content" analyses to show that required organic, biodiversity-conserving, and socially just production methods were used, and "proof-of-location" studies to ensure that certified products are not mixed or contaminated along the farm-to-consumer chain.

3. Three types of comparisons are possible. (1) Degree of conformance: Are ISO/IEC Guide 65, EN 45011, EU 2092/91, and NOP rules substantially in agreement? (2) Genealogy: What is the relation between standards such as the EU 2092/91, enacted in 1993, the 1996 ISO/IEC Guide 65 rule (ISO 2000b), and 2002 US NOP rules? (3) Social histories: Are certification standards linked to, for instance, the EU food safety crisis and/or regulatory dynamics?

4. Beyond structural constraints imposed by the ISO, additional standards and economic dynamics affect organic coffee certifications. To pursue the causes of this intensification further, I undertook a review of archived certification reports from 31 regional certified organic coffee producers' organizations over the period 1994–2002. Though a detailed discussion of changes to certification since 1994 is beyond the scope of this paper, my review indicated that requirements have progressively increased in stringency since at least 1996.

5. *Cargos* represent a form in indigenous governance that is widespread throughout Mesoamerica, ranging from Guatemala in the south through Jalisco in western Mexico. The greatest concentration of *cargo*-governed communities is found in Oaxaca and Chiapas, also the principal Mexican coffee-producing areas (Chance 1989). However, this governance form varies through time and space, with differing degrees of administrative labor associated with it, particularly at its geographic margins. This system is thought to have arisen in the sixteenth century and to have gone through significant changes (Cohen 2002 and 2001; Chance 1989; Chance and Taylor 1985; Cancian 1992). *Cargo* participation is generally thought to produce a "leveling" dynamic that dampens economic differentiation by forcing them to divert their labor power to community service, although the generally applicability of this principle is disputed (Chance 1989). Since the mid 1980s the "brokering" function of *cargos* has been on the ascendancy as *cargo* posts have provided a means of liaising with the Mexican government and with NGO funding sources (Rus 1994; Nigh 1997; Cancian 1992). More recently, as governmental funding has become increasingly tied to projects, the need for *cargo* holders with education and networking skills has increased. During the same

period, however, migration reshaped *cargo* work, in some cases encouraging the entry of women who remain resident in villages and who can manage networks to tap NGO support (Mutersbaugh 2002; McDonald 1999; Fox and Aranda 1996; Hernández 1991).

6. Oaxacan smallholder coffee producers are uniformly organic: certification failures result from a failure to document the origins and movement of coffee through the organization. In a consumer-health sense, peasant farmers are fully organic in that they do not use chemicals in production; in the environmental sense, studies find that coffee farmers build a high degree of biodiversity into their plots (Beaucage 1997; see also Padoch et al. 1998) except when induced not to, as in villages where the coffee parastatal INMECAFE (Mexican National Coffee Institute) convinced producers to cultivate simplified low-biodiversity monocultural "sun" and commercial polyculture (coffee with leguminous trees) production systems (see Nestel 1995; Moguel and Toledo 1999).

7. Sistema Nacional de Capacitación y Extensión Rural Integral (SINDER).

References

Barrett, H., A. Browne, and P. Harris. 2001. Smallholder farmers and organic certification: Accessing the EU market from the developing world. *Biological Agriculture and Horticulture* 19, no. 2: 183–199.

Beaucage, P. 1997. Integrating innovation: The traditional Nahua coffee-orchard (Sierra Norte de Puebla. Mexico). *Journal of Ethnobiology* 17, no. 1: 45–67.

Bray. D., J. Sánchez, and E. Murphy. 2002, Social dimensions of organic coffee production in Mexico: Lessons for eco-labeling initiatives. *Society and Natural Resource* 15, no. 5: 429–446.

Brown, S. 2002. Monitoring, certifying and commercializing sequestered carbon in hillside areas with indigenous rural communities. Paper given at Carbon Sequestration in Hillside Areas conference, Oaxaca.

Burawoy, M. 1979, *Manufacturing Consent: Changes in the Labor Process under Monopoly Capitalism*. University of Chicago Press.

Burawoy, M. 2000, *Global Ethnography*. University of California Press.

Busch L 2000, The moral economy of grades and standards. *Journal of Rural Studies* 16, no. 3: 273–283.

Cancian, F. 1992. *The Decline of Community in Zinacantán*. Stanford University Press.

Carney, J., and M. Watts. 1991. Disciplining women? Rice, mechanization, and the evolution of Mandinka gender relations in Senegambia. *Signs* 16, no. 4: 651–681.

CEPCO. 1996. Coordinadora Estatal de Productores de Cafe Oaxaqueños Annual Report.

Certimex. 2000. Normas para la Produccion y Procesamientos de Productos Ecologicos.

Chance, J.1989. Changes in 20th century mesoamerican *cargo* systems. In *Class, Politics, and Popular Religion in Mexico and Central America*, ed. S. Dow. American Anthropological Association.

Chance, J., and W. Taylor. 1985. Cofradias and cargos: An historical perspective on the Mesoamerican civil-religious hierarchy. *American Ethnologist* 12, no. 1: 1–26.

Chari, S. 2000. The Agrarian origins of the knitwear industrial cluster in Tiruppur, India. *World Development* 28, no. 3: 579–599.

Chayanov, A. [1925] 1986. *The Theory of Peasant Economy*. University of Wisconsin.

COAB. 2000. Factsheet: Standards & ISO. http://www.coab.ca.

Cohen, J. 2001. Transnational migration in rural Oaxaca, Mexico: Dependency, development and the household. *American Anthropologist* 103, no. 4: 954–967.

Cohen, J. 2002. Migration and "stay at homes" in rural Oaxaca, Mexico: Local expression of global outcomes. *Urban Anthropology and Studies of Cultural Systems and World Economic Development* 31, no. 2: 231–259.

Contreras Murphy, E. 1995. La Selva and the Magnetic pull of Markets: Organic coffee-growing in Mexico. *Grassroots Development* 19, no. 1: 27–34.

Crewe, L. 2001. The besieged body: Geographies of retailing and consumption. *Progress in Human Geography* 25, no. 4: 629–640.

Daily, G., P. Ehrlich, and G. Sánchez-Azofeifa. 2001. Countryside biogeography: Use of human-dominated habitats by the avifauna of southern Costa Rica. *Ecological Applications* 11: 1–13.

Daniels, R., M. Hegde, N. Joshi, and M. Gadgil. 1991. Assigning conservation value: A case study from India. *Conservation Biology* 5, no. 4: 464–475.

de Janvry A. 1981. *The Agrarian Question*. Johns Hopkins University Press.

Dudley, N., C. Elliott, and S. Stolton. 1997. A framework for environmental monitoring. *Environment* 39: 16–20, 42–45.

England, P., and N. Folbre. 1999. The Cost of caring. *Annals of the American Academy of Political and Social Science* 561: 39–51.

Fantasia, R. 1988. *Cultures of Solidarity*. University of California Press.

Fox, J., and J. Arranda. 1996. Decentralization and Rural Development in Mexico: Community Participation in Oaxaca's Municipal Funds Program. Center for US-Mexican Studies, University of California, San Diego.

Freidberg, S. 1997. Contacts, contracts, and green bean schemes: Liberalization and agro-entrepreneurship in Burkina Faso, *Journal of Modern African Studies* 35: 101–128.

Gereffi, G. 1999, International trade and industrial upgrading in the apparel commodity chain. *Journal of International Economics* 48: 37–70.

Gobi, J. 2000. Is biodiversity-friendly coffee financial viable? An analysis of five different coffee production systems in western El Salvador. *Ecological Economics* 33: 267–281.

Gomez Tovar, L., M. Gomez Cruz, and R. Schwentesius Rindermann. 1999. *Desafios de la agricultura organica: Comercializacion y certificacion*. Grupo Mundi-Prensa.

Greenberg, R., P. Bichier, and J. Sterling. 1997. Bird populations in rustic and planted shade coffee plantations of eastern Chiapas, Mexico. *Biotropica* 29: 501–514.

Griswold, D. 2000. How much is the market potential for sustainable coffee? *Gourmet Retailer*, November.

Hart, G. 1991. Everyday resistance: Gender, patronage and production politics in rural Malaysia. *Journal of Peasant Studies* 19, no. 1: 93–121.

Hartwick, E. 1998. Geographies of consumption: A commodity-chain approach. *Environment and Planning D* 16: 423–437.

Hernández, L. 1991. Nadando con los Tiburones: la Coordinadora Nacional de Organizaciones Cafetaleras. In *Cafetaleros: La Construcción de la Autonomía*. ed. F. Celis et al. Coordinadora Nacional de Organizaciones Cafetaleras.

Hernandez-Castillo, R., and R. Nigh. 1998. Global processes and local identity among Mayan coffee growers in Chiapas, Mexico. *American Anthropologist* 100: 136–147.

Hoffmann, O. 1992. Renovatión de los actores sociales en el campo: Un ejemplo en el sector cafetalero en Veracruz. *Estudios Sociológicos* 30: 523–554.

Hochschild, A. 1983. *The Managed Heart*. University of California Press.

ISO (International Organization for Standardization). 2000a. ISO in Figures: January 2000.

ISO. 2000b. WTO, ISO, and world trade, the Agreement on Technical Barriers to Trade (TBT). http://www.iso.ch.

ISO/IEC. 2002. Guide 68 (E): Arrangements for the Recognition and Acceptance of Conformity Assessment Results. Copyright office, Geneva.

Johnson, M. 2000. Effects of shade-tree species and crop structure on the winter arthropod and bird communities in a Jamaican shade coffee plantation. *Biotropica* 32: 133–145.

Kaplinsky, R. 2000. Globalization and unequalization: What can be learned from value chain analysis? *Journal of Development Studies* 37: 117–146.

Katz, D. 1997. Eco-friendly coffee farming. *Science* 275: 12–13.

Kearney, M. 1996. Migration, the new indígena, and the formation of multi-ethnic autonomous regions in Oaxaca. Paper distributed at annual meeting of American Anthropological Association.

Kunda, G., and J. Van Maanen. 1999. Changing scripts at work: Managers and professionals. *Annals of the American Academy of Political and Social Science* 561: 64–80.

Latour, B. 1999. *Pandora's Hope*. Harvard University Press.

Lee, C. 1998. *Gender and the South China Miracle*. University of California Press.

Leidner, R. 1993. *Fast Food, Fast Talk: Service Work and the Routinization of Everyday Life*. University of California Press.

Leidner, R. 1999. Emotional labor in service work. *Annals of the American Academy of Political and Social Science* 561: 81–95.

Lohr, L. 1998. Implications of organic certification for market structure and trade. *American Journal of Agricultural Economics* 80: 1125–1129.

MacDonald, K. 2005. Global hunting grounds: Power, scale and ecology in the negotiation of conservation. *Ecumene* 12: 259–291.

MacVean, C. 1997. Coffee growing: Sun or shade? *Science* 275: 1552.

Marsden, T., and N. Wrigley. 1995. Regulation, retailing, and consumption. *Environment and Planning A* 27, no. 12: 1899–1912.

Marx, K. [1867] 1906. The Buying and Selling of Labour-Power. In *Capital: A Critique of Political Economy*, volume 1. Random House.

Meaney, J., J. O'Neil, D. Onsi, and D. Corzilius. 1992. Conserving biological diversity in agricultural forestry systems. *BioScience* 42: 354–362.

Michelsen, J. 2001. Organic Farming in a regulatory perspective. *Sociologia Ruralis* 41, no. 1: 65–84.

Moguel, P., and V. Toledo. 1999. Biodiversity conservation in traditional coffee systems of Mexico. *Conservation Biology* 13: 11–21.

Mountz, A., and R. Wright. 1996. Daily life in the transnational migrant community of San Agustín, Oaxaca, and Poughkeepsie, New York. *Diaspora* 5, no. 3: 403–428.

Murdoch, J., T. Marsden, T., and J. Banks. 2000. Quality, nature, and embeddedness: Some theoretical considerations in the context of the food sector. *Economic Geography* 76, no. 2: 107–125.

Mutersbaugh, T. 1998. Women's work, men's work: Gender, labor organization, and technology acquisition in a Oaxacan village. *Environment and Planning D* 16: 439–458.

Mutersbaugh, T. 2002a. Migration, common property, and communal labor: Cultural politics and agency in a Mexican village. *Political Geography* 21: 473–494.

Mutersbaugh, T. 2002b. "The number is the beast": A political economy of organic-coffee certification and producer unionism. *Environment and Planning A* 34: 1165–1184.

Mutersbaugh, T. 2002c. Building cooperatives, constructing cooperation. *Annals of the Association of American Geographers* 21: 756–776.

Naturland. n.d. Manual de garantía de calidad: la producción ecológica en organizaciones de pequeños agricultures.

Naturland. 1999. Desarrollo de Sistemas de Control internos en Organizaciones de Productores (Organizaciones de Pequeños Agricultores). UC/12/95.

Nestel, D. 1995, Coffee in Mexico: International market, agricultural landscape and ecology. *Ecological Economics* 15: 165–179.

Netting, R. 1993. *Smallholders, Householders: Farm families and the Ecology of Intensive, Sustainable Agriculture*. Stanford University Press.

Nigh, R. 1997. Organic Agriculture and Globalization: A Maya Associative Corporation in Chiapas, Mexico. *Human Organization* 56, no. 4: 427–435.

NOP (National Organic Program, USDA Agricultural Marketing Service). 2003. Accreditation of Certifying Agents. http://www.ams.usda.gov.

OFRF (Organic Farming Research Foundation). 2000. Organic Certifier's Directory. http://www.ofrf.org.

Ong, A. 1987. *Spirits of Resistance and Capitalist Discipline*. SUNY Press.

Padoch, C., E. Harwell, and A. Susanto. 1998. Swidden, Sawah, and In-Between: Agricultural transformation in Borneo. *Human Ecology* 26, no. 1: 3–20.

Paré, L. 1991. Adelgazamiento del INMECAFE o de los Pequeños Productores de Café? In *Cafetaleros: La Construcción de la Autonomía*, ed. F. Celis et al. Coordinadora Nacional de Organizaciones Cafetaleras.

Perez-Grovas Garza, V. A. 1998. Evaluación de al sustentabilidad del sistema de producción de café orgánico en la Unión de Ejidos Majomut en la Región de Los Altos de Chiapas. Master's thesis, Universidad Autónoma Chapingo, Mexico.

Perfecto, I., R. Rice, R. Greenberg, and M. VanderVoort. 1996. Shade coffee: A disappearing refuge for biodiversity. *BioScience* 46: 598–608.

Piñón Jiménez, G., and J. Hernández-Díaz. 1998. El Café: Crisis y organización, los pequeños productores en Oaxaca. Universidad Autónoma Benito Juárez de Oaxaca.

Pred, A. 1990. *Lost Words and Lost Worlds*. Cambridge University Press.

Pringle, R. 1989. *Secretaries Talk: Sexuality, Power and Work*. Verso.

Rasker, R., M. Martin, and R. Johnson. 1992. Theory versus practice in wildlife management. *Conservation Biology* 6, no. 3: 338–349.

Raynolds, L. 2000. Re-embedding global agriculture: The international Organic and Fair Trade movements. *Agriculture and Human Values* 17: 297–309.

Raynolds, L. 2002. Consumer/producer links in Fair Trade coffee networks. *Sociologia Ruralis* 42, no. 4: 404–424.

Renard, M.-C. 1999. The interstices of globalization: The example of fair coffee. *Sociologia Ruralis* 39: 484–500.

Rice, P., and J. McLean. 1999. Sustainable Coffee at the Crossroads. Consumer's Choice Council.

Rice, R. 2001. Noble goals and challenging terrain: Organic and fair trade coffee movements in the global marketplace. *Journal of Agricultural and Environmental Ethics* 14: 39–66.

Rice, R., and J. Ward. 1996. *Coffee, Conservation and Commerce in the Western Hemisphere*. Smithsonian Migratory Bird Center.

Roberts, S. 1998. Geo-governance in trade and finance and political geographies of dissent. In *An Unruly World? Globalization, Governance, and Geography*, ed. A. Herod. Routledge.

Roseboro, K. 2002. Certification seen as a key to success in non-GMO markets. *Natural Products Industry Insider*, September 4.

Rus, J. 1994. Comunidad Revolucionario Institucional. In *Everyday Forms of State Formation: Revolution and the Negotiation of Rule in Modern Mexico*, ed. G. Joseph and D. Nugent. Duke University Press.

Saxenian, A. 1994. *Regional Advantage: Culture and Competition in Silicon Valley and Route 128*. Harvard University Press.

Sayer, A., and R. Walker. 1992. *The New Social Economy: Reworking the Division of Labor*. Blackwell.

Schroeder, R. 1999. *Shady Practices: Agroforestry and Gender Politics in The Gambia*. University of California Press.

Scott, J. 1976. *The Moral Economy of the Peasant: Rebellion and Subsistence in Southeast Asia*. Yale University Press

Scott, J. 1985. *Weapons of the Weak: Everyday Forms of Peasant Resistance*. Yale University Press.

Sosa Maldonado, L., E. Escamilla Prado, and S. Díaz Cárdenas. 1999. Café orgánico: Producción y certificación en Mexico. *El Jarocho Verde* 13–25.

Soto-Pinto, L., I. Perfecto, J. Castillo-Hernández, and J. Caballero-Nieto. 2000. Shade effect on coffee production at the northern Tzeltal zone of the state of Chiapas, Mexico. *Agriculture, Ecosystems & Environment* 80: 61–69.

Steinberg, R., and D. Figart. 1999. Emotional labor since *The Managed Heart*. *Annals of the American Academy of Political and Social Science* 561: 8–38.

Stephen, L. 1993. Weaving in the fast lane: Class, ethnicity, and gender in Zapotec craft commercialization. In *Crafts in the World Market*, ed. J. Nash. SUNY Press.

Talbot, J. 1997. The struggle for control of a commodity chain: Instant coffee from Latin America. *Latin American Research Review* 32: 117–135.

UNDP (United Nations Development Programme) et al. 1999. Environmental Strategies for Sustainable Development in Latin America and the Caribbean: 1999 Regional Action Plan for the Period 2000–2001.

Urry, J. 1987. Some social and spatial aspects of services. *Environment and Planning D 5*, no. 1: 5–26.

Waridel, L., and S. Teitelbaum. 1999. *Fair Trade*. EquiTerre.

Watts, M. 1992. Living under contract. In *Reworking Modernity: Capitalisms and Symbolic Discontent*, ed. A. Pred and M. Watts. Rutgers University Press.

Watts, M., and D. Goodman. 1997. Agrarian questions: Global appetite, local metabolism. In *Globalising Food: Agrarian Questions and Global Restructuring*, ed. D. Goodman and M. Watts. Routledge.

Whatmore, S. 2002. *Hybrid Geographies*. Sage.

Whatmore, S., and L. Thorne. 1997. Nourishing Networks. In *Globalising Food: Agrarian Questions and Global Restructuring*, ed. D. Goodman and M. Watts. Routledge.

Willer, H., and M. Yussefi. 2000. *Organic Agriculture Worldwide*. Stiftung Ökologie & Landbau.

World Bank. 2001. Mesoamerican Biological Corridors Project. www.worldbank.org.

Wright, M. 1997. Crossing the factory frontier: Gender, place and power in a Mexican maquiladora. *Antipode* 29, no. 3: 278–302.π

12

Organic and Social Certifications: Recent Developments from the Global Regulators

Sasha Courville

Consumers and producers alike can be overwhelmed by the choice of options when it comes to the social and environmental credentials of commodity products: organic, Fair Trade, Rainforest Alliance Certified, Utz Certified, shade-grown, Bird Friendly, individual company codes of conduct, and so on. Nowhere is this trend more pronounced than in the international production, trade, and marketing of coffee. While on average representing less than 2 percent of consumption in major markets, what we call sustainable coffees have seen significant growth from a very small base, to total global sales for coffee with an ethical claim to fame in 2002 estimated to be in excess of 1.1 million bags (Giovannucci and Koekoek 2004).

The original social and environmental certification systems of organic and Fair Trade have continued to strengthen their systems and their market share, along with eco-friendly certification dominated by Rainforest Alliance. However, they now face competition from a new generation of sustainable coffee claims, many with lower social and environmental standards. They also face the possibility, or reality in the case of organic agriculture, of government involvement in regulatory activities that have up to now been the domain of civil society. While government involvement can bring certain benefits in terms of building credibility and support for such systems as mechanisms for positive change, there are also new challenges, including contested claims over the control and ownership of the meaning behind these initiatives. They are also being called to improve the professionalism of the services that they provide, including compliance to relevant ISO guidelines. While this has certain advantages in terms of building credibility with government and the private sector, there is a need to ensure that certification is accessible to those who are supposed to benefit from it the most: smallholder producers in developing countries.

This chapter provides an overview of the social and environmental certification systems involved in coffee before addressing three challenges that face the private global regulators. The last section outlines how the global private regulators are beginning to address these challenges.

Current Status of Certification in Coffee

As was mentioned above, the sustainable coffee sector is replete with a range of different standards, labels and claims and there are tremendous differences among these in terms of their standards, the organizations involved, their credibility and their impacts. For example, not all are backed by independent third-party certification systems. Certain initiatives use labels to convey information to consumers, while other initiatives are more focused on behind-the-scenes issues of supply-chain management or harmonization within the coffee system.

This section provides a rough overview of the main certification systems in the sustainable coffee sector, outlining their standards, their organizational structures and operational scope, their verification activities including costs for certification as well as their markets. In this chapter I will focus on multiple-stakeholder-based certification systems for end consumers, as opposed to corporate supply-chain management programs and harmonization initiatives.

The initiatives included in this comparison are organic certification (spearheaded by the International Federation of Organic Agriculture Movements, IFOAM), Fair Trade certification (managed by Fairtrade Labelling Organizations International, FLO), and Rainforest Alliance certification (which uses the standards and systems of the Sustainable Agriculture Network, SAN). In addition to these, the Utz Certified initiative is reviewed as an example of a new-generation sustainable coffee certification system.

Mission, Scope, and Organizational Structure

The International Federation of Organic Agriculture Movements (IFOAM) was founded in 1972 to bring the organic movement together. Today it is a federation of more than 750 member organizations in more than 100 countries whose membership ranges from producers, retailers, non-government organizations, and educators. Membership with full voting rights is open to anyone with a primary interest in organic agriculture. Its highest decisionmaking body is its General Assembly that meets every three years and elects the World Board to govern the implementation of broad strategic goals set by the movement through the General Assembly and various consultations. IFOAM's world board is currently drawn from Argentina, Sweden, Germany, the United States, India, Australia, Senegal, and Japan and includes representatives from a range of sectors involved in organic agriculture, including one member with a background in Cascadian Farms, an organic company owned by the multinational food giant General Mills. At the 2002 General Assembly, there was considerable discussion about the potential repercussions of such a board member, with the membership finally accepting the importance of having the corporate sector represented within the movement, especially in the face of new challenges involved in managing the growth experienced by the organic sector in recent years. This example highlights IFOAM's recognition of the need for different constituencies to come together to debate the future of organic agriculture in ways that safeguard the movement's diversity.

IFOAM's mission is "leading, uniting and assisting the organic movement in its full diversity," and its goal is the worldwide adoption of ecologically, socially and economically sound systems that are based on the principles of Organic Agriculture (IFOAM 2004b). In terms of scope, IFOAM's standard-setting and accreditation activities directly cover or indirectly influence most organic agriculture and certification activities around the world.

Fairtrade Labelling Organizations International (FLO) was founded in 1997 to bring together the various fair trade labeling initiatives in consumer countries, the first of which was Max Havelaar in the Netherlands in 1988. Currently, there are 20 such initiatives operating in 21 countries, the newest of which are in Mexico, Spain and Australia/New Zealand. FLO focuses on internal stakeholder relations, serving the national initiatives in consumer countries, the 422 producer groups certified in the system, and trading companies (FLO International 2004). FLO's highest decisionmaking body is the FLO Board, which includes representatives from national consumer countries, representatives of groups of producers in developing countries, and representatives of traders. One of the two trader representatives is from a traditional fair trade organization (currently Oxfam Wereldwinkels of Belgium); the other is from a more conventional trade background (the current representative is from Green Mountain Coffee Roasters, based in Vermont). With two seats on the board, traders have an important voice but little power.

FLO is revising its governance structure to move toward more balanced stakeholder representation, in response to calls from producers for greater participation in decisionmaking structures. Such calls are in line with the evolution of the Fair Trade certification movement, including increased capacity over recent years on the part of developing country producer groups to engage in FLO policy processes through the development and strengthening of regional Fair Trade producer assemblies, particularly the Latin American and Caribbean Producer Assembly, CLAC (Coordinadora Latinoamericana y del Caribe de Pequenos Productores de Comercio Justo).

FLO Cert was recently established to ensure independence of certification activities. Given its focus on smallholder producers with limited activities in hired labor situations, FLO's scope is self-limiting with respect to production types that are eligible for certification. At the same time, the product range is rapidly expanding from a few commodities focusing heavily on coffee to over 50 product categories ranging from wine, fruits and juices to nuts, spices and cereals. FLO's mission is to improve the position of the poor and disadvantaged producers in the developing world, by setting the Fair Trade standards and by creating a framework that enables trade to take place under conditions respecting their interest (FLO 2005).

The Rainforest Alliance (RA) is the secretariat for the Sustainable Agriculture Network, a coalition of environmental organizations in Latin America created in 1991 with a watchdog group in Denmark. Through a recent process of consolidation, various certification activities undertaken through the Rainforest Alliance, including the SAN agriculture program, have been re-branded as "Rainforest Alliance Certified." The highest decision making body in the SAN is an executive committee comprising the executive directors of each network member group—all environmental NGOs. SAN's mission is to integrate productive agriculture, biodiversity conservation and human development (Rainforest Alliance 2005a). Unlike FLO, SAN certifies both large and smallholder producers in tropical countries and works with the following crops: coffee, bananas, cocoa, citrus, ferns, and cut flowers.

Utz Certification was developed (beginning in 1997) by the retail giant Ahold with cooperation from Guatemalan coffee producers. The latter provided the name Utz Kapeh, which means "good coffee" in the Mayan Quiché language. Utz Kapeh sees itself as a "partnership between coffee producers, distributors and roasters" whose mission is "to enable coffee producers and coffee brands to show their commitment to responsible coffee production in a credible, responsible, and market-driven way" (Utz Kapeh 2004). Although at first it was corporate driven, Utz Kapeh, a not-for-profit organization registered in the Netherlands, has begun to move toward more of a stakeholder-based initiative. The current chair is trader Christian Bendz Wolthers, with fellow board members, including Hans Perk, the manager of Dutch NGO Solidaridad's coffee program, Jeff Hill, president of Java Trading Co., Carlos Murillo, the managing director of a coffee cooperative in Costa Rica, Coopelibertad, and Jan Bernhard, General Manager of Pronatur in Peru (Utz Kapeh 2005). In the past, board members have been appointed by invitation of the existing board; however, there is now an initiative to establish a more formalized nomination process (Rosenberg 2005). Utz Kapeh's scope is limited to coffee production though, it covers both hired labor and smallholder production types in Africa, Latin America and Asia.

Overview of Standards Content

IFOAM's Basic Standards should be considered to be "standards for standards" in that they can be seen to be baseline reference standards for organic agriculture worldwide. In order for them to be used for certification, they need to be fleshed out. Although many people understand organic agriculture as the prohibition of synthetic agrochemicals, the organic standards also include including nature conservation through the prohibition of clearing primary ecosystems, biodiversity preservation, soil and water conservation, a prohibition on the use of genetically modified organisms, diversity in crop production, maintenance of soil fertility and biological activity, among others (IFOAM 2004a). There has also been a basic chapter on social justice since 1996 but this is only beginning to be implemented in earnest by IFOAM accredited certification bodies.

FLO standards are divided into smallholder producer standards and hired labor standards. Smallholder producer standards include social development criteria, such as the ability for Fair Trade to add development potential and for producer groups to have a democratic structure and transparent administration. Economic development criteria include the capacity to transparently administer the Fair Trade premium and a commitment to work toward organizational strengthening. In addition, there are standards on labor conditions that include prohibitions on forced labor and child labor, requirements for freedom of association and rights to collective bargaining, as well as occupational health and safety, applicable if the producer group employs a considerable amount of workers. FLO developed generic environmental standards for all product categories in 2007. Registered coffee importers need to comply with FLO's trade standards that include minimum price and Fair Trade premium requirements, pre-financing if requested by the producer group and a commitment to long-term trading relationships. FLO producer standards are strict in that they contain minimum and progress requirements with the former being required before certification is granted and the latter being requirements on which the producer groups much show permanent improvement over time (FLO 2004).

Rainforest Alliance SAN standards were originally focused on environmental requirements though in recent years the social component has been considerably strengthened. Standards include requirements for ecosystem and wildlife conservation, waste management, water conservation, soil conservation, community relations, as well as fair treatment and good conditions for workers, including compliance with relevant ILO conventions and national laws. Unlike organic standards where synthetic agrochemicals are prohibited, in SAN standards there is a requirement for integrated crop management that includes a prohibition on certain types of agrochemicals, strict control of those allowed and a commitment to their reduction in use over time. SAN standards also include a planning and monitoring component to demonstrate compliance to certification standards and to allow for continual improvement.

The Utz Kapeh standard also covers a broad range of social and environmental requirements, in addition to food safety requirements. This latter component was incorporated specifically to allow Utz Kapeh to be benchmarked as equivalent to the EurepGAP standard of European retailers for food safety (Utz Kapeh and EurepGAP 2004). Specific issues covered include soil management, appropriate use of fertilizers, hygienic procedures for harvesting, post harvest product handling, processing and storage, waste and pollution management, worker health safety and welfare, including worker rights and provisions for education and potable water, conservation policies in place, and energy use. As in the SAN standards, there is a detailed section on the appropriate use, application, handling and storage of agricultural agrochemicals though the term "crop production product" is used instead (Utz Kapeh Foundation 2004).

The Costs of Verification

The organic verification system is the most complicated given duplications between private and governmental regulatory systems. In the private IFOAM system, accreditation is run by the International Organic Accreditation Service (IOAS) that accredited 32 certification bodies operating in 70 countries as of December 2004 (IOAS 2004; IOAS 2004b). For the main markets such as the United States, the European Union and Japan, there are also government regulations and conformity assessment systems in place requiring that organic products be certified by an organic certification body recognized either directly by the importing government as in the case of the United States or Japan, or by another country's government that has an equivalency agreement with the import country (Joint Working Party on Trade and Environment 2002; Swedish National Board of Trade 2003). In terms of the costs of organic certification, these vary widely depending on the certification body. Inspection costs of IFOAM accredited certification bodies typically cost 100–300 euros per day plus travel with the number of days depending on the size and structure of the operation and economies of scale. In addition to this, license fees are generally charged at 0.1–1 percent of turnover. Depending on the certification body, there may also be application and annual fees (Kalus 2003). Given travel costs and differentiated daily rates, local certification bodies or foreign certification bodies using local inspectors are generally much cheaper.

In the Fair Trade system, up until 2003 producer groups did not pay for the costs of inspections and certifications. In 2004, producer group fees were introduced in order to professionalize operations and improve service, including expanding the number of producer certifications per year as resources had been a severe bottleneck. FLO Cert manages the inspection and certification activities, mainly using local inspectors to keep costs down. Initial inspection fees are all-inclusive and range from 2,000 to 5,200 euros, depending on the number of producers and complexity of the organization. For certification renewals, a flat fee of 500 euros is charged plus a percentage fee based on Fair Trade sales, ranging from 0.005–0.01 euros per kilogram (FLO 2005).

Rainforest Alliance certification is managed by the Sustainable: Agriculture Network partner organizations in each country and costs vary slightly among them. Audit costs range from US$150–200 per day plus expenses and overhead (source: letter to author from T. Divney, Technical Operations Manager, SAN, Rainforest Alliance, 2003). There has also been an annual fee per hectare of certified production, though this is currently being revised.

Certification to the Utz Kapeh standard is carried out by a number of largely local approved certification bodies, with a few large multinational certification companies. Fees vary with an average of $250/day for inspection costs in competitive markets where multiple approved certification bodies operate. Certification costs average around $0.15 per bag of coffee for big farms or $0.45 per bag for an average cooperative structure. There are no annual fees (van Heeren 2005).

Markets

Organic products are the fastest-growing segment of the global foods sector, with an estimated value of US$25 billion in 2003 (Rundgren 2004). While growing from a small base, the global market for organic coffee has grown on average about ten times faster than the overall coffee industry, which has a growth rate of 1.6 percent (Giovannucci and Koekoek 2004). In 1999–2000, global organic coffee exports amounted to 15–18 million pounds with a retail value of US$223 million (Kilcher et al. 2002). Figures for Fair Trade certified coffee are more reliable and easier to find, given the traceability system managed by FLO, with Fair Trade coffee reaching 19,895 million tons in 2003, up 26 percent from 2002 (FLO International 2004), the most significant markets being the United States, Germany, France, the Netherlands, and the United Kingdom. As for the Rainforest Alliance, certified coffee volumes grew 174 percent, from 10,000 metric tons in 2002–03 to 27,000 metric tons in 2004–05. A number of large coffee industry players, including Kraft, Procter & Gamble, Drie Mollen Holding, and Lavazza, have launched Rainforest Alliance certified products, with availability in more than 20,000 retail outlets and corporate offices in North America, Europe, and Japan, including the UN's headquarters in New York (Vigilante 2005). While a new initiative, Utz Kapeh certified coffee is now purchased in ten countries; the Netherlands, Norway, Belgium, France, Sweden, the United Kingdom, and the United States are the biggest markets. In 2004, a total of 21,000 tons of Utz Kapeh certified coffee was purchased and it is expected that this will grow by at least 50 percent in 2005 (Rosenberg 2005). Table 12.1 provides for a quick comparison of the schemes.

Table 12.1
Comparative overview of coffee certification systems.

Issue	IFOAM	FLO	RA-SAN	Utz Kapeh
Scope	Global	Developing countries	Tropical agriculture	Global production
	All crops	Expanding range of crops, including coffee	Limited number of crops, including coffee	Only coffee
	All production systems	Only smallholder groups in coffee	All production systems	All production systems
Standards content	Mainly environmental; weak social-justice coverage	Mainly social and economic; new environmental standards	Both social and environmental	Social, environmental, food safety
Verification	Accreditation and cerification systems, both private and governmental	Certification through FLO Cert; separate organization from FLO	Certification done by RA-SAN network members	Certification carried out by approved bodies, both local and international

Challenges of Growth

The coffee certification pioneers have come a long way from their initial development. They have used certified coffee to carve out a space in international trade for consideration of the social and environmental impacts of production. Together they provide a unique vehicle for hundreds of thousands of smallholder coffee producers in Latin America, Africa and Asia to communicate with supply-chain companies and end consumers about the social and environmental benefits of the coffee that they produce. Their markets have seen phenomenal growth rates that dwarf those of the conventional coffee market and they now have the attention of major traders, brands, and retailers.

However, with rapid growth come new challenges. These include the proliferation of new sustainability initiatives following on the success of the pioneers, the challenges of government involvement in a regulatory process previously led by civil society and by industry, and the logistics of managing such growth while still being able to meet the needs of farmers and consumers across continents. This section describes these challenges facing the certification world with the next section outlining ways in which the initiatives are working to address them.

Proliferation of Initiatives

Based on the success of the pioneer systems, a new generation of sustainability initiatives has recently emerged, particularly within the coffee sector. These include the Utz Kapeh system outlined above and benchmarked to the EurepGAP food safety certification program, the Sustainable Agriculture Initiative (SAI) Platform Guidelines on coffee and the Common Code for the Coffee Community (CCCC), as well as individual corporate initiatives, such as Starbucks preferred supplier program and the Neumann coffee group's sustainability standards.

The proliferation of initiatives can be explained by the lack of one system that meets all needs. It is impossible to set standards thresholds at levels that will meet the needs of all stakeholders, with some wanting more stringent standards and others asking for lower bars. For example, it is claimed by those who support competing schemes that organic agriculture is too hard to do and is not possible in all geographic regions. It is said that Fair Trade costs too much for the major brands to take it on in anything more than a token way or that it completely ignores larger coffee farmers who are also in need of market-based systems that can demonstrate and recognize strong social performance. There has also been a lack of integration of social, economic and environmental dimensions of agricultural production, with no system satisfactorily addressing all issues. Furthermore, large corporate players are in need of specific additional features to help them manage complicated supply chains, including quality control and traceability systems. Finally, there are issues of costs and control, with most of the pioneers being slow to react to new market pressures and demands due to their NGO stakeholder-based structures. There has also been a lack of recognition on the part of many players in this arena that robust certification systems are actually quite expensive to build and maintain given the significant information and credibility requirements.

While industry actors are driving the development of competing initiatives, industry is not a monolithic category in the voluntary certification arena. There is a range of possible types of engagement, from participation within the pioneer initiatives inside the decisionmaking structures based on strong value alignment, such as Green Mountain Coffee Roasters' participation in FLO, commitment to the pioneer systems evidenced by significant percentages of certified product in the market, such as Caribou Coffee's commitment to purchasing 50 percent of their coffee under Rainforest Alliance certification by 2008 (Rainforest Alliance 2005b), use of a limited percentage of pioneer initiative certified product as part of a larger product range, such as Costa Coffee shops in the United Kingdom, to active participation in developing alternative initiatives, such as Nestlé's involvement in the Sustainable Agriculture Initiative Platform. These are not mutually exclusive; some companies are engaging with the pioneer certification systems and participating in the development of other initiatives, such as Starbucks' organic and Fair Trade lines parallel to its own preferred supplier program, or Kraft's purchases of Rainforest Alliance certified coffee alongside its participation in the Common Code for the Coffee Community.

As seen from the discussion on organizational structures and leadership, both IFOAM and FLO include corporate representation in their highest decision making structures. Many companies see value in participating in the pioneer initiatives, either based on core social and environmental values held by senior management, expanding markets for certified products or through a need to demonstrate corporate social responsibility commitments in ways that can be independently verified. In the pioneer initiatives, the corporate voice is just one constituency that is balanced by other constituencies in ways that ensure that corporate players do not dominate decision making. Furthermore, the cultural dynamics are such that those corporate actors that do participate in the decisionmaking structures have gained the trust of a wider constituency. This is one point of difference from many of the newer initiatives where industry clearly drives the process even if a semblance of stakeholder representation is assembled.

Although it is easy to understand why such proliferation is occurring, it is hard to tell with any certainty what the impacts will be. One major fear comes from the pioneer Fair Trade, organic, and Rainforest Alliance initiatives. These systems have built up a degree of "credibility and goodwill" from consumers and supporters and there are concerns that with the proliferation of initiatives, it will be very hard for consumers to differentiate among the sustainable coffee offerings (Giovannucci and Koekoek 2004). Debates continue about how the proliferation of certification and labeling systems affects consumers. Some fear it will confuse consumer; others fear that their frustration will cause them to turn away from socially and environmentally preferable products. It is clear that differentiating among the various systems will become increasingly difficult, particularly with respect to such subtleties as lower vs. higher standards in a particular issue area, such as the degree of rigor of a system in evaluating freedom of association or the actual premium that producers receive.

The new initiatives have tended to proclaim themselves as "mainstream," relegating the pioneers to the "niche" category. As an example, Utz Kapeh states that its "ambition is to become the world's leading program for mainstream certified responsible coffee" (Utz Kapeh 2004). Given that Fair Trade can be found as an option in most coffee chains in the United Kingdom and that organic products are now stocked by supermarkets from Copenhagen to Cairo, it is difficult to define what mainstream actually means or to determine whether it is a useful distinction to make. A possible danger is that if consumers choose the lower bar standards over the higher bar standards with or without fully understanding the differences among them, then the most credible and stringent systems may indeed be relegated to the niche. This could, in turn, take the pressure off the powerful actors and new initiatives to perform and improve, to the detriment of the entire certification movement.

While it will take some time to fully understand the market impacts of the proliferation of initiatives, producers are already feeling the effects as they are asked to jump through more and more hoops in order to access high-value markets. The support available for producers to build capacity for compliance varies depending on the initiative and on the region in which the producers are located. For example, while in many regions organic agriculture leads the way in terms of support available through networks, including non-government organizations and extension agencies, local certification bodies implementing training programs, government support programs and successful examples of certified producer groups, there are still regions where such support is difficult to find, particularly in West Africa.

There are also a number of overlaps in the requirements of the various initiatives, requiring producer groups to duplicate systems and forms, adding to the financial and human resource burdens of certification. Without coordination among the sustainability initiatives, such problems will worsen.

In order to ground the discussion, an example will be used to highlight the threat that the proliferation of largely corporate driven initiatives poses to the credibility of the field of voluntary certification initiatives. The German Coffee Association (Deutscher Kaffeeverband) and the German government's international development agency (Deutsche Gesellschaft für Technische Zusammenarbeit, GTZ) initiated the Common Code for the Coffee Community "to foster sustainability in the 'mainstream' green coffee chain and to increase the quantities of coffee meeting basic sustainability criteria" (CCCC 2004). The code is supposed to be a "base reference standard applicable to mainstream coffee, aiming at cooperating with existing standards" (ibid.). Participating in the initiative are producers, represented mainly by national coffee associations, trade union representatives and non-government organizations, as well as a host of industry players, including Nestle, Sara Lee/Douwe Egberts, Tchibo, Volcafe, Kraft, and Hamburg Coffee Company, as well as the European Coffee Federation (CCCC 2005).

Despite repeated claims by the secretariat that the multi-stakeholder based code process has been transparent and participatory (see CCCC 2004; CCCC Management Unit 2005), the process has been widely criticized by coffee producers and NGOs alike, with some NGOs, such as Greenpeace, withdrawing their participation, while others, such as Oxfam International, have been deliberating on whether to continue to participate on the basis of possible benefits to coffee producers (Potts 2004). Producer representatives have been frustrated by the fact that most discussions are held only in English and that producers, in contrast to industry, have not had the time to discuss the potential implications of the code with other producers (de Leon 2005). Oxfam International has also raised concern about the lack of participation by groups that truly represent smallholder farmers and estate workers, rather than just representatives from producing nations (Bloomer 2004). There is a sense shared by many civil-society participants and producer representatives, that the Secretariat, in line with industry, is sacrificing "comprehensive, open debate in the name of pressure for an immediate and premature launch of the Code" (Beekman 2005).

Apart from concerns about the process to develop the code, there are also concerns by both civil-society and producer representatives about the code's content, including the lack of any strong commitment from industry to the code process (Bloomer 2004; de Leon 2005). Oxfam has noted that the "obligations that the Code places on the industry are disproportionately small compared to those placed on producers" (Bloomer 2004). Producers have yet to see any real benefits that would result from their participation and requests for a buyer code have largely been ignored (Potts 2004; de Leon 2005). Although the code espouses the importance of economic viability and "reasonable earnings for all in the coffee chain" (CCCC 2004), no agreement has been reached about economic incentives for producers, with industry actors claiming that agreement on a price premium would constitute anti-trust behavior, bringing legal experts to meetings to support their case. In such a climate, it is no wonder that producer and NGO representatives have raised questions about nature of the multi-stakeholder dialogue.

Of significance to the wider debate about the proliferation of initiatives and the threats that new industry dominated ones pose to credible certification systems, Oxfam has pointed out that "the value of the Code lies in its potential to ensure sustainability in the mainstream of the coffee market" and that "without industry commitment to purchase significant quantities under the Code's criteria, the Code will simply create another niche market, but this time with only minimal benefits to those at the bottom of the supply chain" (Bloomer 2004). Likewise, NGOs are concerned that without any safeguards yet to be put in place, participation in the code process could be used by companies as a marketing tool, misleading consumers as to the real benefits of the process to producers. Without concrete steps taken to address these issues, it is foreseen that the "low cost of participation will undermine sustainable coffee initiatives with higher standards" (ibid.).

The experience of the Common Code for the Coffee Community to date highlights a disturbing new trend in the area of corporate social responsibility. By coming together within a broad based sustainability initiative and sharing the "burden" as

a whole, corporate players can deflect their individual responsibilities and account-
abilities to the larger "multi-stakeholder" initiative (Potts 2005).

The Increasing Role of Government

Given the proliferation of initiatives and consumer concerns about their credibility,
governments have increasingly taken an active role in monitoring and regulating.
The strongest case for this is with organic certification where in the past decade
60 governments have been involved in developing organic regulations and accreditation programs (Commins 2003). This has created a situation where the private
(IFOAM) organic certification system runs largely parallel to the governmental
system, leading to increased costs of certification as certification bodies are forced
into multiple accreditation systems. This cost occurs when market requirements
dictate the use of the IFOAM accreditation system and governments require compliance to their own systems. As an example, it is the policy of UK retailer Sainsbury's
to only accept organic products from IFOAM accredited certification bodies for its
own label organic products (J Sainsbury plc 2005). The issue is further complicated
by the lack of multilateral coordination among governments, requiring a series of
lengthy and rigorous assessment processes at the bilateral level.

While most government activities have focused on organic agriculture, other
certification systems cannot afford to be complacent. A report delivered to the government of France recommended the development of a Fair Trade Regulation to
be undertaken by the French standardization organization AFNOR. The results of
this research have included guidelines to create a National Fair Trade Commission
and a law that establishes Fair Trade as an important component of the French
National Sustainable Development Strategy (Ministre de l'écologie et du développement durable, 2007).

In view of the current nightmare of organic regulation across international trade
and the potential for government involvement in other spheres, there is an urgent
need to better understand the different roles that private initiatives and governments
can play with respect to social and environmental certification and how they could
coordinate to maximize sustainability outcomes.

Accessibility and Cost

One of the biggest barriers to certification is cost—both the cost of certification and
the cost of compliance (which involves capacity building, extension, staffing, capital
investments, and management systems). Costs are particularly high where there is
an absence of local certification bodies in the countries of production, requiring that
overseas inspectors be flown in for short periods of time. In many cases, this may
also mean an absence of technical support services to help producer groups under-

stand and meet compliance requirements. It is also very difficult for local organic certification bodies in developing countries to complete the government conformity assessment systems necessary for sales to the European Union, the United States, and Japan.

Groups of producers of organic-certified coffee have always borne both costs, as was mentioned earlier. Only since January 2004 have groups of producers of Fair Trade certified coffee had to pay for their own certification. Interestingly, certified producer groups themselves agreed to the introduction of certification fees, if it would mean improvements in the quality and accessibility of the service provided. As clients of the certification system as opposed to non-paying beneficiaries, they would be in a better position to demand quality service. There is also the intent by FLO to establish a fund that would be used to support the poorest producer groups who are initially unable to pay for certification costs to gain access to the system.

Many of the challenges facing coffee certification initiatives are compounded at the producer level. The proliferation of initiatives and government regulation in organics, have both combined to increase the participation costs of producer groups where markets are requiring multiple systems compliance. Addressing these challenges will require innovation and cooperation on the part of coffee certification initiatives with initial responses outlined in the next section.

Addressing the Challenges

Initial Responses to the Proliferation of Initiatives

In response to the proliferation of coffee sustainability initiatives, new initiatives have been formed to provide platforms for discussion among the various stakeholders to work toward cooperation and harmonization. One such example is the Sustainable Commodity Initiative of the United Nations Conference on Trade and Development (UNCTAD) and the International Institute for Sustainable Development (IISD). The Common Code for the Coffee Community (CCCC or 4Cs) has also been promoted as a multi-stakeholder platform for coordination.

Most stakeholders recognize that platforms for discussion and coordination are urgently needed, but there is little consensus on how they are to be developed and what actors are facilitating the processes are critical in ensuring a useful space that will yield long-lasting results. For time-poor NGO-based initiatives, such platforms require significant investments and given the number of harmonization platforms, just keeping track of them is a daunting task. It is also not yet clear whether such coordination platforms will result in greater cooperation and harmonization of initiatives or whether they will further polarize the various initiatives into industry-based and NGO-based camps given the dominance to date of industry actors, such as in the Common Code for the Coffee Community process.

While there is a need for cooperation across a broad range of systems, there is also a need for cooperation among the higher standards systems to strengthen their collective position vis-à-vis a more powerful industry lobby within multi-stakeholder platforms. The International Social and Environmental Accreditation and Labelling Alliance (ISEAL) was formed in 1999 to do just that, with a mission to support "members" standards and verification systems to attain a high level of quality and to gain public credibility, political recognition and market success" (ISEAL 2002).

The ISEAL Alliance also serves as a common platform for collaboration of its members. The most significant example of this to date is the Social Accountability in Sustainable Agriculture (SASA) project undertaken from 2002 to 2004 among the ISEAL members involved in agriculture including IFOAM, SAI, FLO and the Rainforest Alliance's SAN. The objectives of the SASA project were to improve social standards setting and auditing in sustainable agriculture for a wide range of agricultural production systems and to develop closer cooperation between the participating initiatives (SASA secretariat 2002). Recommendations on convergence and cooperation include the implementation of mechanisms to facilitate coordination at the inspection and certification level, reducing financial and human resource costs involved in multiple certifications. Through cooperation, the various strengths of the different initiatives can be combined to cover a wider range of social, environmental, and economic issues in a cost-effective manner. Through demonstrated coordination, the ISEAL members would address many of the criticisms levied against them by industry actors that have led to the development of additional sustainability initiatives, thereby placing themselves in a stronger position in multi-stakeholder discussions.

For the organic movement, which has been seeking strategies for addressing social issues in organic standards since 1990, the project provided an opportunity for IFOAM to better understand social standard-setting and auditing methodologies as they relate to agricultural systems around the world. The recommendations also encourage IFOAM to work with its SASA partner organizations with expertise in social certification in facilitating multiple inspections, to avoid reinventing the wheel and duplicating efforts. Furthermore, IFOAM representatives realized the significant difficulties of developing a single set of globally applicable social standards for organic agriculture because of the significant differences in agricultural production systems around the world. As a result, it is likely that in the next few years IFOAM will work with its accredited certification bodies on the effective implementation of the minimal existing social standards within chapter 8 of the IFOAM Basic Standards and will cooperate more closely with complementary social initiatives. The case of Comercio Justo Mexico (a Fair Trade body affiliated with FLO) working in coordination with the Mexican organic certification body CertiMex to provide cost-effective and joint inspections for Fair Trade and organic is a positive example.

Managing an Increasing Role for Government

One reason why governments have become involved in the regulation of social and environmental certification systems is to ensure the credibility of the programs in operation for consumer confidence and protection. In order to address this important goal and to actively participate in shaping a common understanding of the meaning of credibility in this arena, the ISEAL Alliance has been working on various programs to demonstrate the credibility and quality of its member systems to governments and other stakeholders. ISEAL launched its Code of Good Practice for Setting Social and Environmental Standards in April 2004 that includes a requirement for facilitating the participation of developing country stakeholders and small to medium sized enterprises or ensuring that they have influence over the process (ISEAL Alliance 2004a). All ISEAL members including FLO, IFOAM, and the Sustainable Agriculture Network of the Rainforest Alliance, will need to revise their standards setting procedures to ensure that they comply with the ISEAL code.

Furthermore, ISEAL is supporting its accreditation members to improve their systems with peer and external reviews. External reviews of all accreditation system members to the new ISO/IED Guide 17011, the definitive guide for accreditation processes, will take place in 2005 (ISEAL Alliance 2004b). This will demonstrate to government stakeholders the professionalism of the accreditation systems within the ISEAL Alliance.

With these accomplishments, the ISEAL Alliance will be in a position to begin to lobby governments to recognize ISEAL members as credible systems who are worthy partners in cooperation in moving toward sustainable production, trade and consumption.

Specifically within the arena of organic regulation, an important platform that IFOAM, UNCTAD and the FAO have developed to bring the various private and governmental regulatory actors together is the International Task Force on Harmonization (ITF). Since 2003, ITF has enabled periodic meetings of the key regulatory actors to discuss how best to move forward on cooperative strategies to reduce the duplications and the high costs of accessing organic markets. All players have acknowledged the problem in organic regulation and are committed to working toward a resolution. However, it is expected that this will be a long process beginning with an examination of a range of models of cooperation between private and public regulatory agencies and of harmonization among governments

(Courville and Crucefix 2003) as well as comprehensive comparisons of the main organic standards and conformity assessment systems (ITF 2004). From continued discussions aimed at gaining a better understanding of the strengths and weaknesses of the existing regulatory systems, it is hoped that trust can be built among the players, leading to longer-term harmonization.

As can be seen from the above examples with the ISEAL Alliance and the ITF, initial steps to improve relationships between the main certification systems active in coffee and governmental agencies have begun. However, these are only the beginning of much longer-term processes that will take many years to reach end results that will concretely benefit smallholder coffee producers and others participating in such programs.

Working to Increase Accessibility and Reduce Costs

In order to reduce costs and improve accessibility of their initiatives, the global private regulators are working at a number of levels. These include moves to regionalize activities, the use of internal control systems to reduce costs of external inspections and increased coordination among certification systems, based in part on the implementation of SASA recommendations. In view of the urgent need on the part of producers and the concrete nature of these problems, it is in this area where the most direct and timely improvements can be made.

Both FLO and IFOAM have made efforts to decentralize some of their activities, allowing for a stronger regional presence in developing countries. FLO Cert, for example, has opened a regional office for Latin America in Costa Rica, FLO has appointed regional liaison officers in Latin America, Africa, and Asia, and IFOAM has opened a regional office for Africa in Uganda. FLO Cert has also moved to training local inspectors and IFOAM supports the development of local certification bodies in developing countries to keep costs down.

For all certification systems working with smallholders, a great deal of hope has been placed on internal control systems (ICS), originally developed by the organic movement to reduce certification costs. According to IFOAM, an ICS is "a documented quality assurance system that allows the external certification body to delegate the annual inspection of individual group members to an identified body/ unit within the certified operator," and "as a consequence, the main task of the certification body is to evaluate the proper working of the ICS" (Agro Eco 2003). Instead of having to inspect 100 percent of smallholder producer plots within a given association, an external inspector from a certification agency could evaluate the integrity of the internal control system, which itself would inspect all producer members within a given year. As of 2003, 350 producer groups representing about 150,000 farmers were certified organic through internal control systems (Agro Eco

2003). Most of these would be coffee producer groups, in view of the dominance of coffee in smallholder organic production and trade.

The SASA project also focused on the potential of ICS to reduce the costs of certification for groups of smallholder producers. The SASA project found that the establishment of an ICS could also lead to stronger ownership on the part of producers, greater knowledge and control of inputs. The internal review allows for learning through adaptation and improvement and given that people are brought together more regularly, there are opportunities to discuss other issues not specific to ICS goals, strengthening the social capital of the community and the organization. However, these potential benefits are seen when internal control systems develop endogenously as opposed to being imposed through exporter-led processes, seen more often in the African context. SAN has since developed is own internal control system based in large part on the organic model and FLO Cert is currently working on integrating ICS into Fair Trade systems (Pyburn 2004).

At the same time, a number of challenges in the implementation of ICS were also found, including tensions (perceived by producers and extension staff alike) having to do with the dual extension support and internal inspector roles required and the additional documentation burdens placed on producers. The SASA project found that ICS needed to be understood as a process that develops over time. It must also be acknowledged that initial costs may be quite high because of the need to invest in human resources to develop the ICS while covering the costs of 100 percent inspections until the ICS is strong enough to support external inspections and certifications. Also, there is no commonly accepted set of indicators for determining when a producer group is ready for certification via an ICS (Pyburn 2004). These debates came to the fore in early 2007, when the US Department of Agriculture's National Organic Program announced plans to enforce a rule that would prohibit the use of Internal Control Systems as a substitute for inspecting every farm every year. This policy would have undermined ICS and dramatically increased the costs for certified organic production, effectively undermining the efforts of smallholder organic coffee growers throughout the South. However, responses from coffee companies, such as Equal Exchange, and international development NGOs convinced the USDA to reassess these policies (Fromartz 2007).

Given the role of the ICS as a kind of management system for internal organization, monitoring and feedback, initial work is underway to look at using the ICS model as a wider organizational framework. The SASA project recommended the development of a generic management system for smallholder groups that would provide baseline structures to enable efficient and effective management of multiple certification system and other market requirements. A common framework could meet core management system requirements of all initiatives with program-specific modules added on to that. This can also be conceptualized as a total quality system framework for smallholder producers.

The most concrete area of SASA project recommendations to reduce costs of certification focused on ways to avoid duplications and reduce costs of multiple inspections and certifications. These included auditor templates for integrating compliance checks of two or more systems into one physical audit and auditor training guidelines to support the development of "super" auditors who would be able to carry out multiple standard inspections, among others. As an example, given that 45 percent of FLO Certified producer groups (on average across product categories) are organic certified, with 70 percent of Fair Trade sales by volume (average) also being organically certified, there is strong overlap between these two systems and coordination at the inspection level would help to reduce inspection costs, representing a significant proportion of the total certification cost.

Conclusions

As this chapter has demonstrated, the significant growth experienced by sustainable coffee certification systems since the early 1990s has brought a number of new challenges. With respect to the proliferation of certification and supply-chain initiatives, it is too early to fully understand the impacts that this will have on the market for sustainable coffee. Although these systems may enable broader access by smallholder coffee producers to sustainable programs and markets, they may also dilute the benefits that producers gain through their participation. Likewise, whether the distinction between "niche" and "mainstream" coffee initiatives is maintained and how these categories are defined may weaken the position of the pioneering systems, to the detriment of the entire certification movement. However, if a real commitment of cooperation for the benefit of coffee farmers and consumers emerges on the part of all actors, there may be ways of looking at how the various initiatives can complement each other and work constructively to support sustainable coffee production, trade and consumption. Since producers generally have a range of coffees with different quality grades, and since there are many coffee consumer market segments with different price points, with careful coordination there may be room for a number of different types of initiatives.

Without government support or coordination among the initiatives, it will become very difficult for consumers to differentiate between the various claims in terms of standards thresholds and producer benefits. At the same time, government involvement needs to be carefully considered to avoid duplication with private regulatory initiatives. There needs to be a strong commitment on behalf of IFOAM and government regulatory initiatives to work toward the harmonization of organic regulation in international trade.

One role for government could be to find ways to promote credible private initiatives. The high-quality initiatives need to find ways to improve their systems and demonstrate to governments and other stakeholders that they do provide benefits to smallholder producers and consumers alike. ISEAL Alliance initiatives such as the code of good practice in setting social and environmental standards and external reviews of accreditation processes are a good start.

Finally, immediate and concrete action aimed at reducing the cost of certification and at improving the accessibility of sustainability initiatives through supporting local inspectors and certification bodies, providing training materials, capacity building and widespread use and improvement of innovative tools such as internal control systems or even total quality management systems is urgently needed to ensure that smallholder producers actually do benefit from the sustainable coffee movement in which they are key constituents. While they are indeed moving in the right direction, whether the certification pioneers of Fair Trade, organic and eco-friendly coffee systems move fast enough to address these challenges remains an open question.

References

Agro Eco. 2003. Smallholder Group Certification: Compilation of Results. International Federation of Organic Agriculture Movements, Tholey-Theley.

Beekman, B. 2005. Oxfam International Delegate to CCCC. Secretariat, C. Oxfam, Oxfam International: Open Letter to Project Secretariat, CCCC.

Bloomer, P. 2004. Oxfam International, Make Trade Fair Campaign Coordinator. Project Secretariat, C. Oxford, Oxfam International: Open Letter to the Project Secretariat, Common Code for the Coffee Community.

CCCC. 2004. Common Code for the Coffee Community. GTZ, Eschborn.

CCCC. 2005. www.sustainable-coffe.net.

CCCC Management Unit. 2005. Reply of the Common Code for the Coffee Community Management Unit to the letter of OXFAM, sent December 22, 2004.

Commins, K. 2003. Overview of Current Status of Standards and Conformity Assessment Systems—Draft. International Task Force on Harmonization, Geneva.

Courville, S., and D. Crucefix, D. 2003. Existing and Potential Models and Mechanisms for Harmonisation, Equivalency and Mutual Recognition. IOAS.

de Leon, G. 2005. REF: Common Code of the Coffee Community. letter. Guatemala, ANACAFE, FEDECOCAGUA, El Frente Solidario.

FLO (Fairtrade Labelling Organisations International). 2004. Fairtrade Standards for Coffee.

FLO. 2005. www.fairtrade.net.

FLO International. 2004. Annual Report 03/04: Shopping for a Better World.

Fromartz, S. 2007. Is this the end of organic coffee? http://salon.com.

Giovannucci, D., and F. Koekoek. 2004. The State of Sustainable Coffee: A Study of Twelve Major Markets.

IFOAM (International Federation of Organic Agriculture Movements). 2004a. IFOAM Basic Standards—Final Revision Draft.

IFOAM. 2004b. www.ifoam.org.

IOAS (International Organic Accreditation Service). 2004a. www.ioas.org.

IOAS. 2004b. List of IFOAM Accredited Certification Bodies.

ISEAL (International Social and Environmental Accreditation and Labelling). 2002. ISEAL Alliance Vision Document.

ISEAL Alliance. 2004a. Code of Good Practice for Setting Social and Environmental Standards.

ISEAL Alliance. 2004b. ISEAL members FSC and MSC complete peer review against ISO/IEC 17011. *ISEAL Gazette* 2, no. 3.

ISEAL Alliance. 2005. French government leads European regulation of fair trade sector. *Policy Watch* 1: 3–4.

ITF (International Task Force on Harmonization and Equivalence in Organic Agriculture). 2004. Short-Term Actions Towards Harmonising International Regulation of Organic Agriculture -DRAFT.

Joint Working Party on Trade and Environment. 2002. Government regulations affecting trade in products of organic agriculture. In *The Development Dimension of Trade and Environment: Case Studies on Environmental Requirements and Market Access*. OECD.

J Sainsbury plc. 2005. Organic—Our Sourcing Policy www.j-sainsbury.co.uk.

Kalus, D. 2003. Personal communication, SASA Workshop.

Kilcher, L., M. Shaefer, T. Richter, P. van den Berge, J. Milz, R. Foppen, A. Theunissen, S. Bergleiter, M. Stern, F. Staubl, and M. Scholer. 2002. Organic Coffee, Cocoa and Tea. Swiss Import Promotion Programme (SIPPO), Research Institute of Organic Agricultre (FiBL) and Naturland, Zurich.

Ministre de l'écologie et du développement durable. 2007. http://www.ecologie.gouv.fr.

Potts, J. 2004. Provisional Steering Committee for Sustainable Coffee Partnership teleconference minutes.

Potts, J. 2005. Personal communication.

Pyburn, R. 2004. SASA Final Report on Internal Control Systems & Management Systems. ISEAL Alliance.

Rainforest Alliance. 2005a. Agriculture Program Mission.

Rainforest Alliance. 2005b. Caribou Coffee Company Takes a Leadership Role in Backing Sustainable Coffee Growing Practices. http://www.rainforest-alliance.org.

Rosenberg, D. 2005. Personal communication.

Rundgren, G. 2004. IFOAM, a platform for discussion and exchange. Workshop on Alternatives on Certification for Organic Production, Torres, Brazil.

SASA secretariat. 2002. Social Accountability in Sustainable Agriculture: an ISEAL project with FLO, IFOAM, SAI and SAN. Project Outline. SASA project, Australian National University, Canberra.

Swedish National Board of Trade. 2003. Market access for organic agriculture products from developing countries; analysis of the EC Regulation (2092/91). Kommerskollegium, Stockholm.

Utz Kapeh. 2004. Brand/Roaster/Trader Registration Form.

Utz Kapeh. 2005. www.utzkapeh.org.

Utz Kapeh and EurepGAP. 2004. EurepGAP Launches New Coffee Code for European Retail Sector. Utz Kapeh and EUREPGAP.

Utz Kapeh Foundation. 2004. Utz Kapeh Code of Conduct.

van Heeren, N. 2005. Personal communication.

Vigilante, S. 2005. Personal communication.

13

From Differentiated Coffee Markets toward Alternative Trade and Knowledge Networks

Roberta Jaffe and Christopher M. Bacon

Many of the chapters in this book have examined the negative consequences of the coffee crisis and commercially traded coffee commodities. They have also analyzed the multiple responses and explored the tensions and contradictions that have emerged as Fair Trade, organic and other certified production and trade relationships grow and help buffer the consequences of the coffee crisis (Bacon 2005; Petchers and Harris 2006; Goodman 2006; Mutersbaugh 2006). Although these quickly growing differentiated coffee markets represent only 1–2 percent of the global coffee supply (Bacon et al. 2005), emerging empirical research shows that they can provided a buffer for some producer organizations and small-scale farmers (Raynolds et al. 2004).

This chapter asks "What will come next?" What if we take certification as a starting point instead of a finish line? Is it possible to maintain the "alterity" within these certified markets as they rush to mainstream? What if, instead of comparing these differentiated coffee market segments to the corporate controlled conventional coffee markets, we look to the many examples of more local and alternative food networks? What can be learned from farmers' markets, community-supported agriculture, and other local attempts to connect producers and consumers in more direct relationships that are socially just and ecologically restorative and that promote mutual learning and positive change?

This chapter presents a case study of an effort to create an alternative trade and knowledge network. This case study focuses on the Community Agroecology Network (CAN), a young organization in the process of creating a model to help expand and deepen alternative trade networks. Action-oriented researchers and activists in the United States and Mesoamerica started CAN with the idea that a network that linked producers and consumers, as well as producer organizations, could benefit social development and conservation efforts in the producers' communities. In the conclusion, we use scholarship from agro-food studies to examine CAN's experience in a larger context.

Alternative Agro-Food Networks

Alternative agro-food networks are generally defined in opposition to the conventional food system. "What they share in common is their constitution as/of food markets that redistribute value through the network against the logic of bulk commodity production; that reconvene 'trust' between food producers and consumers; and that articulate new forms of political association and market governance." (Whatmore et al. 2003, p. 389). In this chapter, we will touch on a few of the defining characteristics of "alternative" agro-food networks, including issues of trust, closer producer-consumer relationships, empowerment, and strategies to promote a more equitable distribution of the benefits from trade.

The Fair Trade coffee movement started with close social relationships as pioneer cooperatives, clergy inspired by liberation theology, and Northern civil-society organizations searched for strategies that supported indigenous and peasant farmers in their collective struggle for more equality, liberty and a democratic economy (VanderHoff 2002). These relationships have become thinner, stretched and occasionally have disappeared as advocates of Fair Trade adopted a certification-based model, introduced more coffee-industry actors, and grew into a global market involving more than 800,000 farmers and rural workers and accounting for retail sales exceeding a billion dollars in 2005 . In a similar fashion, the pioneering efforts of farmers and environmentalists to launch organic certification can also be seen as an attempt to create an alternative agro-food network. These networks continue to emerge, grow, and change as they mainstream and engage people who find the conventional food system increasingly unpalatable (Whatmore et al. 2003; Murdoch et al. 2000; Goodman 1999). In addition to the global trade networks considered in this chapter, alternative food networks include many local, regional and national initiatives that promote closer relationships and alternative forms of production and consumption *within* the geopolitical North and South (Jaffee et al. 2004; Allen et al. 2003).

Rapid commercial growth as certified products enter the mainstream and more transnational companies, such as Nestlé and Procter & Gamble, try to take a share of the certified organic and Fair Trade market, has caused many to ask what happened to the "alterity" in these alternative agro-food networks. While some search for "alterity" in alternative food systems, other scholars have considered more alternative and conventional notions of sustainability (Orr 1992). This question should be considered with attention to issues of scale in coffee markets. The bulk coffee commodities trade accounts for roughly 90 percent of all the coffee traded, slightly less in the United States and more in selected European countries (Ponte 2004). Although 10 percent of the global coffee trade can be differentiated on various attri-

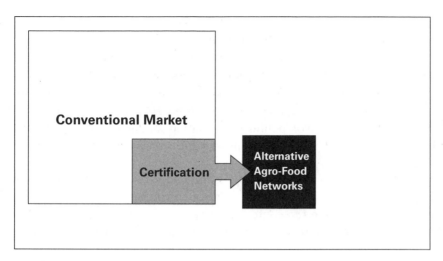

Figure 13.1
Transitioning from the conventional certified toward AAFNs.

butes of quality, including taste, origin, and certifications, only about 2 percent of the global coffee trade carries some type of eco-label or certification (ibid.).

These recent changes have provoked many of those involved in forming the fair trade and organic foods movements to re-converge as they seek to recover the "alterity" within these so-called alternative trade and production networks. Recently, more than 650 activists, civil-society organizations, businesses and others joined together in Chicago at the Fair Trade Futures conference (September 2005) to share ideas and practices about "deepening" the fair trade movement. This search for more alternative approaches is what provoked CAN's founders to ask if developing a knowledge network, where both the producer and consumer are educated about each other, could be a strategy for transforming conventional relationships in the coffee market toward alternative agro-food networks. (See figure 13.1.)

Some scholars suggest that alternative agro-food networks can meet Karl Polanyi's challenge and begin reversing a market centered society as they advance a process of re-embedding the open-market within a series of more progressive social values and societal relationships (Raynolds 2000; Polanyi 1944). Others remain more skeptical regarding the degree of "alternativeness" within the current and expanding range of alternative agro-food networks, and remind scholars that critical evaluative criteria, such as social justice, are absent from such networks (Allen and Kovach 2000; Goodman 2004).

This chapter will only begin to touch upon the many inconsistencies and contradictions surrounding the mainstreaming of market-based fair trade, the alliances between elite Northern consumers and poor Southern farmers, and the complex negotiations concerning issues of price, quality, and market governance. We focus on the still-incipient processes of constructing an organization called the Community Agroecology Network, which exemplifies alternative coffee trade and production networks. Organizations such as CAN posit the question "Can more direct relationships be developed within the differentiated coffee market that move beyond certification and further develop the alternative agro-food network?" As certification fosters public awareness around social issues yet positions itself within the mainstream market, this becomes an increasingly pressing question. This chapter explores these questions within the context of the coffee crisis, presents the CAN case study and then shares a reflective discussion analyzing CAN's experience, struggles, and aspirations.

Finding Opportunities within the Coffee Crisis

As stated elsewhere in this book, the coffee price to producers plummeted to its all time low in 2001, at $0.45 per pound of green coffee (IADP 2002; Bacon 2005). This price decline to levels below harvest costs unleashed a series of adverse consequences among rural workers and small-scale farmers, including hunger, disintegration of families and communities, and migration to cities (CEPAL 2002). This crisis put a human face on a familiar product—coffee—that many hold dear as a part of their daily routine.

Although the prices of green bean coffee have recently increased, the dynamics of the system that set the crisis in motion have not changed (Ponte 2002; Talbot 2004). Thus, it seems predictable that the cycle will repeat itself and a similar crisis will return. At the same time that coffee is a major player in the world commodities market and is controlled by several transnational corporations, it represents an ideal product for social change. It is farmed in 85 countries and exported by more than 50 countries in Central and South America, Africa, and Asia (ICO 2005). Researchers at Oxfam have suggested that 125 million families, of which an estimated 70 percent are small-scale producers, are involved in coffee production and processing (Brown et al. 2001). Most exported coffee finds its way to coffee drinkers in the United States and Europe, where coffee has become an integral part of the cultural way of life. It is produced in environmentally sensitive areas, often on mountainsides that were formerly cloud forests; yet, as an understory plant, it can be grown in ways that protects watersheds and promote biodiversity (Toledo and Moguel 1996; Méndez 2004).

Advocates of sustainable coffee hope that the morning cup offers good-tasting coffee and also appeals to values of social justice and environmental protection. The Fair Trade and certification market has done much to educate the consumer and promote awareness about social issues in the agro-food systems of Europe and more recently in the United States (Raynolds 2000; Renard 1999). This effort focuses on marketing and building label-awareness among coffee drinkers and returning a steady standard price to coffee cooperatives. It has increased coffee drinkers' awareness, and created the opportunity for farmers, their cooperatives, researchers, and university students to engage in innovative networks inspired by social justice and environmental conservation.

One example of these newly forming networks is the rapidly expanding student Fair Trade movement. Although isolated student groups have long been involved in hosting Fair Trade farmers, conducting field internships, and running campaigns to convince their local café to carry Fair Trade certified coffee, these interactions lacked coordination among campus groups. Launched at an event held in Santa Cruz, California in 2004, United Students for Fair Trade (USFT) now links more than 80 affiliated campus-based groups promoting Fair Trade principles and practices throughout the United States (Curnow 2006). A limited map of the emerging global student Fair Trade movements includes the Canadian Student Fair Trade Network, the Student Fair Trade Coalition in the United Kingdom, Hooked in Australia, and a continent-wide network of student Fair Trade activists in Africa called PEACE (Plate-forme d'Étudiants Africains pour un Commerce Équitable / African Student Platform for Fair Trade) (Hussey 2005). Many of these campus-based student activist groups originated with student-led attempts to change what they perceive as exploitative practices in their campus food system. A common entry point for these groups is a campaign to convert their university dining commons and cafés to 100 percent Fair Trade certified coffee. USFT's campus-based member groups have won important campaigns in universities, including Northwestern, Cornell, Harvard, the University of California at Los Angeles, and the University of California at Santa Cruz. The national student movement facilitates communication among otherwise isolated campus groups, encouraging students to join forces in collective efforts to promote Fair Trade principles, practices and products. During a recent student Fair Trade movement gathering, student organizers shared their stories about campus conversion campaigns. However, the successes were qualified by a shared concern that the social justice values and organizing potential that accompany Fair Trade certification are threatened as corporate providers such as Nestlé or Procter & Gamble begin to use the label.

During the 2006 National Convergence, United Students for Fair Trade adopted a focus on anti-oppression work. Through a series of trainings, discussions, and programmatic changes, USFT's leaders have attempted to deepen their understandings about social difference, including the hierarchies and resulting inequalities that accompany the ways that society has historically assigned more power and privilege to specific social groups and organizations. Student leaders and advisors believe that anti-oppression work is an essential strategy to avoid the trap of reproducing the same structures they are organizing to change.

Within the Fair Trade movement, alternative trade supporters believe that by forming grassroots networks, using information technology, and developing long-term relationships globalization can be transformed into a trade system that respects diversity and links sustainable livelihoods with environmental protection (IFAT 2004). Many believe that "another world is possible" through alternative globalizations that could provide opportunities to create more empowering relationships among people that are producing, trading and consuming (Nigh 1997; World Social Forum 2005). To accomplish this, there is a need for people who are producing and consuming to work together in their efforts to develop more sustainable livelihoods and support environmental health. These ideas are the inspiration for many activists, companies, producer organizations,, and others to develop an ever-growing diversity of alternative agro-food networks. These were among the key ideas that prompted both of us to become involved in creating the Community Agroecology Network. The following case study will detail our experience to date.

Case Study of an Alternative Trade and Knowledge Network

History of the Community Agroecology Network (CAN)

In July 2001, six researchers with more than 65 collective years of experience working with communities in Latin America became partners in a network to support the farming communities where they had developed long-term relationships. Each researcher liaison was actively involved in farm-based research and in supporting farmer groups in their organizational development processes. The formation of CAN occurred as the coffee crisis deepened. While each researcher was involved in on-going studies, the communities where they worked were facing increasingly difficult livelihood challenges. The often negative influences of a corporate-centered globalization that has promoted unconstrained competition and profited from lower prices for basic commodities have continued to undermine the economic, social, and environmental fabric of each community. In hopes of providing external support to strengthen local empowerment processes and biodiversity conservation, CAN's founders sought to build a network and cultivate alliances among different communities. We hoped to play a role in developing opportunities to share experiences and develop common projects that would integrate environmental protection with viable livelihoods. Together, this group would work to engage the farm communities and North American consumers in a process that could support agroecology-based sustainable development.

The Community Agroecology Network (CAN) is now a vibrant US-based not-for-profit organization, affiliated with the University of California at Santa Cruz, that links five farming communities in Central America and Mexico with each other and with consumers in the United States. CAN is an international network committed to sustaining rural livelihoods and environments by integrating research, education and trade innovations. The organization's name further describes three concepts around which its goals are developed:

Community: CAN has worked to build a network of community partners. Within each partner community, relations are developed with various organizations and individuals who contribute to the development and work of the network. In the producer communities, this includes farmers and their families; farmer cooperatives; social groups such as women's organizations, churches, schools, etc. (See table 13.1.) The CAN network collaborates in strategic ways to support the partner organizations in implementing their vision of how to integrate sustainable livelihoods with conservation practices. In North American communities, CAN works with universities and with Fair Trade and Alternative Trade organizations. It is also expanding a membership of individuals interested in more sustainable development and more conscientious consumption.

Agroecology: Conservation efforts are especially important in mountainous tropical rainforest ecosystems and they cannot just be relegated to areas peripheral to the human sphere of habitation (Pimentel et al. 1992). Farm systems have the potential to integrate ecological practices that protect watersheds, soils, and the health of their communities. By looking at their farm as integral to a whole landscape, farmers can increase their knowledge base of how to apply ecological principles on their farm (Gliessman 1998). Farmer organizations can become a center for agroecology-based workshops and community-based tree nurseries, composting systems, etc.; farm-based research can focus on the farmers' needs and interests (Pretty 2002; Pretty and Smith 2004). CAN's role is to help facilitate this agroecological approach through research and education, as well as to develop direct links between an increased economic return to the farmers and their protection of environmental resources on their farms and in their communities.

Network: We hope CAN will become a network that promotes a more egalitarian exchange among students, farmers, cooperatives, consumers, and landscapes throughout the Americas. Our goal is to educate participants in and through this network and to encourage them to take an active role in efforts to conserve biodiver-

Table 13.1
CAN's principal international network partners.

	CAN partner communities with PAR researcher liaisons					
	Agua Buena, Costa Rica	Huatusco, Vera Cruz, México	Matagalpa, Nicaragua	Tacuba, El Salvador	Yucatan, México	Santa Cruz, California
First-level partners[a]	Local cooperative	Regional branch of national university	Farmers' cooperative union	Local NGO	Regional university	Univ. of California CAN student organization; agroecology research group
Second-level partners	Internship program run by the cooperative	Farmers' cooperative	National NGO internship program run by the cooperative	Local farmers' cooperative union	Department of Tropical Natural Resources at same regional university	CAN network of supporters and coffee subscribers
Third-level partners	Women's group	Local NGO	Second-level farmers' cooperative	First-level cooperatives	Community-based school gardens program	Univ. of California campus organizations

a. highest representational organization for CAN activities in community sity and to support farmers' livelihoods. Our goal is to build relationships between farmers and consumers so that the coffee drinker understands the individuals and the ecosystems that produce the coffee, and farmers learn about the people drinking their coffee. These more conscious linkages will foster an active and less exploitative participation in the ecology of the food system (Francis et al. 2003).

Addressing the Needs of Small-Scale Farmers, Students, and Northern Consumers

To address the needs of the small-scale farmer and to create viable alternative trade networks, people need to be aware of where their coffee comes from, and who produced it using which practices. To negotiate for better prices, farmers need information about who drinks their coffee and the functioning of the larger coffee market to determine which ways they can influence it. A few decades ago, the isolation of these coffee-producing communities was difficult to overcome. Travel to them was difficult, and global communication was close to impossible. It is not a surprise that these limitations produced a system of control by a few large transnational corporations. It was a system where farmers were not aware of the quality of their coffee beans and the consumer believed what the advertising told them about Juan Valdez. Today, the Internet and increased investments in rural infrastructure have improved communications and access. This increased access offers an opportunity to address some of the exploitive relationships entrenched in the commodity system.

CAN's response to the coffee crisis and farmers' needs is an explicit effort to move beyond the certification markets. The network focuses on improving market return to the farmer, their organization and community; developing sustainable livelihoods; and improving environmental sustainability. In order to improve market return, farmers need to know how their product is judged in the marketplace and learn ways to improve its quality (Daniels and Petchers 2005). They need to understand the multiple levels of transactions that take place to deliver the cup of coffee to the consumer so that they can find ways to improve their position in the market. CAN works with farmers and their organizations as we grapple with the question of how we can make the transition from a vertical hierarchy with small-scale producers at the very bottom to a more egalitarian network. Of the five communities involved in CAN, four depend on coffee production for their cash income. CAN's goal is to embed the alternative market in community-based change that develops long-term relationships between the farmers and consumers, students and researchers. The alternative market is not developed in isolation, but instead is part of long-term research and educational interactions that are components of CAN's emerging program (figure 13.2).

Figure 13.2
Activities within the CAN network.

The direct coffee market is one part of ongoing inter-community relationships. CAN fosters two additional programs that promote this long-term exchange. (1) Participatory Action Research, which engages farmers, their cooperatives and researchers in investigations that seek to both generate information that supports farmers' empowerment processes and promote the transition toward more sustainable farm management practices. (2) Field Internships, which involve university students in action education where they spend a semester in the farm community in a cross-cultural exchange that engages them in community-based projects to promote sustainability. In this integrated structure, the market is just one component of an interconnected network that promotes the education supporting transitions toward sustainability among all participants: farm families and their organizations, researchers, students and consumers.

Participatory Action Research: Farmer Engagement with Research and Learning

As has been documented through the evaluation of many large governmental programs, throwing money at an issue does not necessarily create change. It is the engagement and empowerment of all of the actors that can bring about a shift in the dynamics of a system. Thus, while direct marketing is an important alternative trade focus for the CAN network, this is closely linked with our Participatory Action Research (PAR) program in all five of the CAN communities. Each counterpart organization affiliated with the CAN Network is closely associated with a researcher engaged in active learning and sharing with the farmers. Bacon, Méndez, and Brown (2005) describe participatory action research as a cyclical approach that attempts to involve a wider diversity of stakeholders as active subjects in a process of research, reflection and community-led action for positive change. PAR provides the researcher with a framework in which to work with a group of stakeholders around a specific issue. This process engages cooperatives and the farmers to help set the research agenda, generate information, analyze results and serve as the principal actors who will decide upon and implement the changes that they find appropriate for their situation.

In Tacuba, El Salvador, in Matagalpa, Nicaragua, and in Huatusco, Mexico, researchers and farmers have engaged in issues related to shade-tree diversity on coffee farms. (Guadarrama-Zugasti 2000; Méndez 2004; Bacon et al. 2008; Méndez and Bacon 2005) This research has led to greater understanding of the multi-use of shade trees by farmers and the potential for conserving native tree species within these farms. Researchers seek to develop knowledge partnerships, linking farmers' local and indigenous knowledge about shade-tree species diversity and management with scientific and market knowledge in search of strategies that could link shade-tree diversity to livelihoods and farming practices. In both Nicaragua and El Salvador, the cooperatives have used coffee biodiversity research as part of a strategy to develop cooperative-led agroecotourism initiatives (Méndez and Bacon 2005). Continued research and presentations through international short courses in agroecology may further engage the farmer organizations as they seek more diverse farms and livelihood options.

In Huatusco, Veracruz, Mexico, the researchers Laura Trujillo and Carlos Guadarrama began a series of farmer workshops on improving coffee quality through a cupping activity that offered a blind tasting of several different coffees following the regimen that coffee importers and roasters use. The farmers predicted that the Colombian coffee would taste the best. After using the scale to taste for acidity, flavor, body, aftertaste and balance, they were surprised to find that their own coffee ranked high. This insight opened the door to their asking questions as to how to improve the taste and quality even more through their farming practices. The researchers continue to work closely with the farmers in a series of workshops and on-farm research. In this way the farmers learn how they can improve the marketability of their coffee, along with improving their agroecosystems.

Similarly, the involvement of CAN researchers in PAR creates a network of researchers working collaboratively to share procedures and explore opportunities to have the farmers learn from each other. CAN in its own organizational development is currently building its participatory action research and training infrastructure. Chris Bacon, in collaboration with CAN, was awarded a Switzer Foundation Leadership Fellowship to develop an infrastructure to promote participatory action research and training in and among CAN partners. During the next year, CAN will develop case studies of research and change in each of the communities. Ultimately, CAN's goal is to support ongoing research projects in the partner communities that engage the farmers in learning about and improving their farming systems to promote sustainable livelihoods linked to environmental conservation.

CAN Field Internships: Fostering Student Involvement

University students affiliated with CAN use their connection to coffee as a tool to educate themselves and other Northern citizens about the issues of globalization, commodity markets and the complex social and ecological relationships in coffee-growing communities. Farm families and leaders of local cooperatives are also outstanding teachers. CAN works with the cooperatives in the CAN network to offer field internships in each community. These internships provide an opportunity for a student to live in the community for 10–12 weeks, engage with families on their farms, and become involved in community-based projects. The internship promotes cultural exchange that gives the university student a new way of looking at the world and often an opportunity to test their academic learning. It also gives the farm community a way to share their knowledge about their community and lifestyles while providing an alternative source of income. While in the community, students engage in projects related to community interests and/or the student's individual pursuit. Sample projects to date have included: development of a vermicomposting system, establishment of a computer center in a cooperative, production of a documentary film made by the interns, and participation in longer-term research projects led by CAN's PAR researchers. Beyond the specific projects and the additional source of income to both farmers and their cooperatives, an interaction takes place that is often life changing for all of those involved. Farmers often talk about how they had never really valued their role as a farmer before and now they see they have a lot to share with young people. Students often return with a whole new understanding of the privilege they have and a commitment to work for social change.

CAN is developing, with Oxfam America and United Students for Fair Trade (USFT), an internship program that will connect returning field interns with the USFT's national student Fair Trade movement on campuses and will provide an opportunity for a summer internship with Oxfam America. By building these connections, students have the opportunity to transform their academic learning experience into one of action education, which includes the opportunity to experience different cultures and to better understand globalization. They can then transfer their individual experience to a systemic level by working with organizations that are engaging people in social change.

The Integration of Participatory Action Research, Field Internships, and the Direct Market

The conventional hierarchy of transactions to bring coffee to market depends on lack of economic power and knowledge on the part of the farmer and their organizations. Without negotiating power or the capacity to add value to their product, small-scale farmers become trapped in a system that can involve as many as eight transactions to bring the coffee to market. CAN partners specifically seek to use education and participatory action research to support small-scale farmers in their attempts to leverage the certified coffee markets to greater advantage.

In Agua Buena, Costa Rica, the farmers' cooperative has developed the capacity to ship roasted coffee directly to North American consumers' doors. This process has given the cooperative direct access to the consumer and ownership of the coffee until it reaches the consumer. The direct market was tested and developed collaboratively between the cooperative and CAN using the model of a farmer's market, with the objective of the farmer and consumer having a direct exchange of goods. Consumers have the opportunity to know who grows their coffee and where their coffee is from. The farmer organization is responsible for taking the product directly to the consumer (harvest, roasting, packaging, order fulfillment) and benefits from the added value gained by controlling the product through these intermediary steps. CAN staff and volunteers in Santa Cruz provide customer service and help explore new sales possibilities. The obvious difference here is between a "local" farmers' market or system of community-supported agriculture and an international exchange of coffee. Coffee delivery depends on the postal service, and direct exchange is more difficult; however, e-mail and Internet chat rooms facilitate these interactions. Exchanges of knowledge between consumers and farmers' organization help to maintain this alternative trade network. The result of the direct market is that more of the responsibilities and value added benefits (roasting, packaging, and shipping) stay in the country of origin.

Assessing the Effects of the Direct Market

We have found, that especially during the coffee crisis, an alternative market can help a small group of farmers transition toward sustainability. An example of this is the direct market established with the Cooperativa Coopabuena in Costa Rica. This 40-year-old cooperative of 600 families had been struggling with the low coffee prices that increased their annual debt load. In spring 2003, CAN began the direct marketing partnership described above with them. Although the direct market returned over $3 per pound net profit to the cooperative, it represented only 1 percent of their 3 million pound harvest. Furthermore, Costa Rican cooperative law requires that the profit be distributed equally among the membership, thus making the increased return to individual farm families negligible. The 50 Coopabuena farm families who were engaged in a process of transition toward more sustainable farming practices and were involved with CAN's research and internship programs did not see an explicit gain from the direct market.

In 2004, as a result of the debt burden for Coopabuena and the continuing coffee crisis, the cooperative did not open its *beneficio* (coffee mill) for the first time in 42 years. Farmers searched for other places to sell their harvest, pulled out their coffee plantings, and also emigrated from the community. As the community searched for viable alternatives, the group of 50 farm families who were working with CAN analyzed the direct sales and discovered that they could sell 10 percent of their harvest in this way. This was enough incentive to create the new cooperative of Coopepueblos that is actively engaged in the direct market, in participatory action research and in hosting field interns. While the direct market did not provide a solution for all 600 members of Coopabuena, it did provide an opportunity to create a smaller farmers' organization committed to broader principles of sustainability. The importance of this interplay between farming practices and economic motivation was stated in an interview with Carlos Yones, Director of Fundación Café Forestal, a Costa Rican non-government organization dedicated to supporting socioeconomic and environmental development projects. In Mr. Yones's view, CAN's direct marketing program is the primary incentive for producers in Agua Buena to transform their agricultural production practices by incorporating sustainable methods. He considered it a very good opportunity, especially because of the excellent payment that they receive for their coffee and that this is the incentive for people to join the new cooperative (García 2004).

While Coopepueblos starts with the ideals and hopes of the direct market, it faces many challenges, including lack of capital and capacity to process the coffee. However, in conjunction with the CAN network, it has completed its second coffee harvest as a new organization and is initiating steps to build a *microbeneficio*. As Coopepueblos organizes, it is not only thinking about coffee production and harvesting. Members are also learning about the market and, in partnership with CAN's Santa Cruz office, they are beginning to take a more proactive role in shaping their market. They are also considering ways to diversify income in their community and they are engaging with an international community that wants to protect and regenerate the tropical rainforest in this region. Its success as an organization depends on building an interdependent relationship with its customers, so that both entities are aware of each other's needs and capacities.

The test will be to see whether a small cooperative can sustain itself through direct interactions with a relatively small network of consumers. And if it does, how will it influence the rest of the Agua Buena community? Is there a parallel here with the niche market created for small-scale vegetable and fruit farmers in the United States with the development of local farmers' markets? In other words, can an alternative agro-food network be developed that can sustain small-scale coffee producers? Only the first steps have been taken.

CAN's other network partners do not have national mail systems in which they can successfully send coffee from their cooperatives to a final destination in the North. In these cases, the farmer organizations and CAN have had to rely on the more conventional commodity chain to provide coffee to CAN's network of consumers. Tacuba, El Salvador and Matagalpa, Nicaragua ship coffee through Fair Trade channels to the Port of Oakland, California. CAN has established a partnership with a local Santa Cruz roaster to custom roast this coffee. However, within the context and constraints of the certified Fair Trade and organic marketing system, CAN strives to create an alternative system that returns more to the farmers' organizations and supports transitions toward sustainability. The farmers are part of the CAN network and have access to CAN sponsored workshops on biodiversity conservation and alternative trade networks. They also receive a small number of interns that bring their ideas and much needed additional income to cooperatives and households. CAN has an agreement with the partner cooperatives in Tacuba to pay them Fair Trade prices upon the shipping of the green beans, and then return half of any profit on top of that once the coffee has been sold. CAN tracks the costs to import, roast, package and deliver the coffee. The cooperatives receive a detailed report explaining all of the costs and how their coffee was marketed and sold. For the 2004 harvest, CAN was able to return $2.00 per pound of green beans to these cooperatives compared with the Fair Trade certified price of $1.26 per pound ($1.41 for organic). Keeping farmers informed about their market, the process, and the finances can support their empowerment and foster better decisions. CAN views this educational step as crucial to engaging the farmers in a more egalitarian trade network.

Information gained through close contact with farmers and their cooperatives serves as a tool for educating people in the North as well. University students have played a central role in building this direct market. Since Spring 2003 a group of students has met weekly at the University of California's Santa Cruz campus to learn together and to plan projects in support of the initiatives led by CAN's network of partner organizations. A major goal of this group has been to build an educated consumer base for the coffee being mailed directly from the farmers in Costa Rica. They have done this by engaging a network of people who in turn bring in others to the network. Some of their initial strategies included tabling, presentations, a quarterly newsletter, and holding an annual Holiday Campaign to encourage members of the network to send coffee as a gift to their friends, relatives and colleagues. As the students developed their own understanding of globalization and the commodity chain, they considered how they could change things within their own educational system. They began meeting with and educating the administrators of the UCSC food system. After students' organizing efforts cultivated broad-based campus support for Fair Trade certified, CAN offered administrators a more direct alternative. Once the administrators understood the potential impacts they could have on this community of farmers and how they could make a direct link to the students' academic learning, they found ways to get direct-market coffee into the campus' dining halls.

The student campaign resulted in over 50 percent of UCSC's coffee being purchased directly from the CAN Network. This contract doubled CAN's direct market volume and gave an economic boost to the three partner communities involved in the sale. Furthermore, it has been an impetus for the campus dining halls to develop a sustainability program and is providing a model for students, campus administration and local farmers to work together to serve local, organic fruits and vegetables in the dining halls. The creation of a direct relationship to the specific communities, reinforced by linkages to academic learning, field internships and research, has deepened student and administration understanding of the connections between globalization and campus food choices.

Challenges and Continued Learning

When farmers and cooperative leaders from El Salvador visited Nicaragua to learn from small-scale farmers in Matagalpa who had successfully developed their own second-level cooperatives capable of exporting coffee and returning higher prices to farmers, they asked the hard questions about getting started: How did they deal with issues like changes in leadership? Where did they get the capital to start? What about competition from the private export houses and larger farmers? CAN is not the only organization seeking strategies to confront the coffee crisis; in fact, many small-scale and some larger roasting companies, international development organizations, governments and others have made much larger efforts. However, we are not aware of any other organization that attempts to mix a primarily educational agenda with participatory action research and a small direct market. Accordingly, we feel that it is important to share some of the challenges of building this still young organization and reflect on both obstacles and possible limitations.

We have encountered both logistical and attitudinal challenges related to the ideas of exchange and empowerment within the existing trade system, in cross-cultural exchange with the communities, and in organizational development. We have been surprised to discover how difficult it is for farmers to export their own coffee, especially in small volumes. Even in countries with closer ties to the United States, such as Costa Rica, many actors have invested significant efforts and thousands of volunteer hours in moving the coffee. Each individual bag of roasted coffee sent from Costa Rica to the United States requires a separate registration code from the United States Department of Agriculture. To be done effectively, high-speed Internet access and capacity to complete US government forms in English are needed. It is the same paperwork completed by the shipping importers for each container of food-related imports. However, CAN and the cooperative have had to complete the same paperwork for each one-pound order mailed to a household in the United States.

We also faced logistical challenges in importing small amounts of unroasted coffee from El Salvador and Nicaragua. The current trade system is set up for importing in container unit sizes. It is complicated for importers to work in smaller allotments. CAN spent nine months researching ways to be able to bring in partial containers. Shippers wanted to charge exorbitant amounts. This hurdle was finally solved for both El Salvador and Nicaragua through contacts with alternative importers who were willing to support CAN's efforts. However, we are still working with the partner cooperative in Huatusco, Mexico to find ways to ship small amounts of their coffee to the United States. Developing these alternative and more direct trade systems also requires meeting the consumer demand for quality and service. From regular, on time deliveries to packages arriving undamaged, CAN and the cooperatives have to engage with the consumer to ensure that the needs of all participants in the network are met. Together we have many business lessons to learn as we strive to educate and develop alternatives to move this to a scale that can market more of the communities' coffee.

We have also encountered organizational challenges with the different partners in the network. While a small group of both paid and volunteer staff work to address the many financial and administrative challenges inherent in establishing a new non-profit organization, farmers and their cooperatives must respond to very different circumstances. During the coffee crisis, coffee cooperatives were primarily concerned with the survival of their organizations. Thus their work centered around the logistics of selling more coffee at better prices, providing pre-harvest credit to their members and making the minimum payments to continue operations. The cooperatives tied into Fair Trade and organic markets have received better prices, but they have faced the challenges of balancing increasingly large investments in their marketing programs and closer integration with the Northern markets with the need to move with the logic of their small-scale farmer members and maintain transparency and accountability. Although, cooperative principles focus on the importance of education, the educational and research agendas promoted by CAN staff are secondary to the core operations for each coffee cooperative.

We have also struggled to reach a collective agreement among all network partners regarding issues of empowerment and exchange. The primary challenge concerns the economic inequality between the North, represented in this case by relatively wealthy coffee drinkers, researchers and students in the United States, and the relatively low-income coffee farmers and their communities in the South. Working across these differences is not a simple task. Even after preparation, many students arrive to their field internships hoping to "help" and to immediately plug into meaningful projects that will support the community. Unconsciously, they seem to expect that the community's time, which is tied to agricultural and religious calendars, will match their academic calendar. At a deeper level, we are recognizing that empowerment processes originate from within groups and individuals and that to build a more egalitarian network takes ongoing communication, listening and building of common goals based on mutual values. We are also realizing that significant investments are necessary with our Latin American counterparts in order to support Northern interns during their visits. We recognize that students and researchers have benefited from the knowledge, research and publications and that the workshops and direct marketing links are only a partial payment. We look forward to the day when it is equally as easy for a coffee farmers' daughter to enjoy a three-month internship in the North as it is for a coffee consumer's son to visit Costa Rica.

As the Community Agroecology Network grows, two questions of scalability arise. First, can CAN's network grow to make an impact in linking improved livelihoods with environmental sustainability in the five communities that are part of the network? Second, can this type of network be expanded to other communities? The early responses to both questions are positive. The recent involvement of Oxfam America and United Students for Fair Trade in supporting and promoting the field internship program provides an opportunity for exploring how this network can incorporate other university campuses. In addition, there are other examples of individuals, organizations and companies that are establishing direct relationships with cooperatives and communities. These efforts that focus on developing direct relationships can be the foundation for an international alternative agro-food network that involves multiple universities, small-scale cooperatives and researchers. CAN plans to partner with an emerging web of networks to further the exchange of research, education, and innovative trade ideas.

Discussion and Conclusion

The main focus of this chapter has been on the merits of this attempt to create an alternative coffee trade network and overcome the limitations associated with mainstream certified coffee markets. The certified markets, while creating consumer awareness of the inequities of coffee production and offering price premiums to the farmer cooperative, often operate within the traditional coffee commodity systems, which continue to be controlled mainly by large-scale roasters and retailers. This chapter examines how the awareness developed through the promotion of certification can be expanded to create an alternative producer-consumer relationship. If farmer's markets and community-supported agriculture have created niche markets for small-scale farmers growing multiple crops (Allen et al. 2003), can international alternative networks develop that support sustainable livelihoods for small-scale coffee farmers and promote biodiversity conservation? The Community Agroecology Network is presented as an example of an organization constructing these alternatives. Its three-fold focus on participatory action research and training in coffee communities, action education for university students, and consumer education and direct market exchange represent an attempt to look beyond the market toward the development of knowledge exchanges and long-term relationships with communities. The following section analyzes CAN's incipient efforts in relation to five of the core elements common to alternative agro-food networks (Kloppenburg et al. 2000).

Empowerment

There is potential empowerment for both the farmer and the coffee drinker in breaking down the barriers of trade. Our ideas about empowerment follow those of Freire, Rowlands and others (Freire 1985; Rowlands 1997; Fals-Borda 1991), that concern the ability of individuals and groups to achieve their self-defined goals. This is a process-based approach to empowerment that considers not only what actions are taken (outcomes), by whom (men, women, youth and groups), but also evaluates how it was done (nature of the action) (Zenz 2000). We have found that providing more information and closer connections between farmers, students and coffee drinkers can be mutually empowering (Wilkins 2005). As discussed previously, in Costa Rica, the farmers have used funds from interns and coffee sales through CAN to launch a new cooperative.

In Nicaragua, a US-based coffee company worked with small-scale farmer cooperatives to build coffee tasting labs. The cooperatives appropriated this project using these labs to improve their coffee, better target specific markets, and earn quality-based price premiums. They now demand higher prices in recognition of the quality of the coffee they produce. The CAN researcher in Nicaragua participated in this project and has encouraged the export cooperative to continue training their farmer members through a series of workshops with farmers and youth from the coffee-growing communities. In addition, coffee drinkers appear eager to learn about the faces behind the coffee and how they can make a difference through their advocacy, donations and purchasing (Wilkins 2005). Where exploitation depends on lack of awareness, empowerment depends on exchange of ideas, building understanding and developing long-term relationships that transform ideas and practices.

Diversity

Coffee certification focuses on a standardized set of minimum requirements that may address issues including production methods, prices and trade practices. Taking certification as a starting point instead of a final goal, CAN uses research and education to reveal multiple relationships that move coffee from crop to cup. Agroecological research investigates the many ecological relationships involved in producing coffee, but also considers the subsistence crops such as corn, beans and other fruits. How do these ecological relationships affect biodiversity conservation and local water quality in shade coffee landscapes? How may diverse trade and international development networks connect farmers' cooperatives and their changing livelihoods and local ecologies? CAN's research and education focuses on strategies for diversifying production and linking farmer livelihoods to biodiversity conservation. Advising and Interdisciplinary Research for Local Development and Conservation, CAN's partner organization in El Salvador, has accompanied more than 150 farmers as they have made the transition to certified organic production practices. In Nicaragua, the PAR research process supported an innovative process that has attracted more than 500 short-term visitors to a cooperative-led solidarity ecotourism program. Interns also pay families for providing home stays and hands-on agricultural education. These small steps are part of an ongoing effort to support alternatives and resist the more chemically intensive coffee monocultures that any market centered program focusing on a single product promotes, even among smallholders.

Interdependence

In the current corporate-centered globalization, it is increasingly important and a viable counter-movement to develop closer international relationships and replace assumptions, generalizations and lack of knowledge with more people-to-people relationships. The integration of research, education and trade fosters this development and provides an important direction to create alternatives. There is a need for international alternative agro-food networks that are based on developing relationships rather than just creating better markets. The North-South exchange programs have the tension of establishing meaningful relationships in the context of a history of inequality, colonialism, and current disparities in socioeconomic development standards. Thus building trust and transparency will take time and will require patience from all participants. CAN's field internship program is a valuable bridge for developing this understanding. It offers the opportunity for students to get a deep, personal understanding of the impacts of globalization on individuals and communities in the global south while initiating cultural exchanges where the farm families in the communities become the teachers.

An interdependent alliance of researchers involved in participatory action research can make research and the learning process more useful for farmers and their cooperatives, and it also greatly contributes to the researchers' own development through academic exchange and linking research agendas in different places. As the CAN research group further develops their knowledge in collaboration with their liaison communities, they meet annually to share their experiences and insights with each other and enrolled participants in an international agroecology short course. In 2004 and in 2006 these short courses were held in partner communities (Huatusco and Matagalpa) and incorporated the farmers' experiences.

Harmony with Nature

Instead of focusing on strategies to maximize the yields from a single crop, the livelihoods and agroecology approach seeks to understand the relationships among rural livelihoods, knowledge systems and the ecological processes farmers manage. CAN uses this analysis to inform a project to develop incentives that will link livelihood improvements with environmental conservation. In connecting these farms with CAN's research program, farmers and researchers explore the larger landscape questions of the region and investigate how to collaboratively design landscape management plans that include both farms and reserves. Thus the opportunity exists to create replicable models of coffee landscape management plans that protect the environment, support farmers' empowerment processes and provide viable income to farm families. CAN is planning a workshop to engage farmer organizations in developing their own cooperative-led livelihood and biodiversity conservation management plans. Through farmers gaining their own understanding and formulating regional plans, steps are being taken to integrate conservation with farming practices.

Community

There are two aspects of community that can be explored in this discussion: (1) confronting the coffee crisis at the community level, and (2) forming a community of producers and consumers through a network. In discussing the former, CAN made a strategic decision to work for change at a community level while engaging with community-based organizations as network partners. CAN's development is occurring at a time when many of the coffee communities are disintegrating due to lack of economic viability. Yet, the community level appears to be the appropriate level to involve network participants. We work with representative organizations within these communities. In most cases, farmers' cooperatives are our primary partners.

The research liaison helps to guide CAN's involvement with a specific farming community. Both the community and CAN depend on the long-term relationships established by the researchers. CAN-affiliated researchers have long-term relationships with the partner communities, most preceding the establishment of CAN as an organization. Three have their primary residence in the region of the community they are working with and five are native to the Latin American country of their liaison CAN partner community. This connection is integral to the goal of developing an egalitarian exchange and attempting to overcome the disparity between South and North. As in any community, change is always occurring and different actors take on different roles. An important aspect of CAN's evolution will be to view the capacity to maintain the community relationship as roles pass from one individual to another whether it be the cooperative leader, the researcher liaison, or changes in CAN's first-level partner organizations. (See table 13.1.)

CAN also strives to build a sense of community within the network. There are multiple pathways for this to be pursued. There are the potential relationships among network partners in different producer communities; the alliance that is being formed among researchers and their liaison communities, the South-North relationships among producers and consumers, and the network of students and consumers who become involved in CAN. In many ways, the foundation is just being formed for these relationships by developing awareness of each other among the individuals and partner organizations. In several cases, this is being extended to ongoing communication, exchanges and joint productive activities that further build this aspect of community. Examples of this include: CAN researcher Ernesto Méndez traveled with farmers from the Tacuba cooperatives to Matagalpa, where they worked with coffee farmers and researcher Chris Bacon in demonstrating how to conduct tree diversity studies on coffee farms. CAN's International Agroecology Short course has been hosted in Huatusco (2004), in Santa Cruz (2001, 2003, 2005, 2007), and in Matagalpa (2006), where CAN researchers, together with network partners, provide outreach and education to course participants. In addition, there is the sense of community that is developed between university interns and their host families and communities. Through all of these network links, CAN hopes to break down barriers so solidarity can be formed. As these exchanges continue to develop, we need to acknowledge the socio-economic and cultural differences and emphasize ways to use these differences as a source of strength. We continually search for strategies to promote more equality. However, as time passes and relationships continue to deepen, we are all learning and in so doing, building community.

We have explored what it might look like to move beyond the Fair Trade market. While Fair Trade certification develops consumer awareness, it is focused on mainstreaming and thus becoming part of the transnational corporate system rather than changing it. What can be learned from the formative steps of CAN as a model for encouraging an international alternative agro-food network? CAN attempts to develop multifaceted interactions among partners where marketing represents just one component. Embedding the market in research and education creates a mutual commitment and exchange among actors. As a network it is made up of many diverse players—both individuals and organizations. Each brings different aspects to the network and takes different components from it to share with their community. While there are infrastructure challenges to overcome from transport to quality control, there is strong commitment among participating organizations, students, universities, the farmers and their organizations to get involved in building an alternative institutional framework. And the results can be seen in the formation of a new cooperative based on the direct market and sustainability, as well as in campus dining halls purchasing through the alternative market. As a network our goal is to break down the isolation between and among farmers, students, coffee drinkers, and universities. We believe that by consciously rebuilding these relationships all actors can support each other in transitions toward sustainability. As international organizations and networks committed to grassroots organizing of producers and consumers continue to develop, there is a possibility to form a movement that uses the communication and technology tools of globalization to further alternative trade. CAN is finding that by taking an interdisciplinary approach that integrates building the local economy with environmental conservation and sustainable farming practices the viability of small-scale farming communities is enhanced. In the words of a Costa Rican coffee farmer named Roberto Jimenez: "If I can get more money for my coffee, then I can plant less coffee and plant more trees."

References

Allen, P., and M. Kovach. 2000. The capitalist composition of organic: The potential of markets in fulfilling the promise of organic agriculture. *Agriculture and Human Values* 17, no. 2: 221–232.

Allen, P., M. FitzSimmons, M. Goodman, and K. Warner. 2003. Shifting plates in the agrifood landscape: The tectonics of alternative agrifood initiatives in California. *Journal of Rural Studies* 19, no. 1: 61–75.

Bacon, C. 2005. Confronting the coffee crisis: Can Fair Trade, organic and specialty coffees reduce small-scale farmer vulnerability in Northern Nicaragua? *World Development* 33, no. 3: 497–511.

Bacon, C., V. Méndez, and J. Fox. 2008. Conclusions: Confronting coffee's paradox to negotiate crisis and cultivate sustainability. In this volume.

Bacon, C., V. Méndez, and M. Brown. 2005. Participatory Action Research and Support for Community Development and Conservation: Examples from Shade Coffee Landscapes in Nicaragua and El Salvador. Research brief 6, Center for Agroecology and Sustainable Food Systems, Santa Cruz, California.

Beus, C., and R. Dunlap. 1990. Conventional versus alternative agriculture: The paradigmatic roots of the debate. *Rural Sociology 55*: 590–616.

Brown, O., C. Charavat, and D. Eagleton. 2001. *The Coffee Market: A Background Study*. Oxfam GB.

CEPAL. 2002. Centroamérica: el Impacto de la Caída de los Precios del Café.

Daniels, S., and S. Petchers. 2005. *The Coffee Crisis Continues: Situation Assessment and Policy Recommendations for Reducing Poverty in the Coffee Sector*. Oxfam America.

Fals-Borda, O. 1991. Some basic ingredients. In *Action and Knowledge: Breaking the Monopoly with Participatory Action-Research*. Apex.

Francis, C., G. Lieblein, S. Gliessman, T. Breland, N. Creamer, R. Harwood, L. Salomonsson, J. Helenius, D. Rickerl, R. Salvador, M. Wiedenhoeft, S. Simmons, P. Allen, M. Altieri, C. Flora, and R. Poincelot. 2003. *Agroecology: The ecology of food systems. Journal of Sustainable Agriculture* 22: 3: 99–118.

García, J. 2004. Interview, July 15.

Gliessman, S. R. 1998. *Agroecology: Ecological Processes in Sustainable Agriculture*. Lewis.

Goodman, D. 2004. Rural Europe redux? Reflections on alternative agro-food networks and paradigm change. *Sociologia Ruralis* 44, no. 1: 3–16.

Goodman, D. 2008. The international coffee crisis: A review of the issues. In this volume.

Guadarrama-Zugasti, C. 2000. The Transformation of Coffee Farming in Central Veracruz, Mexico: Sustainable Strategies? Ph.D. thesis, University of California, Santa Cruz.

IADP. 2002. Managing the Competitive Transition of the Coffee Sector in Central America.

ICO. 2005. Exports by Exporting Countries to all Destinations July 2003 to June 2004. www.ico.org.

IFAT 2004. What Is Fair Trade? www.ifat.org.

Kloppenburg, J., S. Lezberg, K. De Master, G. Stevenson, and J. Hendrickson. 2000. Tasting food, tasting sustainability: Defining the attributes of an alternative food system with competent, ordinary people. *Human Organization* 59, no. 2: 177–186.

Méndez, V. E. 2004. Traditional Shade, Rural Livelihoods, and Conservation in Small Coffee Farms and Cooperatives of Western El Salvador. Ph.D. thesis, University of California, Santa Cruz.

Méndez, V. E., and C. Bacon. 2005. Medios de vida y conservación de la biodiversidad arbórea: las experiencias de las cooperativas cafetaleras en El Salvador y Nicaragua. *LEISA* 20, no. 4: 27–30.

Mutersbaugh, T. 2008. Serve and certify: Paradoxes of service work in organic coffee certification. In this volume.

Nigh, R. 1997. Organic agriculture and globalization: A Maya associative corporation in Chiapas, Mexico. *Human Organization* 56, no. 4: 427–436.

Orr, D. 1992. Two Meanings of Sustainability. In *Ecological Literacy: Education and the Transition to a Postmodern World*, ed. D. Orr. SUNY Press.

Pimentel, D., U. Stachow, D. Takacs, H. Brubaker, A. Dumas, J. Meaney, J. O'Neil, D. Onsi, and D. Corzilius. 1992. Conserving biological diversity in agricultural/forestry systems. *BioScience* 42: 354–362.

Ponte, S. 2002. The Coffee Crisis. Issue Paper, Centre for Development Research, Copenhagen.

Ponte, S. 2004. Standards and sustainability in the coffee sector: A global value chain approach. May. Presented at United Nations Conference on Trade and Development and the International Institute for Sustainable Development.

Petchers, S., and S. Harris. 2008. The roots of the coffee crisis. In this volume.

Pretty, J. 2002. Social and human capital for sustainable agriculture. In *Agroecological Innovations: Increasing Food Production with Participatory Development*, ed. N. Uphoff. Earthscan.

Pretty, J., and D. Smith. 2004. Social capital in biodiversity conservation and management. *Conservation Biology* 18, no. 3: 631–638.

Raynolds, L. 2004. The globalization of organic agro-food networks. *World Development* 32, no. 5: 725–743.

Raynolds, L., M. Douglas, and P. Taylor. 2004. Fair Trade coffee: Building producer capacity via global networks. *Journal of International Development* 16: 1109–1121.

Raynolds, L. 2000. Re-embedding global agriculture: The international organic and fair trade movements. *Agriculture and Human Values* 17: 297–309.

Renting, H., T. Marsden, and J. Banks. 2003. Understanding alternative food networks: Exploring the role of short supply chains in rural development. *Environment and Planning* A 35, no. 3: 393–411.

Rowlands, J. 1997. *Questioning Empowerment: Working with Women in Honduras*. Humanities Press.

Talbot, J. 2004. *Grounds for Agreement: The Political Economy of the Coffee Commodity Chain*. Rowman and Littlefield.

Toledo, V., and P. Moguel. 1996. Searching for sustainable coffee in Mexico. In *Proceedings of the First Sustainable Coffee Congress*, ed. R. Rice, A. Harris, and J. McLean. Smithsonian Migratory Bird Center.

Van der Ploeg, H., G. Brundori, K. Kinkkel, J. Mannion, T. Marsden, K. de Toest, E. Sevilla-Guzman, and F. Ventura. 2000. Rural development: From practices and policies towards theory. *Sociologia Ruralis* 40, no. 4: 391–408.

VanderHoff, F. 2002. Poverty Alleviation through Participation in Fair Trade Coffee Networks: The Case of UCIRI, Oaxaca, Mexico. Fair Trade Research Group, Colorado State University.

Whatmore, S., P. Stassart, and H. Renting. 2003. What's alternative about alternative food networks? *Environment and Planning* A 35, no. 3: 389–391.

Whatmore, S., and L. Thorne. 1997. Nourishing networks: Alternative geographies of food. In *Globalising Food: Agrarian Questions and Global Restructuring*, ed. D. Goodman and M. Watts. Routledge.

World Social Forum. 2005. Call from social movements for mobilizations against the war, neoliberalism, exploitation, and exclusion: Another world is possible. *Latin American Perspectives* 32: 3–8.

14

Cultivating Sustainable Coffee: Persistent Paradoxes

Christopher M. Bacon, V. Ernesto Méndez, and Jonathan A. Fox

Although many coffee-growing communities sustain an inspiring combination of cultural and biological diversity, they have been dramatically impacted by the coffee crisis. The coffee crisis is not the first shock to hit these regions, and many observers find it difficult to separate one crisis from the many natural disasters, economic collapses, and political struggles that smallholders and rural workers continue to survive (Bacon, this volume; Skoufias 2003). Nor is crisis in the global South limited to coffee-growing communities. Studies have estimated that from 1980 to 1999 the Latin America and Caribbean region experienced at least 38 major natural disasters and over 40 episodes when GDP per capita fell by 4 percent or more (IADB 2000). A crisis occurs when preexisting conditions and vulnerabilities are met with a trigger event, such as a hurricane, sudden currency devaluation or a commodity price crash (Blaikie et al. 1994). A close analysis of the impacts and responses to a crisis reveals much about the pre-existing vulnerabilities and unequal power relationships (Wisner 2001).

The publicity and public awareness surrounding the coffee crisis, like that accompanying Hurricane Katrina, creates a "teachable moment." This attention opens windows into the uneven power relationships within the global coffee industry and encourages a closer look at social and ecological relationships in coffee-producing regions. Systematic study can reveal the damages and the responses, and can help to identify more productive avenues for confronting future challenges. The crisis also provides an opportunity to delve deeper into political-economic structures and the underlying tensions that accompany international trade and struggles for more inclusive and sustainable rural development processes.

In this concluding chapter, we synthesize the findings of preceding chapters into a single narrative. First, we review the studies that focused on small-scale coffee farmers' changing livelihoods and landscapes. These authors conducted most of the research in these chapters prior to this most recent crisis, and their findings show the pre-existing diversity, continuity, and change in Mesoamerican smallholders' livelihoods and shade coffee landscapes. The following section moves the focus downstream and into the changing coffee markets and certified trade networks, incorporating findings from the preceding chapters into a narrative that links changing coffee farmers' livelihoods and landscapes to sustainability initiatives within the coffee industry. The discussion then engages the paradoxes that must be addressed to develop longer-term strategies to confront the coffee crisis. Finally, we conclude with a brief assessment of the limited impacts of sustainable coffee efforts so far, as well as their future potential.

Livelihoods and Landscapes in Mesoamerican Coffee Regions: Small-Scale Farmers' Livelihoods and Environmental Conservation

At the time when the world was becoming aware of the dramatic social and economic impacts of the coffee crisis, ecological research was increasingly demonstrating that shade coffee agroecosystems conserve tropical biodiversity and other ecosystem services (e.g. water and soil conservation) (Babbar and Zak 1995; Gallina et al. 1996; Perfecto et al. 1996; Muschler 1997; Beer et al. 1998; Moguel and Toledo 1999; Perfecto et al. 2003; Mas and Dietsch 2004; Somarriba et al. 2004; Philpott et al., forthcoming). These smallholders also continue to conserve high levels of crop diversity in their coffee, corn, beans and other crops (Brush 2004). As Gliessman argues in this volume, small-scale traditional coffee farms have higher conservation potential than do larger-scale agrochemical dependent types of coffee management. The data presented in the empirical studies in this book strongly support this argument.

The defining agroecological characteristic of the coffee producers studied in chapters 4–9 is that shade trees are an integral part of their agroecosystem management. A diverse and abundant shade-tree canopy is widely recognized as the basic foundation for low ecological impact and environmentally friendly coffee farms (Perfecto et al. 1996; Somarriba et al. 2004; Gliessman, this volume). The case studies here provide additional evidence. In most of the coffee farms studied, shade-tree biodiversity was high, demonstrating strong potential for on-farm conservation.[1] The two studies that present greatest detail in this respect refer to smallholders in El Salvador and Nicaragua. In El Salvador, Méndez shows, shade coffee cooperatives have almost as much tree diversity (169 species) as a nearby national park (174 species), although tree species composition is different. In Nicaragua, Westphal documents a trend toward a more diversified shade-tree canopy in two different groups of producers reaching a total of 80 tree species in 62 farms. In both countries, shade-tree products support household livelihood strategies. In addition, Martínez-Torres finds a positive correlation between number of tree species and coffee yields. These studies show the conservation potential within small-scale coffee farms. They also demonstrate that a diversity of shade-tree species provides direct benefits to the environment and to farmers' livelihoods (Moguel and Toledo 1999; Somarriba et al. 2004).

In addition to shade-tree biodiversity, two of the studies document the soil conservation impacts of small-scale coffee farmer strategies. Martínez-Torres's research shows that coffee plantations under a low-intensity, no input management approach have the lowest values for an erosion index, and the highest values for a ground cover index, when compared to transition, organic, and conventional management strategies. The conventional management strategy is associated with the highest erosion index and the lowest ground cover index, while organic and transition index values are in the middle. On the other hand, low-intensity management is associated with low coffee yields and correspondingly low household coffee income, as compared to conventional management. This suggests an inverse relationship between ecological and economic benefits. Organic coffee came out as a good compromise, where yield and income are comparable to conventional management, and its environmental indicators comparable to natural systems. In seeking an alternative that will enhance livelihoods and environment, Martínez-Torres's chapter points to a need to improve organic management's soil conservation attributes, while maintaining its economic advantages. Guadarrama-Zugasti compares management practices related to pesticide use between different types of producers, including both small-scale and larger farms (this volume). He finds that small-scale producers were using much lower levels of pesticides and fertilizers, which resulted in lower soil and water contamination problems than those observed in larger farms.

The case studies reveal contradictory patterns between environmental conservation and small-scale farmers' livelihoods. This point is clearly documented by Guadarrama-Zugasti and Trujillo, who show that the coffee farmers in Veracruz, Mexico that have the lowest impact on the environment are also the most socially marginalized (this volume). In the Nicaraguan and the Salvadoran cases presented by Westphal and Méndez, shade trees not only contribute to landscape biodiversity conservation, but also to household livelihood strategies in the form of fruit, firewood and timber throughout the year. However, like their Mexican counterparts, these growers are not able to overcome their high levels of socio-economic vulnerability. Given this situation, they have oriented their agroecosystem management toward a diversified strategy that seeks to compensate for volatile coffee prices. Martínez-Torres focuses more on farmer cooperative development, arguing that these organizations can use certification as an effective capacity building tool (this volume; see also Martínez-Torres 2005). She also contends that the higher prices paid for certified organic coffee make this strategy an alternative that can benefit both the environment and farmers. Bacon presents data from Nicaragua that shows the benefits that cooperative organizations and smallholders can draw from their participation in both organic-certified and Fair Trade-certified networks. He clearly shows that Fair Trade-certified and organic growers who were members of strong cooperative unions were less vulnerable to the volatility of international coffee prices.

Although sustainable coffee certifications may hold promise to improve farmers' livelihoods, they are not accessible to all small-scale growers, nor will they be able to solve all of their problems. Access to international certification is contingent upon such factors as coffee quality, organizational capacity, links to willing coffee buyers, support from development organizations, and the still-limited size of niche markets. Many growers surveyed in Mexico and Central America commented on their inability to become certified for one or more of these reasons (Méndez 2004; Méndez et al. 2006). In addition, debate continues regarding the level of tangible landscape environmental benefits associated with participation in different certification programs (Rappole et al. 2003; Dietsch et al. 2004; Donald 2004; Mas and Dietsch 2004). In some cases, it appears that certifications are already capitalizing on—and often claiming credit for—many of the existing practices that farmers have maintained often for generations. In other cases, it appears that farmers have made significant on-farm investments and management changes to meet certification requirements. In all cases, farmers and their organizations have needed to develop administrative and monitoring programs to fill out all the paperwork associated with coffee certification programs. Although there is increasing evidence that shade certification can support a better balance between livelihood and conservation goals (Perfecto et al. 2005), it is unrealistic to assume that successful conservation in shade coffee landscapes will be achieved through certification alone. Research in tropical ecology is increasingly supporting the possibility of successful conservation in human-dominated landscapes for a diversity of tropical species (Schroth et al. 2004). However, these initiatives will require considerable efforts to achieve landscape-scale management across different types of habitat and with a diversity of rural institutions and social actors (Daily et al. 2003; Méndez 2004). To take advantage of the conservation potential of small-scale coffee farms, these growers and their chosen partners need to have a stronger voice in environmental policy and initiatives. Specifically, international and national conservation actors, including environmental organizations, the national governments, municipalities, and other farmers and activists, need to develop integrated strategies *with* these farmers, which accomplish both conservation and livelihood goals (Méndez, this volume).

Change, Heterogeneity, and External Factors Affecting Farmers' Livelihoods

The empirical case studies presented in this volume also reveal the great diversity of farmers and livelihood strategies that exist in Mesoamerican coffee territories. They point to the dynamism and resilience of these households and their organizations as they negotiate global change. Trujillo and Guadarrama-Zugasti demonstrate the great heterogeneity in terms of production strategies and grower types in the coffee landscapes of Veracruz, Mexico alone (this volume). These strategies and smallholder systems have evolved over long periods of time, and have survived through a myriad of political and economic "crises."

In El Salvador, Méndez shows contrasting interactions between types of cooperatives, levels of shade-tree biodiversity and the importance of these trees to household livelihoods (this volume). This research finds that small-scale independent farms hold higher levels of tree biodiversity and abundance than the larger collectively managed cooperatives. This is associated with independent farmers' livelihood strategies depending more on a diversity of tree products generated on-farm, instead of wages for their agricultural labor on the cooperatives' collectively managed lands. In a similar vein, Westphal's Nicaraguan case documents how independent farmers with different socio-economic histories chose diversified shade-tree management as the best strategy to meet their needs (this volume). She compares one group of smallholders who had maintained their farms in the same places for several generations with a second group of growers who came into this landscape as part of the Sandinista government's agricultural and agrarian reform policies in the 1980s. In the 1980s, the farm was managed as a single large cooperative landholding; during this time government agricultural strategies promoted high chemical inputs and a simplified shade-tree canopy with relatively few tree species. As a consequence of the agrarian reform policies these smallholders received individual titles to their land in the 1990s. During the last decade, these farmers have transformed their plots into diversified agroforestry systems very similar to those that have been managed by independent growers for several generations.

Bacon and Martínez-Torres both show how small-scale farmers and their organizations can, in certain conditions, take advantage of organic and Fair Trade marketing and international networking opportunities (this volume). In both studies, certifications are mediated by strong cooperatives, which have access to international development and solidarity networks. The cooperatives reflect more than two decades of rural social movements in Nicaragua and Mexico. These struggles and local organizing practices predate organic and Fair Trade marketing initiatives. These successes contrast with the smaller numbers of certified farmers and weaker smallholder export cooperatives observed in El Salvador by Méndez (this volume), underscoring the importance of historic struggles, effective local organizations and the networks they create to take advantage of these alternative markets.

These examples demonstrate the interconnections associated with changing global forces, local organizing practices, smallholders' livelihoods, and shade coffee ecologies. The external market influences range from the generally positive effects exemplified by alternative international markets, to the devastating effects of the international coffee price crisis. The role of the state, although ever present, has changed from periods of great influence, as in the Nicaraguan and Mexican examples of chapters 4 and 8, to times of neglect, as in the recent period of the price crisis in most of the countries. However, given the importance of agriculture in these countries, government policy has affected small-scale coffee farms directly or indirectly for centuries. In both Mexico and Central America, access to land by a majority of landless rural inhabitants has been a highly conflictive issue for decades, fueling several of the revolutions across the region. In this context, the state was forced to undertake different types of land reforms in these countries. Many of the cooperatives that exist today emerged through a combination of autonomous organizing from below and the uneven and partial openings from above provided by the governments' agrarian policies—in Mexico dating back to the 1970s, and in Central America beginning in the early 1980s (Fox 1994, 1996; Porter 2002). Some of these have become successful social enterprises, as in the case of Nicaragua, while others continue to struggle with longstanding internal conflicts (e.g. the El Salvador agrarian reform cooperative) (Méndez et al. 2006; Bacon 2005).

Northern actors have acquired increasingly influential roles in Mesoamerican coffee landscapes. They range from alternative trade organizations, such as Equal Exchange, whose focus is on fair trade and social justice, to conservation institutions, like Smithsonian and Rainforest Alliance, which have launched separate "environmentally friendly" certifications. Behind each of these organizations a network of development actors and donors support different social, economic, and environmental projects within coffee landscapes. While many Northern alternative trade initiatives continue to identify as nonprofits and prioritize partnerships with organized producers, other agencies including Utz Kapeh, Rainforest Alliance, and increasingly TransFair USA, have followed growth strategies that lead them to pursue partnerships with the dominant coffee companies. This practice has increased controversy concerning the potential conflicts of interest as certification agencies' operating budgets become dependent on fees paid by transnational corporations. These tensions have provoked protests from both civil-society actors and Southern producer organizations in a struggle for more voice around issues such as the stagnant price premiums and certification standards.

Certified Coffee Responses and North-South Network Dynamics

From the point of view of coffee households and their organizations, participation in alternative trade networks does not always follow a predetermined script. Because of the huge gaps in power and cultural understandings between North and South, alternative coffee marketing partnerships can both encounter and produce previously unforeseen obstacles and misunderstandings.

While these uneven power dynamics are very clear within the consolidated conventional coffee industry, examples also flourish within the certified sustainable coffee trade and production networks. Mutersbaugh provides an economic, ethnographic, and geographically rich account tracing the complex interactions among transnational certification norms, field-level inspection practices, and village social space (2002 and this volume).[2] His account highlights a paradox. In this case, a few Oaxacan village members must simultaneously attempt to enact their dual imperative as community members bound by tradition and culture to serve their village, and as organic inspectors who need to "inspect" their neighbors according to criteria set by the international certification networks. Their attempts to make the farming practices legible to the global certification bodies often make certification requirements "illegible" to their local communities (Mutersbaugh, this volume).[3] Paradoxically, Oaxacan producer organization efforts to reduce their dependence on foreign certification agencies by building their own certification capacity provoked the unintended consequence of bringing North-South tensions inside the producers' organizations.

Certified coffees are the more recent outcome of a long history of repeated interactions between coffee related institutions, farmer organizations, and agroecological farming practices (Bray et al.; Gliessman, this volume; Martínez-Torres, this volume). Bray, Sánchez, and Murphy (this volume) trace the processes that contributed to Mexico's emergence as the first exporter of certified organic and later shade-grown coffee. Their insightful analysis reminds us that organic coffee did not emerge in Mexico in response to certification agencies; on the contrary, it is the result of decades of grassroots efforts to build autonomous smallholder production and marketing organizations, longstanding shade coffee management practices, and changing government policies in Southern Mexico (Bray et al., this volume; Celis 2003; Fox 1994, 1996; Hernández and Celis 1994; Porter 2002; Snyder 2001). Bray and his colleagues highlight the essential role of strong farmer organizations that have both the willingness and the capacity to participate in certification programs. These authors also caution against expecting dramatic social justice impacts from certified marketing, given volatile coffee markets.

How has participation in different cooperative and trade networks mediated the coffee crisis in terms of the vulnerability of smallholders' livelihoods? In chapter 7, Bacon explores how Nicaraguan farmers who are linked into certified trade and production networks received higher farm gate prices and felt more secure in their land tenure. Although these farmers' livelihoods were less vulnerable, they were by no means completely protected from the crisis. These findings are similar to those found in other parts of Nicaragua and Central America (Mendoza 2003; Utting 2007; Méndez et al. 2006).

Most farmers, including those connected to certified networks, reported that they have experienced difficulties in maintaining food security, keeping their children in school, and improving their livelihoods. A recent study that surveyed almost 500 small-scale coffee farm households in four Mesoamerican countries found that these families continue to be plagued by poverty, including deficient access to education, potable water, and housing (Méndez et al. 2006). Coffee is still the main source of income for these families, but the low volumes produced by most of these farmers generate low yearly returns. Although Fair Trade and organic certifications have resulted in farmers gaining better prices for their coffee, this has not had a significant impact on household income per person. Another recent study, from the Global Development and Environment Institute, also highlights the relatively low prices that even certified farmers receive at the farm gate, demonstrating that these prices are not sufficient to compensate farmers for the multiple environmental benefits generated by their shade coffee production practices (Calo and Wise 2005). Calo and Wise make the convincing argument that many small-scale coffee farmers are providing the rest of the world with a significant—and uncompensated—environmental subsidy.

Much of the discussion of the impacts of Fair Trade coffee ignores the fact that limited international demand prevents certified producers from selling most of their coffee at Fair Trade-certified prices. Remarkably, only 20 percent of the export quality coffee produced by Fair Trade-certified cooperatives is sold under these preferred terms (TransFairUSA 2005). The percentages of their total harvest sold at Fair Trade prices are often even smaller, since up to 20 percent of a farmer's coffee is of lower grade and is not exportable and thus they sell it at very low prices in local markets. For producers, because of insufficient Northern demand, participation in Fair Trade networks therefore falls far short of its potential. These trends hold in many certified markets, although organic coffee farmers are generally able to sell close to 80 percent of their crop at certified prices. These market realities remind us that the Southern producers continue to be much more organized than Northern "conscious" consumers. Moreover, most of the profit in the value chain, even for certified commodities, continues to accrue to Northern intermediaries and retailers. One economist claims that "only 10 percent of the premium paid for Fair Trade coffee in a coffee bar trickles down to the producer" (Harford 2006, cited in *The Economist* 2006b, p. 74). In this context, Northern social movements, socially responsible businesses, and other market makers have a lot of catching up to do in the struggle toward more balanced North-South partnerships for sustainable coffee (Fox 2006).

Most scholars agree that, although the higher standard certifications, such as Fair Trade and organic, have provided opportunities to strengthen smallholders' organizations (Raynolds 2002; Murray, Raynolds, and Taylor 2006; Bray et al., this volume), their household-level rural development impacts remain limited by the relatively low coffee outputs, expanding smallholder households and small volumes of a farmer's coffee sold into these preferred markets (Méndez et al. 2006). Furthermore, the low price premiums received in these markets have not offset rising costs of living or mitigated small-scale farmers' continued cultural and economic marginalization and the larger political-economic inequalities between the North and South.

The Changing Global Coffee Trade

The studies presented in this book remind us that the crisis of international coffee prices is not a homogeneous force spreading across a flat coffee-producing world. Rather, the lower prices ripple through thousands of trade and production networks, including those organized around more alternative (organic and Fair Trade) and conventional marketing principles. These hybrid networks connect—or do not connect—into a heterogeneous landscape of social and ecological relationships formed through decades of local organizing and farming practice (Bacon, forthcoming). In this section, we consider the relationships between the global crisis and restructuring in the coffee industry.

From 1999 through 2003, the price of a pound of green coffee fell from US$1.20 to between US$0.45 and US$0.65 (Bacon, this volume). During this same period retail prices in Northern markets remained largely unchanged and in some cases increased. Although prices paid to producers rebounded to about $1.00/pound, the impacts of the crash and the pre-existing chronic poverty remained. In other words, even though prices began to recover, the coffee crisis continued (Petchers and Harris, this volume). The price crisis overwhelmed vulnerable rural economies and further threatened the biodiversity associated with traditional coffee production (Bacon 2006; IADP 2002; Toledo 1997). From 1999 through 2002, the total monetary value of Central American coffee exports declined from US$1.678 billion to US$700 million (IADP 2002). The low prices increased debt burdens and provoked bankruptcies among coffee exporters, millers, and farmers in producing countries.

The point here is that any assessment of the social and environmental impacts of alternative production and marketing initiatives must take the "big picture" of conventional coffee markets into account. As Goodman notes, "by wide consensus, the origins of the present crisis are to be found in the breakdown of the International Coffee agreement (ICA) in 1989, the ensuing relaxation of supply controls, and the cumulative weight of chronic over-production on "green" coffee prices in world export markets" (this volume). Most authors in this book share a structural and historical perspective that links the coffee crisis to neo-liberal governance and corporate consolidation in the coffee industry. Pronounced structural shifts have occurred in the coffee value chain in favor of Northern retailers (Talbot 1997; Talbot 2004; Pelupessy and Muridian 2005; Kaplinsky 2000). Since the ICA's breakdown (1989), coffee producers' export earnings have fallen from $10–12 billion to less than $5.5 billion, whereas international coffee market revenues have risen from $30 billion to over $70 billion. In short, the share of producing countries in the coffee value chain has fallen from 30 percent to less than 8 percent (Ponte 2004). A similar analysis reveals that many of the sustainability certifications also share a relatively low percentage of their final retail price with producers and their organizations.[4]

Although some of the preceding studies analyzed trends within the conventional coffee markets, this book has given relatively little attention to the strategies of specific companies. The dominant corporate response has been continued "business as usual," including higher profit taking and relatively superficial changes, though some have made symbolic concessions by marketing their own certified brands. Petchers and Harris (this volume) recall the importance of huge segments of the coffee market that are relatively untouched by Fair Trade and organic initiatives, such as instant coffee. They cite a report from an investment bank concerning Nestlé's control of 56 percent of the instant coffee industry: "Nothing else in food and beverages is remotely as good." The report estimates that, on average, Nestlé makes 26 cents of profit for every dollar it received for instant coffee (Deutsche Bank 2000), and notes that Sara Lee, one of the world's largest coffee companies, had a 17 percent profit in 2002.

Yet growth trends and public awareness of Fair Trade drove even Nestlé to make a symbolic gesture, launching a Fair Trade-certified coffee "Partners' Blend" (Beattie 2005). This move has, according to *The Economist*, "convinced activists that the [Fair Trade] movement is caving in to big business. Nestlé sells over 8,000 non-Fair Trade products and is accused of exploiting the [Fair Trade] brand to gain favorable publicity while continuing to do business as usual." According to FLO International, however, "you are winning the battle if you get corporate acceptance that these ideas are important" (2006b, p. 74). In the case of Kraft, which led the transnational corporate incursion with the Rainforest Alliance certification, purchases increased from 5 million pounds in 2004 to a projected 20 million pounds in 2006—but certified coffee still accounts for less than 2 percent of their total coffee purchases (Weitzman 2006a).

Future research should focus more directly on the coffee industry, to reveal the inner workings of the large-scale transnational coffee company responses to the coffee crisis. A framework that analyzes the coffee crisis as a corporate credibility and public-relations problem rather than a farmers'-livelihood struggle could reveal fascinating new information. For example, applying a critical corporate social responsibility lens (Utting 2002) might reveal how the top ten coffee companies that control more than 75 percent of the industry deployed publicity campaigns, charity giving, government lobbying efforts, and self-certification campaigns, supported ethical trade initiatives (Utting 2007), and restructured their supply chains in order to profit from the market opportunities created by the collapse of the quota system within the International Coffee Agreement.

Certified Solutions to a Systematic Crisis?

Critical questions remain. Civil-society organizations, companies, and certifiers celebrate together when Starbucks, McDonald's, and Procter & Gamble sign up to provide certified Fair Trade coffee and sustainable coffee surveys consistently document 20 percent growth rates in certified coffee markets (Giovannucci and Koekoek 2004). However, how do these high growth rates compare to the rest of

Table 14.1
Size of global conventional and sustainable coffee market, 2003–04.

Volume (metric tons)	Market segment	Percent of total	Source
4,659,522	Conventionally traded coffee	90.70	This table
480,000	Estimated exported volume of differentiated coffee	9.30	Lewin et al. 2004
5,139,522	Total green coffee exported	100	ICO 2005
Certified coffee exports			
26,400	Organic	0.51	Ponte 2004
28,283	Fair Trade[a]	0.55	TransFairUSA 2005
660	Shade Grown	0.01	Ponte 2004
10,000	Rainforest Alliance	0.19	Courville, this volume
14,000	Utz Kapeh	0.27	Courville, this volume
65,702	Estimated total[b]	1.28	This table

a. Of which 14,642 is also organic.
b. 13.7% of differentiated coffee is also certified.

Table 14.2
Summary of sustainable coffee certification systems.

Name of certification	Who are the certifiers and what are their criteria?	Who do they certify and where?
Certified organic	Farms are certified organic by third-party inspectors who follow an international code for each crop. IFOAM or USDA and others accredit the inspection and certification agencies. *Certification criteria*: Prohibit the use of synthetic chemicals; encourage farmers to preserve and recuperate soil fertility by managing the ecological processes on their farm. *Price premiums*: Generally range from $0 to $0.40/lb of green coffee depending on quality and demand.	Certified organic coffee production occurs around the world. Many are small-scale farmers but there are some larger operations. Earlier certification occurred primarily in Latin America. Mexico and Peru continue to be leaders in the organic coffee exports. Recently more farms have been certified in Africa.
Certified Fair Trade	FLO sets standards. FLO-Cert. is an independent inspection and monitoring agency. *Certification criteria*: Include participation in a democratically controlled small-scale farmer's organization. Smallholder organizations encouraged to export their coffee directly, the promotion of sustainable agriculture, long term contracts and access to credit. *Price premiums*: Minimum prices paid to smallholder exporters are $1.31 conventional and $1.46 organic arabica coffees. Premiums are high when conventional coffee prices are low and vice versa. Prices can exceed the minimums depending on quality and demand.	Fair Trade certified coffee producer associations must be primarily small-scale farmers. More than 600,000 small-scale farmers belong to over 197 Fair Trade certified cooperatives (Ponte 2004). More than 2/3 of the FT certified coffee comes from Latin America (TransFair USA 2005). According to the standards, importers and roasters are also monitored and certified by national Fair Trade initiatives.
Rain Forest Alliance (RA)	Organized a network of conservation-oriented NGOs to inspect farms and promote biodiversity conservation. The standards were initially written for larger landholdings; however, they are now being adjusted for smallholders. *Certification criteria*: Require following national labor laws, improving	Primarily larger estates but also small-scale farmers' cooperatives. Mostly in Latin American coffee-producing countries.

Table 14.2
(continued)

Name of certification	Who are the certifiers and what are their criteria?	Who do they certify and where?
	conditions for coffee workers, and a number of conservation practices, including minimum levels of shade tree density, water conservation and the elimination of more toxic chemicals. *Price premiums*: Price premiums generally range from $0.0 to 0.15/lb; sometimes higher depending on coffee quality and demand.	
Utz Certified	Utz Kapeh Foundation sets standards based on a set of general "good agricultural practices." *Certification criteria*: Standards intended to reduce environmental damage, and require humane worker treatment, including following national labor laws. *Price premiums*: Price premiums generally range from 0 to $0.15/lb; sometimes higher depending on coffee quality and demand.	Many producers in Latin America but also growing in Asia (India, Indonesia, Vietnam) and Africa. Primarily larger landholdings, but a few small-scale farmer cooperatives.

the global coffee trade? How many of the more than 1.5 billion cups people drink every day are linked to a certified trade and marketing network that makes sustainability claims? After the initial task of naming all these certifications, we begin asking questions: How do they work? Where did they come from? Why have they emerged at this time? What paradoxes do they suggest?

In 2003, countries exported 5,139,522 metric tons of coffee (ICO 2005). Table 14.1 summarizes the volumes of coffee traded through different channels in 2003. About 10 percent of the global coffee supply is differentiated through some specific quality, origin or certification (Ponte 2004). As of 2003, 1–2 percent of the global coffee supply was differentiated by one of the four major certification programs reviewed in table 14.2. Although small, this certified market segment has grown rapidly during the last ten years. Considering the rapid growth rates by 2005 the trade of certified coffees probably accounts for 2–4 percent of the global coffee trade. For, example the data in table 14.2 already show dramatic increases in the total trade of organically certified coffees.

We believe there are many reasons why sustainable coffee initiatives have emerged, proliferated, and rapidly expanded. Media coverage of the coffee crisis and civil-society mobilizations have created pressure to fill the regulatory vacuum following the disintegration of the International Coffee Agreement more than a decade ago. As governments "outsourced regulation" (O'Rourke 2003), sectors of the coffee industry have responded in unanticipated ways. Global civil-society organizations have teamed up with certifiers and progressive coffee roasters in an attempt to exert a degree of social control and "re-regulate the coffee industry from the street" (Utting 2005). Although civil-society campaigns, such as those conducted by Oxfam, Global Exchange, and various religious groups, have clearly used certification to leverage industry actors, these efforts have a long way to go to become part of a new social contract (Giovannucci and Ponte 2005), and specialty coffee companies acting in their own self-interest to ensure the quality and stability of their supply chains have accounted for the larger changes in terms of total coffee volumes. This is not the case with many of the coffee industry's largest actors, represented by the National Coffee Association. The NCA has actively lobbied against congressional legislation on coffee quality and purity standards (i.e., that products labeled as coffee must contain 100 percent coffee) (US House of Representatives 2002). In contrast, the Specialty Coffee Association of America, which represents more than 2,600 mainly small-scale roasters, retailers, and importers, has lobbied Congress in favor of quality standards and funding for coffee sustainability initiatives (Bacon, this volume).

Each star in the current constellation of certification initiatives has a different history (Rice and McLean 1999). These origin stories set in motion many of the processes that continue to play out at the policy and market interface. Social movements, notably those supporting organic farming and trade justice, have played fundamental roles in creating Fair Trade and organic certifications (Goodman, this volume; Moore 2004; Jaffee 2007). Equal Exchange pioneered alternative coffee marketing from Nicaragua as part of the US movement against war in the Central America in the 1980s. Progressive church-based constituencies in Europe linked to liberation-theology-inspired cooperatives in southern Mexico generated the initial demand for Fair Trade and organic coffee in the late 1980s (Porter 2002; Vanderhoff 2002). Global Exchange's threat to campaign against Starbucks after the Seattle World Trade Organization protests of 1999 quickly drove the company to begin buying Fair Trade-certified coffee. Although corporate involvement in certified Fair Trade and organic coffee has provoked much "internal" debate over the potential contribution of transnational companies, the lack of smallholder voices in governance decisions, the certification of larger landholdings, the use of genetically engineered crops within different certification schemes, and, more significant, the active engagement of both the environmental and larger social justice movements in these debates illustrate a sense that these certification schemes continue to hold potential for positive social and environmental change. These debates are largely absent in both Rainforest Alliance and Utz Kapeh certifications due to a general lack of interest from most producer and social-movement organizations.

Yet one challenge for the future growth of Fair Trade coffee demand, at least in the United States, is the lack of sustained campaigning by organized constituencies that could potentially be mobilized to challenge conventional North-South trade relationships (i.e., churches, environmental organizations, labor unions, organized immigrants). Fair Trade's share of the European market is significantly higher and appears to be embedded in social institutions as well as in supermarkets. In contrast, the demand for Fair Trade coffee in the United States appears to be driven primarily by a handful of progressive coffee roasters, by churches and civil-society organizations (such as Oxfam America), by café owners, and by a large number of individuals.[5] Activist engagement may be deterred in part by the proliferation of labels, as well as the perception that some labels involve lower social and environmental standards. The main exception to this general pattern in the United States is on college campuses. Following in the footsteps of anti-sweatshop organizing, United Students for Fair Trade has organized more than 80 campus-based social justice groups promoting Fair Trade principles, practices and polices as part of a larger global justice movement.

Of the authors represented in this volume, only Courville takes us inside the complicated worlds of voluntary multi-stakeholder codes of conduct and certified coffee initiatives. She describes the moves of some dominant industry actors to support the Common Code for the Coffee Community (4-C), which is a set of voluntary sustainability standards intended to make an incremental change that will move the conventional coffee industry toward sustainability. The 4-C steering committee includes corporate actors, international coffee organization members and limited participation from national producer associations and civil society. Both producer organizations and civil-society organizations have considered withdrawing their support. Self-certification initiatives, such as Starbucks' Preferred Supplier Program and Neumann Coffee Group's sustainability standards, are also on the rise. These corporate strategies can promote traceability and coffee quality, provide a managed response to some sustainability demands, and help large companies generate proprietary information. However, the lack of an independent third-party verification system has caused many to question the credibility of these programs.

While the credibility of these self-certification initiatives is under scrutiny, others question the future directions of the higher-bar (organic and Fair Trade) certified coffee programs. In fact, an expanding group of Fair Trade and organic pioneers have started to look at these certifications as a starting point instead of a potential finish line (Jaffe and Bacon, this volume). Some believe that the democratic and ecological principles promoted by the original social-movement actors that created organic and Fair Trade certification are increasingly threatened as transnational companies and larger landholders become active stakeholders and participants in the governance of these certification systems (Mutersbaugh et al. 2005; Renard 2005; González and Nigh 2005). As Jaffe and Bacon observe in this volume, the founding principles that motivated people to create these "alternative" agro-food networks are increasingly contradicted by efforts to adjust as sustainability certifications to fit into the conventional trade systems they initially sought to transform. This effort to prioritize quantity over quality serves the economic interests of certifying agencies and other intermediaries that earn more with higher volumes. At the same time, if successful in increasing still-insufficient demand for certified coffee, this approach does serve the interests of producers—though whether private commercial producers will benefit more than organized smallholders remains to be seen.

Understanding Paradoxes to Confront the Crisis

Attempts to confront the coffee crisis will fail in the long term unless certain paradoxes are addressed. Actors concerned with sustainable rural development in Mesoamerica can ill afford to continue sidelining these challenges in an effort to simplify the message and/or boost sales for their self-defined "solution" to the coffee crisis.

René Mendoza succinctly stated the primary coffee paradox in the title of his 2002 book: *La Paradoja del Café: El Gran Negocio Mundial y la Peor Crisis Campesina* (The Paradox of Coffee: A Great Global Business and the Peasants' Worst Crisis). The central paradox involves the unequal power relationships that have constructed the global coffee commodity chains and markets, leading to booms in coffee consumption and crises in coffee-producing countries (Topik and Pomeranz 1999.). Paul Katzeff, a founder and a twice-elected president of the Specialty Coffee Association of America, says that over 500 years in the coffee trade has made people rich and it has made people poor (Katzeff 2002). However, as Daviron and Ponte note (2005), the paradox of wealth and poverty is not the only paradox accompanying this golden bean in its journey from crop to cup.

Rich Lands, Poor People: Can Environmental Conservation Improve Farmers' Livelihoods?

Will the paradox of rich lands and poor people (Peluso 1994) persist, or can farmers enhance their livelihoods while contributing to environmental conservation? Traditional shade coffee already provides significant environmental benefits.

However, farmers' rural livelihoods are increasingly vulnerable. Rural households adapt with strategies that will provoke changes in land management strategies, cooperative organizations, migration and a host of still poorly understood processes as people struggle for survival. Participation in Fair Trade and organic markets may offer a partial solution. For the smallholders who initially had low yields, the transition to organic production increased yields (Martínez-Torres, this volume; Damiani 2002). Yet organic production has lowered yields for many previously conventional farmers, at least in the short to medium term, and the analysis includes the increased labor investments needed to manage organic farms (Calo and Wise 2005). In principle, the organic price premium can partly compensate for the costs of the transition, as can the price premium from associated increases in quality—but producers can realize the gains only years after making the investment. In most cases, producers are expected to absorb the income loss involved in the transition process, until certification is achieved.

Previous agroecological research has considered the dynamics of this transition process (Gliessman 2000). However, new research is needed to better understand the social and economic tradeoffs that accompany these changes, especially given higher international coffee prices and lower organic price premiums. Although lacking compensation, smallholders' organizations continue to provide valuable social benefits (e.g. local democracy, farmer autonomy, social support systems), and environmental services (e.g. biodiversity and water and soil conservation). Without direct policy and market incentives that better link all local peoples' livelihood improvements with environmental conservation strategies, many coffee territories may soon encounter increased environmental damage or increased poverty, or both.

Hungry Farmers: Production for Subsistence and/or for Sale

Most coffee smallholders are already diversified, producing staple foods for household consumption as well as coffee for the market. Smallholders generally also sell surplus staple crops and fruits into local markets and keep about 10 percent of their coffee for household consumption. These farmers already manage multiple diversification strategies, including different off-farm livelihood activities and the production of fruits, vegetables and animals both inside the coffee agroecosystems and in other production fields. Though their degree of diversification varies, for smallholders to operate in both subsistence and commercial economies involves negotiating very different logics simultaneously.

To generate agricultural income while limiting dependence and risk is easier said than done. The tradeoffs involved are still not well understood by most scholars and development professionals. In Nicaragua, smallholders claimed to grow half or more of the food they ate during the 2001–02 coffee harvest (Bacon, this volume).

Subsequent research conducted throughout Mexico and Central America has found that households on average grew about one third of their food, although Salvadoran smallholders reported higher levels of production for subsistence (Méndez et al., forthcoming). The same comparative study also found that a high percentage of small-scale coffee farm households reported that they experienced periods of hunger during the preceding two years.

One hypothesis is that the continued focus on the producer price premiums and cooperative business development could further undermine local food security efforts, if specific attention is not given to supporting and enhancing the diverse social and ecological relationships that have sustained farmers' livelihoods and agroecosystems for generations. It is also clear that, although pioneering theories addressing the peasant economy (Chayanov 1966; Netting 1996) still provide useful and frequently overlooked insights into changing farmers' livelihoods, new work is needed to apply and adjust these approaches in changing times. Scholars have called for a return to the problematic of the classical "agrarian question" in the context of globalizing foods (Goodman and Watts 1997). This volume uses coffee to explore many of these issues and interrogate future avenues for research and action in order to cultivate sustainability in coffee territories. Among the important trends in Mesoamerican coffee production are changing farmer typologies (Guadarrama-Zugasti, this volume), evolving livelihood strategies as households become more closely engaged with local cooperatives and expanding "alternative" trade networks, and increasing rates of international migration (Lewis and Runsten 2006). Of these three trends, migration has had especially dramatic impacts on smallholder households and coffee-production systems (Benquet 2003; Aranda Bezaury 2006.).

Sustainable Coffee as an Alternative to International Migration: Coffee, Development, and Migration in Mesoamerica

If migration is one response to the coffee crisis, is sustainable coffee an alternative to migration? Is migration a source of capital for sustainable coffee? Campaigners for both immigrant rights and small-scale farmers have noted that some of those migrants who have died attempting to cross the Arizona desert came from communities that had not experienced significant out-migration until the coffee price crisis, such as central Veracruz (Hérnandez Navarro 2004). Many researchers and practitioners have implicitly or explicitly hoped that Fair Trade and organic coffee could be an alternative to migration. Meanwhile, family remittances appear to have helped coffee families to survive. Yet researchers have only just begun to study the relationships between sustainable coffee initiatives and migration. One of the most promising initial studies focused on a community with cooperative members connected to organic and Fair Trade networks that had a prior track record of migration. Lewis and Runsten (2006, p. 18) concluded that "international migration can be a means to better capitalize coffee production for higher yields, quality, and returns. But coffee prices would have to be higher. Nominal [farmworker] wages have doubled in 5 years in Cabeza del Río [Oaxaca], but the [floor] price of Fair Trade coffee has not risen in over 10 years." To the degree that the transition to organic production requires substantial labor investments, the increased cost of local labor (both in terms of cash for day labor and the opportunity cost of family labor associated with increased migration rates and labor shortages) will make the spread of organic production more difficult for those families and communities where migration is an option. Changing migration patterns and the associated impacts on international trade, farmers' livelihoods, and rural landscapes will continue to be sources of research questions.

Struggling for Survival or Sustainability: From Diversified Farms to Diversified Livelihoods

The studies presented here reveal the great diversity among shade coffee livelihood strategies in Central America and Mexico. This heterogeneity can inform efforts to create sustainable livelihoods in coffee farming communities. In this respect, it is important to move beyond the conventional agronomic response to coffee crises, which has sought to support farmers by diversifying the crops within coffee plantations. Examples of this include intercropping bananas, oranges or timber with existing coffee and shade trees. This response has been continuously repeated through cyclical coffee price crises since the 1930s, with very limited success (Trujillo 2001). To move beyond crop diversification and into *livelihood diversification* it is necessary to start with a deeper understanding of the current farm household characteristics and strategies. Farmer typologies as exemplified by Guadarrama-Zugasti and Trujillo in this volume are one way to analyze and synthesize this type of information. This knowledge forms the basis for a process, which is led by farmers and their organizations, to seek diversified livelihood strategies that go beyond coffee production. Depending on the characteristics of farmers and landscapes, these diversified livelihood strategies could include strengthening local food security, developing agroecotourism, handicrafts, community forestry, nontimber forest products, or adding value by increasing the involvement of farmers in the coffee processing and marketing chain, as many regional Mexican and some Nicaraguan coffee cooperatives have done. In Mexico, ISMAM produces their own vacuum-packed canned coffee, and UCIRI even sells small jars of instant coffee, and in Nicaragua a few cooperatives have started selling specialty coffee domestically and even launched a line of all-female-produced coffee (under the name Flor de

Café). A desire for these types of alternatives was explicitly expressed by coffee farmers in El Salvador in a focus group held in 2001, during a critical period in the recent coffee crisis (Méndez 2004). However, it is important to point out that this will require in-depth case studies and context-specific alternatives that will take more time and effort than most conventional development projects.

Successful initiatives to diversify livelihoods will require the creation of strong partnerships and networks that include farmers, researchers, governments, development and conservation organizations, the coffee industry, and engaged consumers. In this volume, Jaffe and Bacon analyze an example of one such initiative: the Community Agroecology Network. Larger-scale efforts include Equal Exchange in Boston and Cafédirect in the United Kingdom. A concerted effort is necessary to take more alternative initiatives to scale. Policy actions are called for at scales ranging from the local to the national and international. This approach will be difficult to apply broadly without the support of state and international development policy makers. The more successful interventions will take process into account as they address farmers' immediate survival needs while working toward longer-term sustainability. In this way, sustainable coffee initiatives hold the potential to connect Northern efforts that seek more meaning through sustainable consumption with Southern sustainability efforts that start with survival as their top priority.

Certification Systems: North-South Collaboration or "Institutionalized Mistrust"
Mutersbaugh's research (this volume and 2002) shows how this paradox of North-South relations unfolds within coffee communities and organizations. Mutersbaugh shows the contradictory nature of how certified organic production requires that village members serve distant institutions, translating organic farming practices for international certifiers and complicating them for their smallholder neighbors. While Fair Trade and organic certification systems are widely presented by their advocates in the North as emblematic of their concern for small-scale farmers, third-party monitoring reflects the need to assure buyers that coffee is indeed produced and traded under specified conditions. Certification systems are based on the principle of submitting to external scrutiny in exchange for a price premium, given that both intermediaries and consumers are understandably reluctant to pay such premiums on faith. As a result, certification processes reflect a system of "institutionalized mistrust." Few examples of the violation of basic standards are needed to damage the credibility of certification more generally. (See e.g. Weitzman 2006b.) Since the system requires scrutiny of producers rather than consumers, there is an inherent imbalance in how it is perceived, in spite of its official discourse in favor of North-South equality and producer-consumer collaboration.

While many of these partnerships are indeed value-driven for some participants, there is no escaping the fact that these are market-based partnerships as well, based at least as much on interests as on ideas. One forum in which these issues play out is the negotiations about price premiums and contract fulfillment. While certification systems impose distinct burdens on producers and their organizations, it is important to recognize that they function in qualitatively different ways. Neither Utz Kapeh nor Rainforest Alliance provides guaranteed price premiums to exporters or producers. Organic certification does not guarantee a premium; nor do organic premiums necessarily cover the additional labor costs required (Calo and Wise 2005). However, high demand and coffee quality have resulted in organic premiums that averaged 0.42/pound above conventional prices during the 2002–03 coffee harvest (Méndez et al., forthcoming). This should not be interpreted as an upward trend for organic coffee price premiums since these data refer to a time period with very low international commodity prices. Whether or not Fair Trade buyers will actually increase their purchase price when the market price rises remains an open empirical question. In principle, the rules state that if the market price is higher than the floor price, the market price plus a Fair Trade (social) premium shall apply (FLO 2004, p. 10). However, in practice, when the market price recently rose above the Fair Trade price and coffee shortages decreased cooperatives' total available production, significant tensions emerged between cooperatives and some importers, leading to a 4 percent contract default rate in 2005 (Camps et al. 2005). Some Fair Trade intermediaries pushed hard to hold producer cooperatives to the pre-season purchase price, converting the price floor into a de facto ceiling. Smallholders may be best off when they combine Fair Trade and certified organic sales. However, declining real price premiums and the fact that this requires significantly more labor lessens the appeal of these celebrated farming and trade practices.

This paradox highlights a few of the increasingly visible North-South tensions associated with expanding certified coffee markets. These incidents indicate a growing culture of mistrust that has accompanied new certified coffee initiatives and mainstreaming strategies that do not involve solidarity and direct people-to-people relationships. These relationships will need investment and conscious nurturing if people seek to maintain the "alternative" principles within food systems that initially shared a common effort to redistribute value and reconvene "trust" between food producers and consumers (Whatmore et al. 2003, p. 389). The ability to reconvene trust and redistribute value speaks to closer alliances among producers and consumers and addresses issues of accountability and transparency within the alternative trade network governance structures.

Accountability and Transparency in the Supply Chain?

Fair trade and organic systems often claim to have fewer intermediaries in the supply chain than conventional trade. However, from the perspective of institutional economics it is important to study the functions that the old intermediaries once performed, since new actors will have to perform many of them (Warning 2006). The actors involved in each stage of certified coffee trade and production often have significant differences among them in terms of their accountability relationships and access to market information. Small producers gain market leverage and institutional capacity to the degree that they scale up to form larger cooperatives and federations. Larger memberships can also increase cooperatives' influence in the policy process, for those that campaign for access to government supports, such as access to credit, infrastructure investments, or payments for environmental services. Indeed, in Mexico producers have been creating grassroots organizations for decades and many have managed to leverage government support programs. (See, e.g., Celis 2003; Snyder 2001; Ejea and Hérnandez Navarro 1991.) Field research shows that without these programs the fraction of producers who lose money on their coffee crop would be substantially higher (Calo and Wise 2005; Lewis and Runsten 2006).

However, pursuit of economies of scale can involve costs for cooperatives as well. In addition to the overhead involved in maintaining larger institutions, additional costs that come with greater size include the greater potential for distance between leadership and membership. Where larger cooperatives include members that produce coffee of varying quality, the leadership's need to be accountable to the membership as a whole may create incentives to use returns from the highest-quality coffee to subsidize others' production. This internal cross-subsidy may generate important advantages for the organization, especially if a larger membership is indeed associated with greater policy influence. Yet if this strategy is not the result of a fully informed democratic decision, the risks of alienating those members with higher-quality coffee go up.[6]

Larger cooperatives, with more ambitious financial and marketing operations, will have a more difficult time keeping the membership fully informed about their activities, unless major investments are made in both financial transparency and grassroots economic literacy. It is difficult for leaders to keep members informed about why various fees and the costs of processing, credit, and transportation costs are discounted from the prices that the cooperatives pay their members for coffee—especially if the leaders themselves lack full information about decision making, risks, and opportunities throughout the rest of the supply chain. This may become especially problematic when market prices rise. Small-scale producers with limited access to market information are easily confused when private buyers offer prices that are similar to or higher than the prices their own cooperatives offer—but only at the end of the harvest, when little coffee is available to be actually traded. If cooperatives sign fixed-price contracts as insurance against price decline and then the market prices rise, the cooperatives must decide whether to break the contracts with international buyers or to take a loss and try to convince their members to sell to the cooperative for lower farm gate prices. The challenges involved in explaining and managing perceptions of prices involve issues of accountability, trust, and transparency between different actors in the supply chain.[7]

Scaling Up or Selling Out: The Role of Producers' Voices in Making Fair Trade Coffee Fairer

The possible tradeoffs between scale and values provoke intense debates in the coffee industry. Since both Utz Kapeh and Rainforest Alliance launched their certifications of large-scale coffee plantations and a market-growth strategy based on transnational corporations, the question of selling out seems to some to be answered from the outset. Scholars, journalists, food activists, and conscientious eaters have wrestled with this same tension for more than a decade as big companies and large landholders threaten to displace the smallholder and local market pioneers in the rapidly expanding organic industry (Goodman 2002; Vos 2002; Guthman 2004; Pollan 2006). Fair Trade coffee's roots in solidarity-based social movements and smallholder cooperatives combined with the current mainstreaming strategy provides more fertile ground for debate (Goodman, this volume; Jaffe and Bacon, this volume; Renard 2005; Jaffee 2007; O'Nions 2006; Murray et al. 2006).

At its core, this debate concerns the extent to which Fair Trade can avoid being co-opted by the corporate centered market system it was set up to challenge and transform. Specific interrelated criticisms have emerged from multiple directions within and surrounding the Fair Trade movement. Southern rural producers' associations, such as Via Campesina and El Movimento Sem Terra (MST), have focused their organizing and farming efforts on food sovereignty, which refers to peoples' right to define the type of food and agriculture they want, including their ability to access sufficient healthy food, and capability to determine the degree of food self-reliance consistent with their cultural values (Rosset 2003; Via Campesina 2006). Although a food-sovereignty-centered strategy does not negate trade, these organizations and their allies have criticized Fair Trade's narrow focus on exports, instead of embedding Fair Trade in an approach that prioritizes increased support and protection for local and national food production, as well as agroecological farming practices (O'Nions 2006, p. 21).

Within the Fair Trade coffee movement, critical debates continue to brew around governance issues including the composition of FLO's Board of Directors, minimum guaranteed price premiums, licensing products sold by transnational companies, and the possible inclusion of large coffee plantations (Renard 2005).[8] Southern producers have little voice in global coffee markets and the four dominant coffee-certification initiatives. However, this is beginning to change in the Fair Trade system, as certified cooperatives in Latin America have organized to form Coordinadora Latinoamericana y del Caribe de Pequeños Productores de Comercio Justo (CLAC). Producers now hold four of the twelve seats on Fairtrade Labelling Organizations' board of directors, including a seat named by CLAC's general assembly. Among the aspects of Fair Trade governance discussed at regional CLAC meetings were the following:

• CLAC called for the inclusion of two consumer representatives on the FLO board of directors, and the restructuring of the board to be accountable to a general assembly.

• CLAC stated resistance to attempts from some national certification initiatives and industry members to lower the Fair Trade minimum price. Later CLAC commissioned a study, which demonstrated that real Fair Trade prices have declined for at least 10 years, while the costs of sustainable production have increased (Bacon 2006). In October 2006, CLAC's general assembly reviewed the study and proposed that by the end of 2008 FLO increase the minimum Fair Trade conventional prices by 12 percent and the combined Fair Trade and organic prices by 21 percent. The FLO Board of Directors responded within 6 months with a 7 percent increase for Fair Trade conventional coffees and an 11 percent increase for Fair Trade organic coffees (FLO 2007). They have also committed to conducting an extensive study to assess the costs of sustainable and develop a proposal Fair Trade coffee minimum prices in 2007.

• CLAC clearly stated its opposition to including large-scale landholders in the Fair Trade system and caution regarding the participation of transnational companies (Renard 2005; CLAC 2006). CLAC also argued against the participation of large-scale transnational corporate exporters, such as Atlantic coffee in this system.

• CLAC stated that it did not want to limit Fair Trade market growth, that it "welcome[s] companies willing to make a serious commitment to Fair Trade," and that it "dislike[s]" what happens when, for example, "a company that dominates 25 percent of the world market for one product . . . decides to buy only 0.002 percent of their annual coffee as [Fair Trade-certified] and in all their propaganda claims that they are now a company that is part of the fair trade system" (CLAC 2006b, p. 4).

Different actors in the system will have different interests and perspectives on this paradox. For example, Starbucks' Fair Trade marketing campaigns have been seen as "greenwashing" by some socially aware consumers and by many small roasters who are more heavily committed to sustainable coffee. Scholars, businesses, and activists have criticized "Starbuckian" behavior for their coffee-purchasing strategies and for their aggressive retail behavior (Utting 2007). Others celebrate Starbucks for their corporate social responsibility, including the provision of health care to temporary employees and relatively transparent social and environmental reporting practices. For many farmer export cooperatives, the fact that Starbucks purchased 11.5 million pounds of Fair Trade coffee in 2004 clearly matters, especially for the cooperatives that otherwise may have had difficulty finding enough Fair Trade buyers. While only 3.7 percent of Starbucks coffee was Fair Trade in 2004, this accounted for more than 25 percent of the Fair Trade coffee sales in the United States (Starbucks Coffee 2005; TransFair USA 2005). In other words, when it comes to assessing Starbucks' role, producer cooperatives and small roasters do not share exactly the same interests. When different actors make specific recommendations for sustainable coffee, a careful analysis of interests reminds us that where one stands often depends on where one sits.

Starbucks' commitment to corporate social responsibility and sustainable coffee is clearly well beyond that of Nestlé. Nestlé launched Nescafé Partners' Blend as its first—and only—Fair Trade-certified product in 2005. This purchase probably represents less than 0.002 percent of their total coffee sales, and only one of Nestlé's 8,500+ products (CLAC 2006a; O'Nions 2006). Fair Trade producers associations and social-movement organizations (including the World Development Movement, which helped create the UK-based Fairtrade Foundation that licensed Nestlé to sell this product) have protested vociferously (O'Nions 2006).

As certified sustainable coffee initiatives enter the mainstream, motivated actors will continue to push the next innovation, taking certification as a starting point, while other actors struggle toward a certified finishing line. Many of the actors (such as Global Exchange and Lutheran World Relief) that are lobbying some of the largest coffee companies to carry Fair Trade and organic coffees are also launching their own sustainable coffee enterprises. Movement-motivated organizations are developing these strategies at different scales ranging from domestic Fair Trade initiatives in the Global South and community-based certifications, to attempts to form global "alternative" food networks and the increasing common joint ventures among the more committed Fair Trade companies, coffee-producing cooperatives, and even churches. This market-based competition is too often measured in terms of total sales, press coverage, and number of participating retails outlets. If sustainable coffee advocates do not soon re-orient their success measures to address issues of social development, empowerment processes, and environmental health, these movements risk participating in a process that sells out in order to scale up.

More than 2,500 years ago, Lao Tzu said "The Great Integrity is a Paradox." (2002, p. 43) He encouraged all those searching for truth to both celebrate and nourish paradox. We follow this tradition by suggesting that people interested in confronting the coffee crisis and sustaining Mesoamerican livelihoods and ecosystems must also, as Diane Rocheleau (1999) eloquently states, "confront complexity and deal with difference." We have found that cooperative (UWCC 2005) and agroecological (www.agroecology.org) principles can serve as effective tools to deepen both dialogue and practice as actors negotiate paradox in search of strategies to confront the coffee crisis without reproducing the same structures that created it. These principles can serve as evaluative concepts that guide interdisciplinary analysis and international development interventions intended to support dynamic transitions toward sustainability in coffee territories (Méndez and Gliessman 2002).

Conclusions

What will the future look like in Mesoamerican coffee territories? It is clear that most smallholder households have developed three primary livelihood survival strategies: diversification, migration, and attempts to increase their total income from coffee sales. This book reveals the heterogeneity and interconnections among changing farmers' livelihoods, shade coffee production, and sustainable coffee initiatives. This understanding should inform any intervention intended to sustain the peoples, cultures, communities, and ecological processes in these coffee territories. Paradox will also accompany strategies for change. The preceding chapters have addressed strategies related to increasing opportunities through participation in certified coffee programs and diversification.

The field-based evidence available so far does not support hopes that these certified coffee markets will be a "magic bullet" cure for structural poverty and crisis. However, some of these programs, notably Fair Trade and organic networks, have played important roles in supporting smallholders' organizations, biodiversity conservation, reducing vulnerability to the coffee crisis, increasing international awareness of the social and environmental costs of the current coffee system, and creating a savvy group of smallholder coffee producers, now actors on the international stage. The constraints include price volatility, North-South power imbalances, declining and in some cases non-existent price premiums received at the farm gate, the many certification costs producers pay, and a general lack of effort to seek support from the state.

While the more likely assessment suggests the persistence of the same imbalanced global coffee economy (Topik and Clarence-Smith 2003), the chapters in this volume suggest that a few political changes during the second half of 2006 may carry with them seeds of change. Election results throughout Latin America show that voters are increasingly rejecting the consequences of neo-liberal economic policies. One example of these changes is evident in Nicaragua, where the government is negotiating plans to shift some government support to cooperatives and other organizations promoting a more social economy. Second, the Latin American and Caribbean network of Small Fair Trade Producers' research and lobbing efforts convinced Fairtrade Labelling Organizations International to increase minimum prices paid to cooperatives by 4–7 percent and to commit to studying the costs of sustainable production. This move is contrary to the general tendency among all certifiers, which is to generally increase quality standards and requirements, while real price premiums remain stagnant. If the Fair Trade system is able to overcome a host of other problems, and other certifications follow suit, this could represent a step in a race to the top instead of scaling up and selling out. However, what happens in the coffee territories will depend on how producers and their organizations negotiate these risks and opportunities.

In addition to these evolving relations with states and markets, producers' survival and sustainable development processes will depend on their ability to build accountable and efficient organizations, increase yields, prioritize food sovereignty and diversify. While many coffee cooperatives have increased business capacity during the last decade, even within these organizations opportunities to improve accountability and efficiency exist (Méndez et al. 2006). A well-planned place- and livelihood-specific diversification strategy could simultaneously increase food sovereignty, sustain agroecological processes, and improve yields in optimum coffee-growing regions. These efforts will require increased investment from producers as well as from their allies in government, business, and civil society. If certified coffee initiatives are to become an important component in this strategy, they will have to be scaled up.

Indeed, the fact that only about 2–4 percent of the global coffee supply is sold through any kind of certified markets and the fact that only 20 percent of the Fair Trade-certified coffee (about 1 percent of the world supply) is sold through these "alternative" channels suggest that one of the main imbalances in the world coffee system is between organized producers in the South and the largely individualistic environmentally and socially aware consumers in the North. Just as the environmental movement over the past several decades has made gains that support public access and collective ownership, often in contradiction with the neo-liberal economic models, coffee certification must expand its base to link social and environmental goals that bring consumers and growers into closer and more equal relationships.[9] Scholars can play an important role in networking knowledge in service of these processes, and out of paradox they can create opportunity for change that cannot be easily co-opted and concentrated in the hands of a few (Freire 1985; Sevilla-Guzmán and Woodgate 1997; Fals-Borda and Rahman 1991; Fox 2006; Prechtel 2003).

When Northern citizens and consumers have "caught up" and are able to hold businesses, certification agencies, and governments accountable to their public claims and responsibilities to support human rights and sustainable development, while simultaneously working through markets by purchasing at least as much coffee as certified smallholder cooperatives can produce, then we will be better positioned to assess the degree to which sustainable coffee initiatives really offer a long-term alternative. In the meantime, there are plenty of opportunities to serve a little less milk and a little more social and environmental justice (Zabin 2006).

Notes

1. Note that this evidence challenges the simple dichotomous assumption that farming inherently undermines biodiversity. *The Economist* editorialized "buy organic, destroy the rainforest" (2006a).

2. This account is also applicable to agricultural certification programs, including organic and the Smithsonian Bird Friendly certification.

3. Another example of the institutionalized mistrust that frequently accompanies attempts to use certification as a tool to overcome North-South divides concerns the specific requirements to comply with certification efforts. González and Nigh address a few of these tensions in their 2005 paper, which highlights increasing tendencies toward larger landholder takeover in Mexico's certified organic agricultural sector. They also question organic inspectors' requirements for composting as unnecessary in many tropical soils. While we disagree with this from an agroecological perspective since farmers will probably be interested in compost to increase their coffee yields, we agree with their effort to highlight the general North-South tensions. Perhaps a better example is the individual terraces that many inspectors have required on Mexican certified organic farms. An interview conducted during a recent field study in Mexico found that one of the primary justifications for this practice was so that neighbors and visitors to the community could "see" the difference between certified organic farms and uncertified ones (Trujillo 2006).

4. A calculation evaluating the fairness in the distribution of coffee rents compares the percentage of the final retail price returned to farmer. Ponte has estimated that a cup of specialty coffee sold at Starbucks returns about 1 percent to the farmer, whereas a higher percentage is returned if the coffee is sold roasted but not brewed. The estimates for organic coffee are about the same. A Fair Trade whole bean espresso roast sold by the pound and on sale at the supermarket returned 21 percent to the export cooperative (about the same as the target under the International Coffee Agreement) and about 11 percent when sold at a specialty coffee shop (Ponte 2005).

5. The *Boston Globe* recently reported that "according to the National Coffee Association, more than half of coffee drinkers who have heard of Fair Trade buy it—nearly twice the rate of organic" (Dicum 2006).

6. Only large cooperative organizations with complex internal redistributive pricing structures are able to use the income from premium prices to partly compensate the losses to transitional producers by offering them intermediate prices, as in the case of Oaxaca's CEPCO, where costs of certification reached an estimated 3.4 percent of total sales (Calo and Wise 2005). In the CEPCO case, however, the complex internal price system may have contributed to some members' loss of trust in the central leadership, leading eventually to a split in the organization.

7. Another example of these North-South tensions concerns debates about the governance and standard setting within these certified initiatives (Muradian and Pelupessy 2005; Renard 2005). Although Southern producers have traditionally had little or no voice within all four of the dominant sustainable coffee-certification initiatives, this is beginning to change. In fact, producers have used Fair Trade networks as a tool to strengthen the ties among smallholder coffee cooperatives and, after intense lobbying, win four of the twelve seats on FLO's board of directors. From one perspective, these represent continued South-North tensions; from another angle, they show that the movement has moved beyond romantic notions of solidarity with relatively small rural development impacts to clearly political debates and business negotiations. While this may be healthy, the challenge is to simultaneously recover strong solidarity ties amidst the processes of expanding the market.

8. In 2004, committed coffee companies and civil-society organizations lobbied to have Transfair USA list all licensees by pounds of Fair Trade coffee sold and by the percentage of their total sales that number represented. This movement was led by Equal Exchange, Dean's Beans, Peace Coffee, and Cooperative Coffees, which at the time represented about one-half of all Fair Trade coffee sold in the United States. Transfair USA refused to provide the information or even to ask the licensees to reveal it, citing "corporate confidentiality." According to the same editorial, "it seems incongruous that in a movement that demands transparency by the farmers, a similar demand is not made of the companies—especially when those figures would give consumers a fair and complete picture of a company's commitment to fair trade thereby strengthening their capacity to make a reasoned choice" (Cycon 2005).

9. Coffee also offers a powerful medium that could allow socially responsible businesses including many actors within the Specialty Coffee Association of America to play a role in efforts to unite environmental and global social justice movements. Of course, the foundations of these links would be value alignment outside of the markets. Some of these connections are already visible in the close ties between organic with fairly traded coffees, but more effort is necessary if these two movements are going to build stronger alliances and then create "alternative" food networks that emerge in a more participatory way that keeps the important components of the power and the rewards more firmly in the hands of the least empowered—including the farmers coffee-producing communities, and the coffee drinkers. As this journey develops, it will become increasingly important to understand and promote the state's public policy interventions, since these policies can either create or undermine the social institutions that will facilitate fairer trade and a more inclusive sustainable rural development process (Miranda 2003). Scholars with a more historic perspective are also quick to point out the fact that many of the most promising certification initiatives are very insignificant in their efforts to reclaim the benefits that the state provided, such as minimum prices to producer countries, through the quota system within the International Coffee Agreement (Talbot 2004). However, others have found promise within the new flexibility and closer relationships between roasters and producers found within the current international market.

References

Aranda Bezaury, J. 2006. El impacto de la migración en los hogares cafeterleros: El nuevo rol de la mujer campesina. Presented at XXVI International Congress of Latin American Studies Association, San Juan, Puerto Rico.

Babbar, L. I., and D. R. Zak. 1995. Nitrogen loss from coffee agroecosystems in Costa Rica—Leaching and denitrification in the presence and absence of shade trees. *Journal of Environmental Quality* 24, no. 2: 227–233.

Bacon, C. M. 2006. Estudio de costos y propuesta de precios para sostener el café, las familias de productores y organizaciones certificadas por Comercio Justo en América Latina y el Caribe. Coordinadora Latino Americana y del Caribe de Pequeños Productores de Comercio Justo.

Bacon, C. M. forthcoming. A spot of coffee in crisis: Nicaraguan smallholder cooperatives, fair trade networks and gendered empowerment processes. *Latin American Perspectives.*

Bacon, C. M. 2005. Confronting the Coffee Crisis: Nicaraguan Smallholders Use of Cooperative, Fair Trade, and Agroecological Networks to Negotiate Livelihoods and Sustainability. Ph.D. dissertation, University of California, Santa Cruz.

Beattie, A. 2005. Nestlé embraces "Fair Trade" label. *Financial Times,* October 7.

Beer, J., R. Muschler, D. Kass, and E. Somarriba. 1998. Shade management in coffee and cacao plantations. *Agroforestry Systems* 38, no. 1–3: 139–164.

Benquet, F.M. 2003. Crisis cafetalera y migración internacional en Veracruz. *Migraciones Internacionales* 2, no. 2: 121–148.

Blaikie, P., T. Cannon, I. Davis, and B. Wisner, eds. 1994. *At Risk: Natural Hazards, People's Vulnerability, and Disasters.* Routledge.

Brush, S. 2004. *Farmers' Bounty: Locating Crop Diversity in the Contemporary World.* Yale University Press.

Calo, M., and T. Wise. 2005. Revaluing Peasant Coffee Production: Organic and Fair Trade Markets in Mexico. Globalization and Sustainable Development Program, Tufts University.

Camp, M., S. Flynn, A. Portalewska, and T. Tidwell Cullen. 2005. A cup of truth. *Cultural Survival Quarterly* 29, no. 3.

Celis, F. 2003. Nuevas formas de asociacionismo en la cafeticultura mexicana: El caso de la CNOC. In *Sociedad civil, esfera pública y democratización en América Latina: México,* ed. A. Olvera. Fondo de Cultura Económica/Universidad Veracruzana.

Chayanov, A. V. 1966. *The Theory of the Peasant Economy.* University of Wisconsin Press.

Coordinadora Latinoamericana y del Caribe de Pequeños Productores de Comercio Justo (CLAC). 2006a. II Asamblea General (VI Asamblea Regional) del CLAC, Dominican Republic.

Coordinadora Latinoamericana y del Caribe de Pequeños Productores de Comercio Justo (CLAC). 2006b. Informe de la Junta Directiva (2004–2006). II Asamblea General (VI Asamblea Regional) del CLAC, Dominican Republic.

Cycon, D. 2005. Confessions of a Fair Trader. *Cultural Survival Quarterly* 29, no. 3.

Damiani, O. 2002. Organic Agriculture in Guatemala: A Study of Producer Organizations in the Cuchumatanes Highlands. Office of Evaluation Studies, IFAD.

Daviron, B., and S. Ponte. 2005. *Global Markets, Commodity Trade and the Elusive Promise of Development*. Zed Books.

Deutsche Bank. 2000. Soluble Coffee: A Pot of Gold?

Dicum, G. 2006. Fair to the last drop? *Boston Globe*, October 22.

Dietsch, T. V., S. M. Philpott, R. A. Rice, R. Greenberg, and P. Bichier. 2004. Conservation policy in coffee landscapes. *Science* 303: 625–626.

Donald, P. F. 2004. Biodiversity impacts of some agricultural commodity production systems. *Conservation Biology* 18, no. 1: 17–37.

Economist. 2006a. Ethical food: Good food? December 9.

Economist. 2006b. Special report: Food politics, voting with your trolley. December 9.

Ejea, G., and L. Hérnandez Navarro, eds. 1991. Cafetaleros: La construcción de la autonomia. Servicio de Apoyo Local, Mexico City.

Fals-Borda, O., and M. A. Rahman, eds. 1991. *Action and Knowledge: Breaking the Monopoly with Participatory Action-Research*. Apex.

FLO (Fairtrade Labelling Organizations International). 2004. Criterios de Comercio Justo para Café. June draft.

FLO. 2007. FLO announces increase in Fairtrade premium and organic differential for coffee. www. fairtrade.net.

Food First (Institute for Food and Development Policy). 2006. http://www.foodfirst.org/.

Fox, J. A. 1994. Targeting the poorest: The role of the National Indigenous Institute in Mexico's National Solidarity Program. In *Transforming State-Society Relations in Mexico: The National Solidarity Strategy*, ed. W. Cornelius, A Craig, and J. Fox. Center for US-Mexican Studies, University of California, San Diego.

Fox, J. A. 1996. How does civil society thicken? The political construction of social capital in rural Mexico. *World Development* 24, no. 6 : 1089–1103.

Fox, J. A. 2006. The impact of migration on coffee production in southern Mexico. Discussant comments, XXVI International Congress of the Latin American Studies Association, San Juan, Puerto Rico.

Fox, J. A. 2006. Lessons from action-research partnerships. *Development and Practice* 16: 27–38.

Freire, P. 1985. *The Politics of Education : Culture, Power, and Liberation*. Macmillan.

Gallina, S., S. Mandujano, and A. Gonzalez-Romero. 1996. Conservation of mammalian biodiversity in coffee plantations of central Veracruz. *Agroforestry Systems* 33: 13–27.

Giovannucci, D., and S. Ponte. 2005. Standards and a new form of social contract? Sustainability initiatives in the coffee industry. *Food Policy* 30: 284–301.

Giovannucci, D., and F. Koekoek. 2004. The State of Sustainable Coffee: A Study of Twelve Major Markets. CENICAFÉ.

Gliessman, S., ed. 2000. *Agroecosystem Sustainability: Developing Practical Strategies*. CRC Press.

Goodman, D. 2000. The changing bio-politics of the organic: Production, regulation, consumption. *Agriculture and Human Values* 17: 211–213.

Goodman, D., and M. Watts, eds. 1997. *Globalising Food: Agrarian Questions and Global Restructuring*. Routledge.

González, A. A., and R. Nigh. 2005. Smallholder participation and certification of organic farm products in Mexico. *Journal of Rural Studies* 21: 449–460.

Guthman, J. 2004. Back to the land: The paradox of organic food standards. *Environment and Planning A* 36: 511–528.

Harford, Tim. 2006. *The Undercover Economist*. Oxford University Press.

Hérnandez Navarro, L. 2004. To Die a Little: Migration and Coffee in Mexico and Central America. International Relations Center, Americas Program. www.americaspolicy.org.

Hérnandez Navarro, L., and F. Celis. 2004. Solidarity and the new campesino movements: The case of coffee production. In *Transforming State-Society Relations in Mexico: The National Solidarity Strategy*, ed. W. Cornelius, A. Craig, and J. Fox. Center for US-Mexican Studies, University of California, San Diego.

Inter-American Development Bank (IADB). 2000. Social Protection for Equity and Growth.

IADB. 2002. Managing the Competitive Transition of the Coffee Sector in Central America. Background report for conference.

ICO (International Coffee Organization). 2005. Exports by Exporting Countries to All Destinations July 2003 to June 2004.

ICO. 2005. Organic Coffee Export Statistics.

Jaffee, D. 2007. *Brewing Justice: Fair Trade Coffee, Sustainability and Survival*. University of California Press.

Kaplinsky, R. 2000. Spreading the Gains from Globalization: What Can Be Learned from Value Chain Analysis? IDS Working Paper 110.

Katzeff, P. 2002. The Cuppers Manifesto. Thanksgiving Coffee Company.

Lao Tzu. 2002. *Tao Te Ching*. Barnes & Noble.

Lewis, J., and D. Runsten. 2006. Does Fair Trade Coffee have a future in Mexico? The impact of migration in a Oaxacan community. Presented at XXVI International Congress of Latin American Studies Association, San Juan, Puerto Rico.

Lewin, B., D. Giovannucci, and P. Varangis. 2004. Coffee Markets: New Paradigms in Global Supply and Demand. World Bank.

Martínez-Torres, M. E. 2006. *Organic Coffee: Sustainable Development by Mayan Farmers*. Ohio University Press.

Mas, A. H., and T. V. Dietsch. 2004. Linking shade coffee certification to biodiversity conservation: Butterflies and birds in Chiapas, Mexico. *Ecological Applications* 14, no. 3: 642–654.

Méndez, V. E. 2004. Traditional Shade, Rural Livelihoods, and Conservation in Small Coffee Farms and Cooperatives of Western El Salvador. Ph.D. thesis, University of California, Santa Cruz.

Méndez, V. E., and S. R. Gliessman. 2002. Un enfoque interdisciplinario para la investigación en agroecología y desarrollo rural en el trópico Latinoamericano. *Manejo Integrado de Plagas y Agroecología (Costa Rica)* 64: 5–16.

Méndez, V. E., C. Bacon, S. Petchers, D. Herrador, C. Carranza, L. Trujillo, C. Guadarrama-Zugasti, A. Cordón, and A. Mendoza. 2006. Sustainable Coffee from the Bottom Up: Impacts of Certification Initiatives on Small-Scale Farmer and Estate Worker Households and Communities in Central America and Mexico. Research report, Oxfam America.

Mendoza, R. 2002. La Paradoja del Café: El Gran Negocio Mundial y la Peor Crisis Campesina. Nitlapán-UCA.

Mendoza, R., and J. Bastiaensen. 2003. Fair trade and the coffee crisis in the Nicaraguan Segovias. *Small Enterprise Development* 14: 36–46.

Miranda, B. A. 2003. Captial Social, Institucionalidad y Territorios: El Caso de Centroamamérica. San Jose, Costa Rica: Instituto Interamericano de Cooperación Agrícola.

Moguel, P., and V. Toledo. 1999. Biodiversity conservation in traditional coffee systems of Mexico. *Conservation Biology* 13, no. 1: 11–21.

Moore, G. 2004. The Fair Trade movement: Parameters, issues and future research. *Journal of Business Ethics* 53: 73–86.

Muradian, R., and W. Pelupessy. 2005. Governing the coffee chain: The role of voluntary regulatory systems. *World Development* 33, no. 12: 2029–2044.

Murray, D., L. Raynolds, and P. Taylor. 2006. The future of Fair Trade coffee: Dilemmas facing Latin America's small-scale producers. *Development in Practice* 16, no. 2: 179–192.

Muschler, R. G. 1997. Shade or sun for ecologically sustainable coffee production: A summary of environmental key factors. In Proceedings of the 3rd Scientific Week. CATIE.

Mutersbaugh, T. 2002. The number is the beast: The political economy of organic coffee certification and producer unionism. *Environment and Planning A* 34: 1165–1184.

Mutersbaugh, T., K. Daniel, M.-C. Renard, and P. Taylor. 2005. Certifying rural spaces: Quality-certified products and rural governance. *Journal of Rural Studies* 21: 381–388.

Netting, R. 1996. *Smallholders, Householders: Farm Families and the Ecology of Intensive, Sustainable Agriculture.* Stanford University Press.

O'Nions, J. 2006. Fairtrade and global justice. *Seedling,* July: 18–21.

O'Rourke, D. 2003. Outsourcing regulation: Analyzing nongovernmental systems of labor standards and monitoring. *Policy Studies Journal* 31: 1–29.

Pelupessy, W., and R. Muradian. 2005. Governing the coffee chain: The role of voluntary regulatory systems. *World Development* 33, no. 12: 2029–2044.

Peluso, N. 1994. *Rich Forests Poor People: Resource Control and Resistance in Java.* University of California Press.

Perfecto, I., R. Greenberg, and M. Vand der Voort. 1996. Shade coffee: A disappearing refuge for biodiversity. *BioScience* 46, no. 8: 598–609.

Perfecto, I., A. Mas, T. Dietsch, and J. Vandermeer. 2003. Conservation of biodiversity in coffee agroecosystems: A tri-taxa comparison in southern Mexico. *Biodiversity and Conservation* 12: 1239–1252.

Perfecto, I., J. Vandermeer, A. Mas, and L. Soto-Pinto. 2005. Biodiversity, yield, and shade coffee certification. *Ecological Economics* 52: 435–446.

Philpott, M., P. Bichier, R. Rice, and R. Greenberg. Forthcoming. Field testing ecological and economic benefits of sustainable coffee certification: Do organic and fair trade do enough? *Conservation Biology.*

Pollan, M. 2006. *The Omnivore's Dilemma: A Natural History of Four Meals.* Penguin.

Ponte, S. 2004. Standards and Sustainability in the Coffee Sector: a Global Value Chain Approach. United Nations Conference on Trade and Development and International Institute for Sustainable Development.

Ponte, S. 2005. The coffee paradox: The commodity trade and the elusive promise of development. Presentation to International Coffee Organization.

Porter, R. 2002. *The Coffee Farmers' Revolt in Southern Mexico in the 1980s and 1990s.* Edwin Mellon.

Prechtel, M. 2003. *The Toe Bone and the Tooth.* HarperCollins.

Rainforest Alliance. 2006. Products from Farms Certified by the Rainforest Alliance—Coffee Producers. Accessed March 3,2006. Available at: http://www.rainforestalliance.org/programs/agriculture/shop/coffee-producers.html

Rappole, J. H., D. I. King, and J. H. Vega-Rivera. 2003. Coffee and conservation. *Conservation Biology* 17, no. 1: 334–336.

Raynolds, L. T. 2002. Poverty Alleviation through Participation in Fair Trade Coffee Networks: Existing Research and Critical Issues. Ford Foundation.

Renard, M.-C. 2005. Quality certification, regulation and power in fair trade. *Journal of Rural Studies* 21: 419–431.

Rice, P., and J. MacLean. 1999. Sustainable Coffee at a Crossroads. Consumers Choice Council

Rocheleau, D. 1999. Confronting complexity, dealing with difference: Social context, content, and practice in agroforestry. In *Agroforestry in Sustainable Agricultural Systems*, ed. L. Buck et al. CRC Press.

Rosset, P. 2003. Food sovereignty: Global rallying cry of farmer movements. *Backgrounder* [Institute for Food and Development Policy] 9, no. 4: 1–4.

Schroth, G., G. A. B. da Fonseca, C. A. Harvey, C. Gascon, H. L. Vasconcelos, and A. M. N. Izac, eds. 2004. *Agroforestry and Biodiversity Conservation in Tropical Landscapes.* Island.

Sevilla-Guzmán, E., and G. Woodgate 1997. Sustainable rural development: From industrial agriculture to agroecology. In *The International Handbook of Environmental Sociology*, ed. G. Woodgate and M. Woodgate. Elgar.

Skoufias, E. 2003. Economic crises and natural disasters: Coping strategies and policy implications. *World Development* 31: 1087–1102.

Snyder, R. 2001. *Politics after Neoliberalism: Reregulation in Mexico*. Cambridge University Press

Somarriba, E., C. Harvey, M. Samper, F. Anthony, J. Gonzalez, C. Staver, and R. Rice. 2004. Biodiversity in coffee plantations. In *Agroforestry and Biodiversity Conservation in Tropical Landscapes*, ed. G. Schroth et al. Island.

Starbucks Coffee. 2005. Beyond the Cup: Corporate Social Responsibility. Fiscal 2005 Annual Report.

Talbot, J. M. 1997. Where does your coffee dollar go? The division of income and surplus along the coffee commodity chain. *Studies in Comparative International Development* 32: 56–91.

Talbot, J. M. 2004. *Grounds for Agreement: The Political Economy of the Coffee Commodity Chain*. Rowman and Littlefield.

Toledo, V. M., and P. Moguel. 1997.Searching for sustainable coffee in Mexico: The importance of biological and cultural diversity. In *Proceedings of the First Sustainable Coffee Congress*, ed. R. Rice, A. Harris, and J. McLean. Smithsonian Migratory Bird Center.

Topik, S., and W. Clarence-Smith. 2003. *The Global Coffee Economy in Africa, Asia, and Latin America*. Cambridge University Press,

Topik, S., and K. Pomeranz. 1999. *The World Trade Created: Culture, Society and the World Economy, 1400 to the Present*. M. E. Sharpe,

TransFair USA. 2005. 2004 Fair Trade Coffee Facts and Figures.

Trujillo, L. E. 2000. Political Ecology of Coffee. Ph.D. thesis, University of California, Santa Cruz.

Trujillo, L. E. 2006. Resultados Finales Mexico. Final Case Study Report for the Project Sustainable Coffee from the Bottom Up. Universidad de Chapingo, Huatusco, Mexico.

University of Wisconsin Center for Cooperatives (UWCC). 2005. History and Theory of Cooperatives. http://www.wisc.edu.

US House of Representatives Commitee on International Relations. 2002. The Coffee Crisis in the Western Hemisphere.

Utting, K. 2007. Assessing the Sustainability Impacts of Ethical and Fair Trade Coffee: Cultivating Opportunities in Northern Nicaragua? Doctoral thesis, University of Leeds.

Utting, P., ed. 2002. *The Greening of Business in Developing Countries: Rhetoric, Reality and Prospects*. Zed Books and UNRISD.

Utting, P. 2005. Rethinking Business Regulation: From Self-Regulation to Social Control. UNRISD Programme Paper TBS-15.

Utz Kapeh. 2006. Certified Producers. http://www.utzkapeh.org.

VanderHoff, F. 2002. Poverty Alleviation through Participation in Fair Trade Coffee Networks: The Case of UCIRI, Oaxaca, Mexico. Fair Trade Research Group, Colorado State University, Fort Collins. http://www.colostate.edu.

Via Campesina. 2006. What Is La Via Campesina? What Are Its Priorities? http://www.viacampesina.org.

Vos, T. 2000. Visions of the middle landscape: Organic farming and the politics of nature. *Agriculture and Human Values* 17: 245–256.

Warning, M. 2006. Microeconomic considerations about Fair Trade coffee: The Impact of Migration on Coffee Production in Southern Mexico. XXVI International Congress of Latin American Studies Association. San Juan, Puerto Rico.

Weitzman, Hal. 2006a. Coffee with a conscientious kick. *Financial Times*, August 16.

Weitzman, Hal. 2006b. The bitter cost of "fair trade" coffee. *Financial Times*, September 8.

Whatmore, S., P. Stassart, and H. Renting. 2003. What's alternative about alternative food networks? *Environment and Planning A* 35, no. 3: 389–391.

Wisner, B. 2001. Risk and the neoliberal state: Why post-Mitch lessons didn't reduce El Salvador's earthquake losses. *Disasters* 25: 251–268.

Zabin, C. 2006. Coffee and migration in the Mixteca and the Sierra Norte de Oaxaca: A discussion. XXVI International Congress of Latin American Studies Association. San Juan, Puerto Rico.

About the Contributors

Christopher M. Bacon (christophermbacon@gmail.com) is an S.V. Ciriacy-Wantrup Postdoctoral Fellow in the Geography Department at the University of California at Berkeley. He finished his Ph.D. work in Environmental Studies at UC Santa Cruz and continued to teach in the departments of Sociology, Engineering, and Latin American and Latino Studies. Previously he served in Peace Corps Nicaragua and worked for the World Resources Institute. His work investigates linkages among globalization, alternative trade networks, changing landscapes, and struggles for rural survival and sustainability.

David B. Bray (brayd@fiu.edu), a professor of Environmental Studies, is Director of the Institute for Sustainability Sciences in Latin America and the Caribbean at Florida International University. He served as Foundation Representative at the Inter-American Foundation from 1986 to 1997. He has received research funding from the Ford, Hewlett, and Tinker Foundations and from the US Agency for International Development. He co-edited *The Community Forests of Mexico*.

Sasha Courville (sasha@isealalliance.org) is Executive Director of the ISEAL Alliance, a membership-based organization of leading social and environmental standards and conformity assessment initiatives. When she wrote her chapter for this volume, she was a research fellow at the Regulatory Institutions Network at the Australian National University. Her expertise is in social and environmental certification systems and in new forms of regulatory relationships between voluntary instruments and governments.

Jonathan A. Fox (jafox@ucsc.edu), a professor in the Department of Latin American and Latino Studies at the University of California at Santa Cruz, has done field research in rural Mexico since 1982. His books include *Indigenous Mexican Migrants in the United States* (co-editor, 2004), *Demanding Accountability: Civil*

Society Claims and the *World Bank Inspection Panel* (co-editor, 2003), and *Cross-Border Dialogues: US-Mexico Social Movement Networking* (co-editor, 2002). He has published articles in *Perfiles Latinoamericanos*, in *Annual Review of Political Science, Latino Studies*, and in *Development in Practice*. In 2004 he was awarded the LASA/OXFAM Martin Diskin Memorial Lectureship for his contributions to action research. His current projects focus on how accountability and transparency have affected rural development policies in Mexico.

Stephen R. Gliessman (gliess@ucsc.edu) is the Alfred E. Heller Professor of Agroecology at the University of California at Santa Cruz, where he is also a member of the faculty of Environmental Studies. He founded the UCSC Agroecology Program and co-founded the UCSC Program in Community and Agroecology and the non-profit Community Agroecology Network. His research interests include agroecology, food system sustainability, farm conversions, action education, agroecological education, and alternative production of wine grapes and olive oil.

David Goodman (hatters@ucsc.edu), professor and former chair of Environmental Studies at the University of California at Santa Cruz, explores social relations and institutional forms at the intersection of food systems, nature, and technological change. Currently he is researching the quality "turn" in food provisioning and the politics of localism in advanced industrial countries. He is a co-author of *From Peasant to Proletarian* (1981), *From Farming to Biotechnology* (1987), and *Re-fashioning Nature* (1991) and a co-editor of several volumes, including *Globalizing Food* (1997).

Carlos Guadarrama-Zugasti (carolusver@yahoo.es) is a research professor at the Centro Regional Universitario Oriente, an extension of the Universidad Autónoma Chapingo based in Huatusco, Veracruz, Mexico. He was a co-founder and the first director of the agroecology major at Chapingo University. His work focuses on the study of traditional agricultural technology, peasant economy, regional agriculture, agroecological education, and sustainable coffee agroecosystems.

Shayna Harris (shayna.harris@gmail.com) is a Fulbright scholar studying Fair Trade coffee cooperatives in Brazil. Previously she was Oxfam America's Coffee Program Organizer. Her work connects US high school, university, and community advocates with issues of social justice and globalization through the Make Trade Fair campaign. Work related to Fair Trade has led her on extensive travels throughout the United States, Ethiopia, Mexico, and Central America. Her senior honors thesis combined field research and analysis to describe indigenous migration patterns from a coffee-growing region of Southern Mexico to the United States. She is a founding member of the advisory board of United Students for Fair Trade.

Roberta Jaffe (rmjaffe@ucsc.edu) is a co-founder and a co-director of the Community Agroecology Network. Her career has been focused on environmental science education. She founded and served as Executive Director of Life Lab Science Program. She is also a lecturer in science education and sustainable agriculture at the University of California at Santa Cruz. An author of environmental science education curricula for elementary schools, she directs a national online mentoring program for beginning science teachers at UCSC's New Teacher Center.

María Elena Martínez-Torres (martineztorres@ciesas.edu.mx) is a faculty member in the Environment and Society Program of the Center for Research and Graduate Studies on Social Anthropology-Southeast Campus (CIESAS-Sureste), located in San Cristóbal de las Casas, Chiapas, Mexico. Also Director of Desarrollo Alternativo, AC, she is the author of *Organic Coffee: Sustainable Development by Mayan Farmers* (2006) and of a number of articles on the Zapatista movement in Chiapas.

V. Ernesto Méndez (emendez@uvm.edu) is an assistant professor in the Environmental Program and the Department of Plant and Soil Science at the University of Vermont. Previously a research associate and a lecturer in the Department of Environmental Studies at the University of California at Santa Cruz, he is a founding member of two sister nonprofit organizations: Advising & Interdisciplinary Research for Local Development and Conservation, in El Salvador, and the Community Agroecology Network, in California. He analyzes interactions between agriculture, rural livelihoods, and biodiversity conservation in tropical landscapes, with an emphasis on small-scale farmers and their organizations in Central America. Since 1999 he has contributed to participatory action research with coffee farmers in the coffee landscapes of Tacuba, El Salvador. More information on his work and publications can be found at http://www.uvm.edu/~emendez/.

Ellen Contreras Murphy (Ellen_Murphy@fws.gov) is the Mexico Program Coordinator for the North American Wetlands Conservation Act in the US Fish and Wildlife Service's Division of Bird Habitat Conservation. While working in the Inter-American Foundation's Office of Learning and Dissemination, she led an evaluation team that followed the progress of the organic coffee producer Union de Ejidos La Selva. She is the author of "La Selva and magnetic attraction of markets: The cultivation of organic coffee in Mexico" (*Grassroots Development* 19, no. 1: 27–34).

Tad Mutersbaugh (mutersba@uky.edu), Director of Latin American Studies and an associate professor in the Geography Department at the University of Kentucky, has been doing research in Mexico since 1988. He has authored many articles on rural development, on gender and economic development, and on organic certification.

Seth Petchers (shp@georgetown.edu) is working with a start-up organization to launch domestic Fair Trade certification in India. Previously he was Coffee Program Coordinator at Oxfam America. In this role he integrates Oxfam's policy, corporate engagement, public education, and development work as it pertains to coffee. Before joining Oxfam America he worked in a joint project with the non-profit organization Chocolate Matters and the World Bank. He has also worked with the International Task Force on Commodity Risk Management and as Certification Manager at TransFair, USA. His contributions include curriculum development for trainers and farmers and implementation of training programs in Nicaragua and Tanzania.

José Luis Plaza-Sanchez (consultordes@aol.com) has been a consultant for the Inter-American Foundation, for the United Nations Development Program, for the Food and Agricultural Organization, and for many other Mexican and international agencies. He served as the Director General of Agriculture in Mexico's Agricultural Ministry (SAGAR) during the presidency of Ernesto Zedillo. He is currently a consultant for the National Commission for the Development of Indigenous Peoples (CDI).

Laura Trujillo (letover@yahoo.com) is a professor in the Centro Regional Universitario Oriente at the Universidad Autónoma Chapingo, based in Huatusco, Veracruz, Mexico. She co-founded the first agroecology major in Latin-America (at Chapingo University). Her research interests include political ecology, the rural development implications of global markets, and environmental risk inequality from a political ecology and gender perspective.

Silke Mason Westphal (silkemw@yahoo.com) is DanChurchAid's Regional Representative in Central America. Her previous experience includes assignments as program adviser, consultant, and researcher. Her chapter in this volume is based on research she carried out as a research fellow at Roskilde University in association with the CATIE-MIP/AF project, a participatory research and extension project on Integrated Pest Management and coffee agroforestry in Nicaragua.

Index

Fertilizers
Place
Co-opt - mainstreaming

Keep them in the loop - Farmer retention